Physiological Adaptations for Breeding in Birds

Physiological Adaptations for Breeding in Birds

Tony D. Williams

PRINCETON UNIVERSITY PRESS
Princeton and Oxford

Published by Princeton University Press, 41 William Street,
Princeton, New Jersey 08540
In the United Kingdom: Princeton University Press, 6 Oxford Street,
Woodstock, Oxfordshire OX20 1TW

press.princeton.edu

Library of Congress Cataloging-in-Publication Data

Williams, Tony D.
Physiological adaptations for breeding in birds / Tony D. Williams.
p. cm.
Includes bibliographical references and index.
ISBN 978-0-691-13982-1 (hardcover : acid-free paper) 1. Birds—Reproduction.
2. Females. 3. Birds—Physiology. 4. Adaptation (Physiology) 5. Phenotype.
6. Birds—Variation. 7. Birds—Ecology. I. Title.
QL698.2.W45 2012
598.13'8—dc23 2011041066

British Library Cataloging-in-Publication Data is available

This book has been composed in Sabon

Printed on acid-free paper. ∞

Printed in the United States of America

10 9 8 7 6 5 4 3 2 1

Contents

Illustrations

Abbreviations

A4, androstendione

BMR, basal metabolic rate

cGnRH-I, chicken gonadotropin-releasing hormone-I

cGnRH-II, chicken gonadotropin-releasing hormone-II

DEE, daily energy expenditure

E2, estradiol-17β

EGF, epidermal growth factor

EGFRL(s), epidermal growth factor receptor ligand(s)

fshr, follicle-stimulating hormone receptor

FSH, follicle-stimulating hormone

GnIH, gonadotropin-inhibiting hormone

GnRH, gonadotropin-releasing hormone

Hb, hemoglobin

Hct, hematocrit

HPG, hypothalamic-pituitary-gonadal

Id, inhibitor of differentiation

IGF, insulin-like growth factor

IGF-I, insulin-like growth factor-I

IGF-II, insulin-like growth factor-II

LH, luteinizing hormone

MAPK, mitogen-activated protein kinase

RBC, red blood cell

RFG, rapid follicle growth

ROS, reactive oxygen species

RYD, rapid yolk deposition

T, testosterone

T3, triiodothyronine

T4, thyroxin

Ta, ambient temperature

VIP, vasoactive intestinal peptide

VLDL, very low density lipoprotein

VLDLy, yolk-targeted very-low density lipoprotein

VO2max, maximum oxygen consumption

VTG, vitellogenin

Acknowledgments

My thanks to Sir Brian Follett, who read the entire draft manuscript of this book for me and who was not afraid to prompt me to improve my writing; I am very grateful for his feedback and for his suggestion for the title of the book. The following read and provided comments on specific chapters: Samuel Caro, Alistair Dawson, and Marcel Visser (chapter 3); Oliver Love and Katrina Salvante (chapter 4); Keith Sockman (chapter 5); Frederic Angelier (chapter 6); and Glenn Crossin, David Green, and Michael Romero (chapter 7). Thanks to all of them for their constructive feedback. Many people generously provided me with original or high-quality figures or even their original data so I could redraw figures: Frederic Angelier, Pierre Bize, Anne Charmantier, Francis Daunt, Alistair Dawson, Maaike de Heij, Dany Garant, Corine Eising, Paul Flint, Gilles Gauthier, Jennifer Grindstaff, Rene Groscolas, Lars.Gustafsson, Alan Johnson, Marcel Lambrecht, Denis Lepage, Oliver Love, Jesko Partecke, Michael Romero, Patricia Schwagmeyer, Charles Thompson, Marcel Visser, and John Wingfield. Joe Williams and a second anonymous reviewer provided very helpful comments on the draft manuscript.

Special thanks to everyone who took part in the E-BIRD network and meetings between 2004 and 2008, and especially to the USA and European coordinators, John Wingfield, Marcel Lambrechts, and Marcel Visser. E-BIRD was a fantastic, stimulating, thought-provoking experience and the genesis of many of the ideas contained in this book. Thanks to the National Science Foundation, European Science Foundation, and National Sciences and Engineering Research Council (NSERC) for funding this network, and thanks to NSERC for continuing to fund my research through their Discovery Grant program. This allowed me the flexibility to think about and then to write this book.

At Princeton University Press I would like to thank Robert Kirk for convincing me to write this book in the first place and Alison Kalett and especially Stefani Wexler for advice and helpful suggestions during the completion of the book. Thanks to Dimitri Karetnikov at PUP for frighteningly efficient help with figures, to Debbie Tegarden for handling production of the book, and to Gail Schmitt for the copyediting.

At Simon Fraser University I would like to thank all the undergraduate and graduate students, including the BISC 807 class, who sat through my rants on different sections of the book and helped refine (and even change) my ideas. All the graduate students and postdocs in my lab help make the science challenging but fun; special thanks to Sophie Bourgeon, Glenn Crossin, Kristen Gorman, Oliver Love, Calen Ryan, Katrina Salvante, Will Stein, François Vézina, and Emily Wagner.

Thanks to Parmindar Singh Parhar and all the staff of Renaissance Coffee at Simon Fraser for keeping me going when I really didn't feel like writing anymore!

Finally, this book is for my wife Karen—who did let me write a second book after all—and for Nicolas and Natasha.

Physiological Adaptations for Breeding in Birds

CHAPTER 1

Introduction

THIS BOOK IS PRIMARILY about physiological mechanisms, but it also addresses the specific question of what we know about the physiological, metabolic, energetic, and hormonal mechanisms that regulate, and potentially determine, *individual*, or phenotypic, variation in key *reproductive life-history traits*, *trade-offs* between these traits, and *trade-offs* and *carry-over effects* between different life-history stages. Initially I will focus on the avian reproductive cycle (from seasonal gonadal development, through egg-laying and incubation, to chick-rearing), and then I will expand this view to consider reproduction in the broader context of the annual cycle and over an individual's entire lifetime (incorporating early developmental effects, maternal effects, and late developmental effects such as senescence). Throughout I will focus on and develop two major themes: that we need to consider reproductive physiology and ecology from a *female perspective* and that we need to consider the causes and consequences of *individual (phenotypic) variation* in reproductive life-history traits.

1.1. Structure of the Chapters

Although this book is about physiology I have rooted the book's organization firmly in the domain of evolutionary biology, or life-history, by asking the question, Which reproductive life-history traits contribute most to individual variation in lifetime fitness? An increasing number of long-term, individual-based population studies of birds and mammals are providing answers to this question, measuring differences in fitness between individuals, and assessing the causes of these disparities (Clutton-Brock and Sheldon 2010). Figure1.1 shows an example for multiple life-history traits for a pedigree of more than 2,100 individual female great tits (*Parus major*) studied across 39 years in Wytham Wood, England (McCleery et al. 2004). This study, and numerous others cited in this book, shows that the traits most strongly correlated with lifetime fitness (estimated as total number of offspring *recruited* to the breeding population) are lifetime fledging success (the total number of young *fledged* from all breeding attempts), longevity, clutch size, and laying date. Traits such as egg size, nestling mass, natal dispersal, and body size do not significantly influence total fitness. This approach is, admittedly, not perfect: only traits for which large amounts of individual data are available can be entered into these models. As an example, a major focus in avian biology in the last 10–15 years has concerned yolk hormones, antioxidants, and antibodies as potential determinants of offspring quality (see chapter 4). Given sufficient data, which currently do not exist, it is *possible* that these traits could prove to be important components of variance in lifetime fitness. Similarly, as I will highlight in chapter 6, it is much easier to obtain simple, reliable metrics for laying date, egg size, and clutch size than it is to measure the multiple, composite traits that characterize individual variation in incubation and chick-rearing effort (e.g., incubation onset, duration, constancy, provisioning rate, load size, and foraging distance), so the latter are not currently included, explicitly, in these analyses of lifetime fitness. Nevertheless, variation in incubation and chick-rearing effort is presumably subsumed within the trait of lifetime number of fledglings, implying that these *are* important components of variance in fitness. Taking this approach also highlights other important, unresolved evolutionary and mechanistic questions; for example, why are other traits, such as egg mass, highly variable among individuals if this variation is *not* related to variation in fitness?

Taking the approach outlined above, the main chapters of this book deal with timing of breeding (chapter 3), egg size and egg quality (chapter 4), clutch size (chapter 5), and parental care (incubation and chick-rearing, chapter 6). Each of these chapters is structured in the same way. I describe variation in the trait of interest (e.g., clutch size) or the composite traits that make up more complex reproductive behavior (e.g., variation in on-nest and off-nest bouts in relation to incubation). I focus on *individual* variation within a population or species, and, with one or two exceptions, this book does not take a comparative approach. Having described how variable these traits are, I ask if, and how, this variation is related to fitness and if there is evidence that selection is acting on these traits. Finally, having provided an evolutionary context for individual variation in these key reproductive life-history traits, I ask what we know about the physiological, metabolic, energetic, and hormonal mechanisms that regulate, and potentially determine, trait variation. Each chapter ends with a summary, and readers might like to start with it to get a quick overview of the main points. Reproduction does not occur in isolation; rather, an individual's fitness is dependent on successful integration of multiple life-history stages (non-breeding, breeding, molt, and migration) across the complete life cycle. In chapter 7 I place breeding in this broader context and consider linkages between different life-history

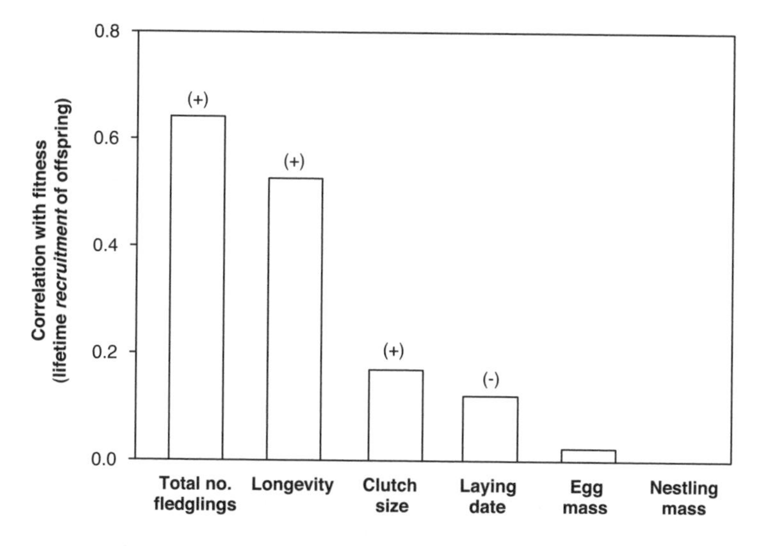

Fig. 1.1. Correlations of different life-history traits with total fitness (lifetime recruitment) in the great tit. (Based on data in McCleery et al. 2004.)

stages, including carry-over effects, costs of reproduction, and other trade-offs. How does an individual's activity during the non-breeding or pre-breeding period influence the subsequent reproductive period, and how, in turn, does reproduction influence late-season events such as post-fledging parental care, molt, migration, and ultimately future fecundity and survival? As in the previous chapters, having considered individual variation in these linkages between life-history stages, I then ask what we know about the physiological mechanisms that regulate potential trade-offs and carry-over effects.

1.2. A Primer on Reproduction in *Female* Birds

The reproductive traits that contribute most to lifetime fitness are either directly (laying date, clutch size) or indirectly (e.g., number of fledglings, where the upper limit is set by the clutch-size decision) related to a female's decision on *when* to breed and what *level* of primary reproductive investment to make during egg-laying. Furthermore, a female's decision on when to initiate a breeding attempt has important consequences for many components of reproductive effort much later in the season, for example, the probability of multiple brooding, the probability of breeding-molt overlap, and perhaps sex-specific survival. Chapter 2 provides a comprehensive overview of the physiological and endocrine control of female reproduction, focusing on the mechanisms underlying egg production, which must be involved in regulating and generating individual variation in these traits. The female perspective that pervades this book, and in particular chapter 2, stems from what I perceive as a disconnect in avian biology between physiological research on free-living or non-poultry species, where much of the focus has been on male birds, avian ecology, which *has* focused on female traits (e.g., egg and clutch size), and poultry studies, which *have* focused on females and egg production. Seasonal breeding provides one clear example. *Female-specific* physiological processes, such as vitellogenesis, yolk deposition, follicle growth, and oviduct development, are critical determinants of the timing of breeding, but in the majority of studies on environmental control of seasonal reproduction and gonadal development, male birds have been the model of choice! This disconnect has been long-standing: more than 40 years ago Farner et al. (1966) stated that "the accumulation of data on the ovarian cycle, both in the field and the laboratory, has been necessarily at a much slower rate [than for testis function]." G. Ball and Ketterson's recent review (2008) pointed out just how little we know about control of seasonal reproduction in *females* and highlighted the fact that "sex differences in relation to the environmental control of reproduction are [still] not a major research focus for the field overall." Therefore, I believe that bringing a female-specific perspective to bear within a life-history framework will help to move the field forward.

1.3. Individual Variation

It is almost axiomatic that heritable, individual variation is the raw material for natural selection. Darwin's *Origin of Species* (1872) is replete with references to *variability* and to *individual differences*, and he highlighted the fact that "these ... afford materials for natural selection to act on and accumulate" (p. 34). Interestingly, Darwin fully recognized that these differences could be physiological in nature, describing studies of individual variation in the branching of nerves and in the structure of muscles in insects. The analysis of individual variation has been embraced by many areas of ecology and evolutionary biology (Biro and Stamps 2008; Clutton-Brock and Sheldon 2010; Dingemanse et al. 2010; Hamel et al. 2009; Nussey et al. 2007; A. Wilson and Nussey 2010), but until recently, physiology and endocrinology have all but ignored individual variation (A. Bennett 1987; T. Williams 2008). A consideration of individual variation forces us to ask a different set of questions with regard to regulatory mechanisms underlying reproductive traits. The question, What is the physiological difference between non-breeding and egg-producing females? might be satisfactorily answered by showing that the former has baseline plasma estrogen levels, whereas the latter has highly elevated plasma estrogen levels. These results would be taken as evidence that estrogens play a role in regulating egg production. However, in understanding individual variation we need to ask the question, Does a female with a high egg-producing phenotype (many, large eggs) have higher plasma estrogen levels than a low egg-producing phenotype (few, small eggs)? In other words, do estrogens, or estrogen-dependent mechanisms, explain *individual* variation in the reproductive phenotype? As a further example, although elevated plasma prolactin levels are generally associated with parental care, is prolactin causally related to individual, phenotypic variation in the *amount* of parental care? This approach therefore gets at the heart of a mechanistic understanding of phenotypic variation.

1.4. What Is *Not* in This Book?

As a consequence of the specific perspective described above, there are many topics that are *not* covered in this book. I have not considered male-specific topics, such as testis structure and function, sperm competition, song, testosterone-immune function trade-offs, etc. I have tried to avoid any extensive consideration of social behavior per se since this topic has been comprehensively covered in Elizabeth Adkins-Regan's excellent book, *Hormones and Animal Social Behavior* (2005). I have also not included behavioral syndromes or avian personalities, although these represent other areas where ecological studies are rapidly outpacing any mechanistic understanding (Biro and Stamps 2008; Dingemanse et al. 2010; Reale et al. 2010). There is relatively little genomics in this book, even though it is considered by some to be one of the "hotter" areas of avian biology (Bonneaud et al. 2008). Sequencing of the zebra finch (*Taeniopygia guttata*) genome is indeed a "major step forward for avian ecology and evolutionary biology" (Balakrishnan et al. 2010), and genomics might indeed "revolutionize our understanding of birds" (Edwards 2007). However, most work to date has focused on comparative genomics, population genetics, speciation, and systematics, and genomics work on natural bird populations is still in its infancy (Ellegren and Sheldon 2008). In particular, application of genomic approaches to questions of plasticity, phenotypic variation, and adaptation has hardly begun, especially for reproductive and life-history traits (but see, e.g., Abzhanov et al. 2004; Abzhanov et al. 2006; Cheviron et al. 2008; Fidler et al. 2007). A main conclusion of this book very much supports Edwards' (2007) statement that "ornithologists [should] maintain focus on the age-old questions in ecology." However, my own view is that the "age-old questions" considered in this book, such as why clutch size, or parental effort varies among individuals, and the consequences of this variation, have *not* remained intractable simply because of a dearth of non-genomic approaches. Rather we have just stopped focusing on certain fundamental questions even though they remain unresolved and can be tackled with traditional, *experimental* approaches (albeit enhanced and aided by genomics knowledge and techniques). Identifying the genes or DNA sequences underlying variation is life-history traits is an important goal (Ellegren and Sheldon 2008), but unveiling the *physiological mechanisms* linking genotypes to phenotypes, and ultimately to fitness, in natural populations is even more important.

1.5. Avian Reproduction in a Changing World

Global climate change is currently impacting all aspects of the annual cycle of birds, from the tropics, through the temperate zone, to the Poles (Gaston et al. 2005; Nevoux et al. 2010; Visser, et al. 2004; S. Williams et al. 2003). Climate change is affecting the phenology of breeding in birds, the timing of migration, demography, population dynamics, and species' distributions and ranges (Carey 2009; Visser 2008; Visser, et al. 2004). Since the timing of breeding determines the timing and pattern of subsequent breeding stages (see chapter 3), climate change affects all components of the reproductive cycle, including incubation, nestling development time, and probability of second clutches (e.g., Husby et al. 2009; Møller et al. 2010; Matthysen et al. 2010) via effects on a suite of correlated life-history traits (Both and Visser 2005; Garant, Hadfield, et al. 2007). Regional differences in climate change (e.g., between breeding and non-breeding areas) and variation in responses to climate change at different trophic levels (e.g., between birds and their insect prey) have led to mismatching of resource availability and arrival dates in migratory birds (T. Jones and Cresswell 2010) and to mismatching of peak food availability and chick-rearing in some breeding birds (Gaston et al. 2005; Visser et al. 1998; Visser et al. 2004). However, population responses to climate change are complex, and shifts in the timing of breeding and migration have varied markedly between species, occurring in some populations but not others (Carey 2009; Parmesan 2006). Several studies have reported or predicted negative effects of climate change on population dynamics of certain species (e.g., Both et al. 2006; Jenouvrier et al. 2009), but some studies have also suggested there might be positive effects of climate change on other populations, at least in the short-term (D'Alba, Monaghan, and Nager 2010; Gaston et al. 2005). Long-term monitoring studies of bird populations provided some of the earliest and clearest evidence for changes in phenology associated with climate change (Crick et al. 1997). However, to date most studies have simply, though importantly, documented change after it has occurred; a critical goal is to develop the ability to predict how *future* climate change will affect avian populations. This capacity, in turn, will require that we obtain a far better understanding of the *physiological response mechanisms* that link environmental cues and environmental conditions to reproductive and life-history decisions, and it is therefore essential that research on phenological, life-history events integrates both mechanistic and evolutionary perspectives (Visser et al. 2010). Physiological or mechanistic hypotheses can themselves provide a valuable approach for predicting future effects of climate change *provided we know enough about the basic, underlying physiology* (Portner and Farrell 2008; Portner and Knust 2007; Sinervo et al. 2010). As this book demonstrates, this is at present not the case for most of the key reproductive life-history traits in birds, a situation that adds urgency to the need to refocus our research efforts. Hopefully, by focusing on the basic physiological, metabolic, energetic, and hormonal mechanisms underlying female-specific reproductive traits within a broader life-history context, this book will contribute to advancing our progress in this area.

CHAPTER 2

The Hormonal and Physiological Control of Egg Production

As I WILL DESCRIBE in the remaining chapters of this book, the reproductive traits that contribute most to variation in lifetime fitness are either directly (e.g., laying date, clutch size) or indirectly (e.g., number of fledglings, where the upper limit is set by the clutch size decision) related to a female's decision on *when* to breed and what *level* of primary reproductive investment to make during egg-laying. Females therefore play a predominant role in the timing of breeding both from a physiological and genetic perspective, a circumstance that should dictate a *female-biased* perspective in physiological studies (Caro et al. 2009; Sheldon et al. 2003; van de Pol, Bakker, et al. 2006). The timing of egg production and egg-laying are dependent on the regulation of *female-specific* physiological processes, such as yolk-precursor production (vitellogenesis), rapid yolk deposition (RYD), follicle growth, and oviduct development, and regulation of these traits represents the final, absolute prerequisite "decision" that must be made by the female before a breeding attempt can be initiated. Many other aspects of reproductive investment, such as hormonally mediated and immunological maternal effects, involve the transfer of substances from the mother to eggs during egg production and probably involve some of the same mechanisms that mediate yolk uptake (see section 2.1.4). Thus, understanding the physiological mechanisms underlying egg production, and the proximate control mechanisms that integrate environmental cues, is of critical importance to understanding a wide range of adaptations to breeding in birds.

Previous reviews of physiological control of avian reproduction have focused on male birds and usually also on mechanisms at the level of the pituitary, hypothalamus, or higher neural centers (Bentley et al. 2007; A. Dawson 2002, 2008; A. Dawson and Sharp 2007; A. Johnson and Woods 2007). Here I will concentrate almost exclusively on the regulation of *female* reproduction and on the physiological regulation of ovary and oviduct function during egg production. Since female reproduction is highly seasonal and occurs in response to seasonal environmental variability, I will consider what we know about how information encoded in

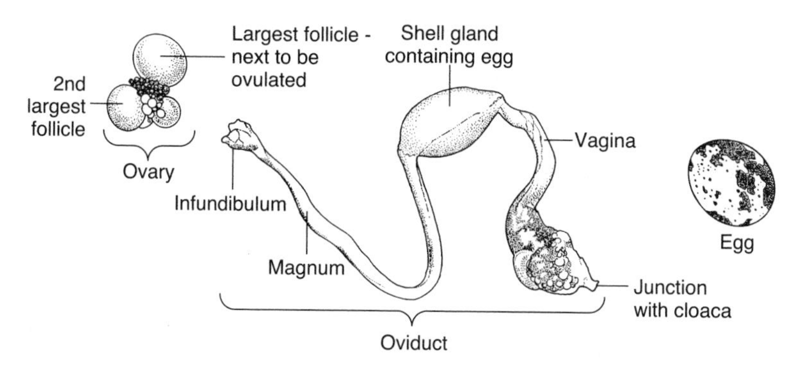

Fig. 2.1. Reproductive system of a breeding female common quail showing the different components of the ovary and oviduct. (Redrawn from Brooke and Birkhead 1991, with permission of Cambridge University Press.)

photoperiodic and non-photoperiodic cues (food, temperature, and social factors) is integrated by the neuroendocrine system and transduced into an (hormonal) output that ultimately determines ovarian function and egg formation. I will briefly review important aspects of the neuroendocrine system that regulates gonadal function (i.e., the hypothalamic-pituitary-gonadal [HPG] axis), but, again, I will focus primarily on control of the ovary. I will compare data for males and females in terms of HPG-axis function and reproductive development to provide evidence that males cannot be considered good models for females, and I will highlight some discrepancies, inconsistencies, and gaps in our knowledge with regard to control of female reproduction. Throughout I will attempt to highlight and describe mechanisms that might generate or constrain *individual* variability in female reproductive traits.

2.1. Overview of the Female Reproductive System

The reproductive system of female birds comprises two organs essential for egg formation and egg-laying: the ovary and the oviduct (fig. 2.1). Two bilaterally symmetrical gonads and oviducts develop in the female avian embryo, and two fully developed ovaries are common in birds of prey (Falconiformes) and the kiwi (*Apteryx spp.*), but in nearly all other species only the ovary and oviduct on the left side are functional in adult life (Kinsky 1971).

The ovary is suspended from the dorsal wall of the body cavity near the top of the kidneys and consists of a large number of oocytes embedded in a stroma of connective tissue. Oogenesis is terminated at hatching, at which time the ovary of the chick embryo contains about 480,000

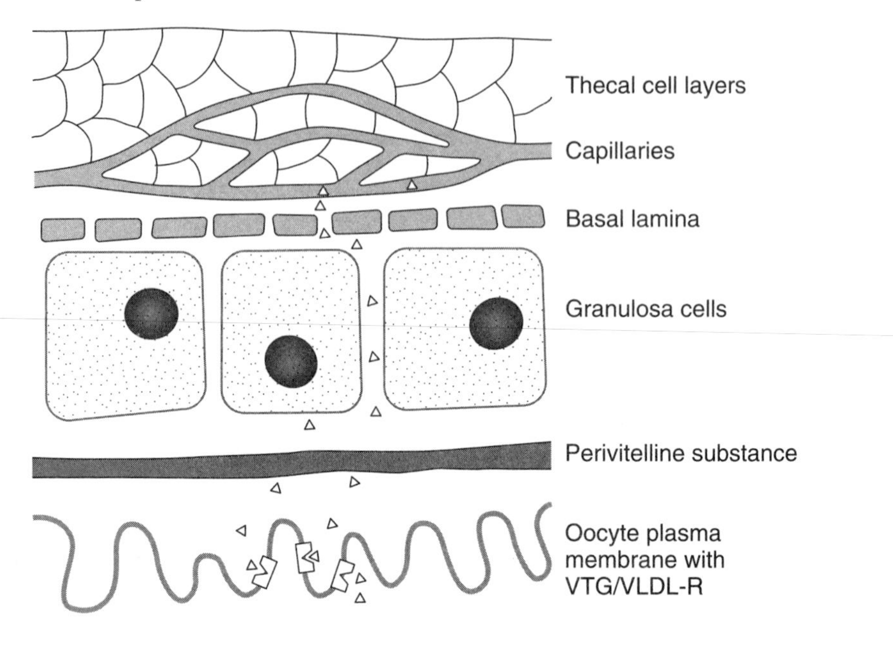

Fig. 2.2. Histological organization of the follicle tissue layers surrounding the developing oocyte, indicating the pathway of yolk precursors (triangles) from the capillaries to receptor-mediated yolk uptake at the oocyte plasma membrane.

primary oocytes. Even in domesticated species selected for high laying rates, only 200–500 oocytes will eventually mature and be ovulated, and in free-living species as few as 1–20 oocytes mature to produce eggs, depending on clutch size and life span. Once an oocyte is selected into the vitellogenic follicle hierarchy, there is a variable phase of rapid follicle growth, or rapid yolk development. During this period the developing ovum is surrounded by an ovarian follicle comprising (a) the oocyte plasma membrane, (b) a perivitelline layer (analogous to the mammalian zona pellucida), (c) a granulosa cell layer, (d) a basal lamina, and (e) theca interna and externa cell layers (fig. 2.2).

The nucleus of the oocyte—the germinal vesicle—that ultimately forms the embryo is located in the germinal disc region of the granulosa cell layer. Granulosa and thecal cells proliferate during follicle maturation and synthesize and secrete the hormones required for yolk development, oviduct function, and control of female sexual behavior. At ovulation the fully developed yolk passes into the oviduct, where the albumen and shell are added to complete egg formation. Fertilization takes place in the infundibulum, the top, funnel-shaped part of the oviduct, within 15–30 minutes of ovulation (Birkhead and Møller 1992). However, in contrast to mammals and most other vertebrates, the female is the heterogametic sex in birds (ZW), and the sex of the offspring is under maternal control (Rutkowska and Badyaev 2008). In the chicken (*Gallus gallus*), offspring sex is determined during the first meiotic division, which occurs within 1–2 hours of ovulation and involves the segregation of the sex chromosomes and movement of either the Z or W chromosome to the ovum and movement of the other sex chromosome to the polar body (Etches 1996; Pike and Petrie 2003). Postovulatory follicles containing the granulosa and thecal cell layers are not homologous with the corpus luteum of mammals and persist for at most a few days (A. Johnson and Woods 2007). Follicles that start to mature but that are not ovulated become atretic, and the oocyte and yolk material are reabsorbed. Atresia is common in pre-vitellogenic follicles, but rare in vitellogenic follicles (A. Gilbert et al. 1983; A. Johnson and Woods 2007), and occurs through the process of apoptosis, or programmed cell death.

2.1.1. Pre-vitellogenic Follicle Development

Follicle development proceeds in two stages: (a) an initial slow stage that lasts from 60 days to several years, resulting in the development of a large number of pre-vitellogenic "white" follicles, and (b) rapid follicle growth (RFG) or rapid yolk development (RYD), which lasts from days to several weeks and in which the majority of the proteins and lipids are added to the yolk, by uptake of yolk precursors from the blood via receptor-mediated endocytosis, to produce large, yolky "yellow" follicles (fig. 2.3). An intermediate stage of pre-vitellogenic growth has been described, which lasts weeks to several months in the chicken and involves incorporation of small amounts of a lipoprotein-rich white yolk (A. Johnson 2000). This process generates a cohort of somewhat larger follicles (8–14 follicles, 6–8 mm in diameter in the chicken), some of which are selected to enter the final maturation stage, but 75%–90% of which become atretic and are reabsorbed (Woods and Johnson 2005). A similar intermediate phase of pre-vitellogenic follicle development may occur in free-living species, but it is not well characterized (e.g., see Astheimer 1986 and section 2.1.2).

Various studies have suggested that follicle-stimulating hormone (FSH), inhibin, and growth factors such as epidermal growth factor (EGF) and insulin-like growth factor (IGF-I) (S. Palmer and Bahr 1992; Woods and Johnson 2005) are involved in regulating follicle selection, and many other putative regulatory factors have been identified in mammals (Fortune 2003; Skinner 2005). However, the mechanism(s) that determines which follicles will be recruited from a pre-hierarchical cohort of oocytes for the final stages of differentiation and RYD remains largely unknown (A. Johnson and Woods 2009). Follicle selection involves functional

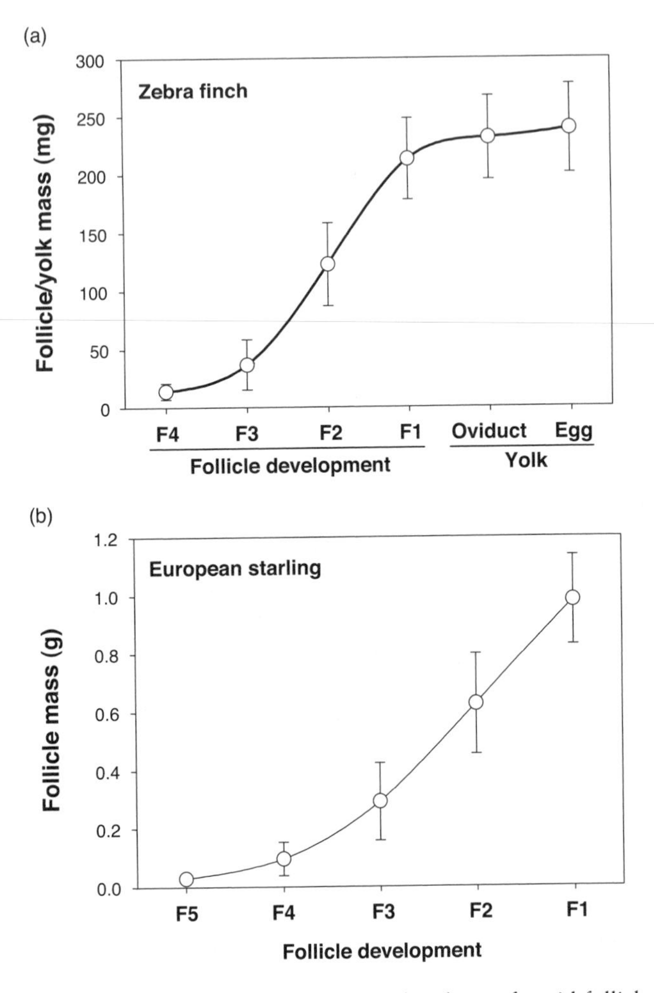

Fig. 2.3. Follicle and yolk development during the phase of rapid follicle growth, or rapid yolk development, in (a) the zebra finch and (b) the European starling.

differentiation of the granulosa and thecal cell layers, and the earliest marker of selection is increased expression of follicle-stimulating hormone receptor (*fshr*) mRNA in the granulosa cell layer (Woods and Johnson 2005). Johnson and colleagues (A. Johnson et al. 2008; A. Johnson and Woods 2009) recently proposed an "internal, follicle autonomous" model whereby autocrine/paracrine factors maintain pre-hierarchical follicles in an undifferentiated state. This model suggests that epidermal growth factor receptor ligands (EGFRLs) initiate mitogen-activated protein kinase (MAPK) signaling and up-regulate several inhibitor of differentiation/DNA binding (Id) isoforms, such as Id1, Id3, and Id4, which function as transcription factors that inhibit *fshr* mRNA expression. Inhibition of MAPK signaling increases the expression of another Id isoform (Id2) that is associated with elevated *fshr* mRNA expression. Progressively increasing FSH responsiveness is then thought to lead to expression of the luteinizing hormone receptor and initiation of follicular steroid hormone synthesis. As A. Johnson and Woods (2009) point out, current studies of avian ovarian follicle growth and differentiation, as described above, are necessarily conducted in vitro and are directed at a cellular and molecular understanding. How these current models of follicle selection are integrated with environmental cues to determine daily and seasonal patterns of follicle development, what A. Johnson and Woods (2009) refer to as an "external model" of follicle selection, remains unknown.

Pre-vitellogenic ovarian development and follicle growth has not been studied in detail in free-living species due to the problem of identifying the reproductive status of females *before* the first egg appears in the nest. As I discuss below, if a *prolonged* developmental phase of pre-vitellogenic growth is typical of all birds, then it will have important implications for understanding how and when critical environmental cues might initiate seasonal development of the female reproductive system (see section 3.6.1 and fig. 3.11). Some early studies of captive birds did investigate early gonadal development in detail, measuring growth constants for the very earliest periods of testis and ovarian growth. In Japanese quail (*Coturnix c. japonica*), which will undergo full ovarian development under artificial photoperiods, the increase in ovarian mass is log-linear (from about 20 mg up to 150 mg) for the first 12 days of photostimulation with 20L:4D, at which point yolky, vitellogenic follicles are evident (B. K. Follett, pers. comm.). If this initial rate of gonadal growth (the log growth constant, *k*) is plotted against day length, it is clear that this process is photoperiodically dependent in quail; that is, *initial* rates of gonadal growth are higher under longer day length, although this effect appears to be stronger in male than female quail (fig. 2.4). However, in white-crowned sparrows (*Zonotrichia leucophrys gambelii*), *k* is generally much lower than in quail, especially in females, and is only weakly photoperiod-dependent at "intermediately" long day lengths, that is, between 12L and 16L (M. Morton et al. 1985), not at day lengths longer than 16L (Farner et al. 1966; Lewis 1975).

It is not known what the *functional* significance of any phase of pre-vitellogenic growth is, or what is actually occurring during this period at the cellular and physiological level, especially in free-living birds. Presumably there is some cell proliferation, and in the chicken there is incorporation of lipoprotein-rich white yolk (A. Johnson and Woods 2007).

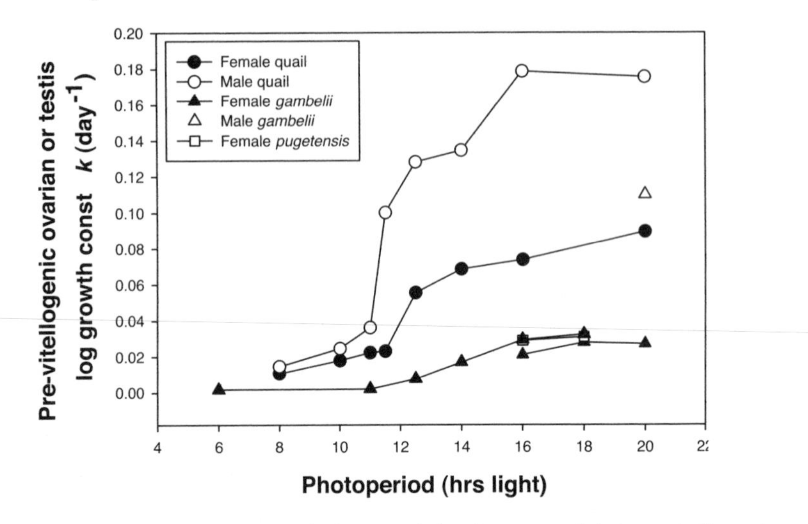

Fig. 2.4. Log of growth constants (k) for initial development of the pre-vitellogenic ovary and for testis development in relation to photoperiod. (Based on data from Brian Follett, pers. comm.; and in Farner et al. 1966; Lewis 1975; and M. Morton et al. 1985.)

However, this process does not involve significant steroidogenesis (e.g., increased plasma estrogen), expression of key regulatory genes for vitellogenin (VTG) or the vitellogenin receptor, or initiation of oviduct growth. In short, is early pre-vitellogenic gonadal growth a time-dependent developmental phase that must be completed *before* vitellogenesis and the final stages of yolk formation can be initiated? Does this initial cell proliferation determine subsequent steroidogenic or yolk uptake capacity during RYD? Or does the slow, pre-vitellogenic growth phase simply represent some "non-functional" process that ticks along until environmental cues are sufficient to initiate the latter stages of ovarian development? The answers to these questions are fundamentally important to identifying the environmental cues and conditions that prevail when the female reproductive system is finally "switched on," that is, the point at which females are committed to egg production.

2.1.2. Rapid Follicle Growth (RFG) or Rapid Yolk Development (RYD)

Mechanisms underlying the final stages of follicle growth after follicle selection are particularly important because they determine the key life-history traits of egg size and number and, potentially, laying date. Ovary size varies markedly with the females' reproductive status and with the number of follicles entering the final rapid growth phase of yolk development. In the ovary of the laying chicken, four or five large yellow follicles increase in size from 0.15 g to 12–14 g (40 mm in diameter) each and form an ovarian follicle hierarchy, along with thousands of smaller undeveloped yellow and white follicles (0.5–6.0 mm in diameter; fig. 2.5a). Similarly, in the zebra finch the ovary of a non-breeding female weighs 20 mg but will increase in size over 5–6 days to 250 mg in breeding females as yolk formation progresses. In species with single-egg or small clutches, fewer follicles may be recruited to the follicle hierarchy (e.g., albatross, fig. 2.5b), but in general the number of recruited follicles is always greater than the clutch size (see chapter 5). The duration of the RYD phase is highly variable among species, lasting 3–4 days per follicle in small passerines, 6–10 days in the domestic chicken, and 14–25 days in large seabirds (Procellariformes and Spheniscidae).

Empirical data on follicle size distributions in developing ovaries of several species support the idea that single oocytes are selected and sequentially initiate yolk uptake at approximately 24-hour intervals (in species that lay one egg every 24 hours). Thus, birds develop a hierarchy of vitellogenic, yolky follicles from the smallest (F5 or F4) to the largest F1 follicle, with the F1 follicle being the next follicle to be ovulated (see fig. 2.5a). Each follicle has a more or less parallel, exponential rate of growth, with the maximum rate of yolk deposition occurring in the last half to one-third of RYD (Astheimer and Grau 1990; Ojanen 1983b). This traditional model predicts that there is little overlap in size between sequentially developing follicles and that the hierarchical order of oocyte selection is maintained at ovulation and oviposition; that is, the first follicle to enter RYD is the first to ovulate and the first to be laid (fig. 2.6a). Badyaev et al. (2005) suggested another much more flexible model of yolk formation in which the growth rates of individual follicles are highly variable (fig. 2.6b). This flexible model appears to be required to underpin proposed mechanisms for adaptive sex differences in oocyte growth (Badyaev et al. 2005; Pike and Petrie 2003) and sex-specific "allocation" of maternal hormones (Badyaev, Acevado Seaman, et al. 2006; see chapter 4). In this latter model, rapidly growing follicles can "overtake" slower-growing follicles and ovulate sooner relative to their timing of recruitment into the follicle hierarchy. For example, in figure 2.6b, the F4 follicle grows more rapidly and overtakes the slow-growing F5 follicle, and similarly the F2 follicle overtakes the F3 follicle. The order of recruitment of follicles into RYD (F5, F4, F3, F2, F1), then, is potentially very different from the order of ovulation (F4, F5, F2, F3, F1; fig. 2.6b).

There is considerable evidence for this alternative model of rapid yolk development in house finches (*Carpodacus mexicanus*), although all of that work is based on the use of lipophilic dyes that stain yolk rings in developing yolks (Grau 1976), rather than on the direct measurement of

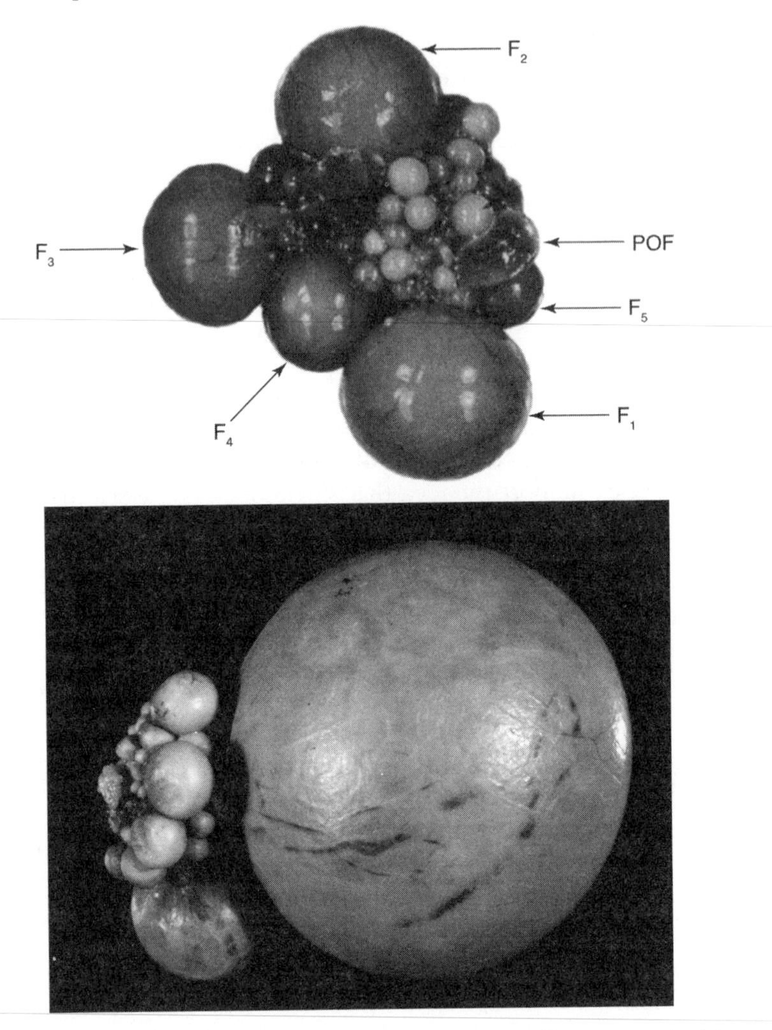

Fig. 2.5. Fully developed, vitellogenic ovary illustrating differences in the hierarchical structure of yolky, ovarian follicles for: (a) the domestic hen (from Johnson 1990, reprinted with permission of Science Reviews 2000 Ltd.) and (b) the wandering albatross (provided by Brian Follett).

follicle size (Badyaev, Hamstra et al. 2006; Young and Badyaev 2004). However, empirical data on follicle size in female zebra finches and European starlings (*Sturnus vulgaris*) from females dissected during laying do not support this flexible model. The two models make very different predictions about the size distribution of developing follicles at any point in time during the laying cycle. According to the traditional model, at the

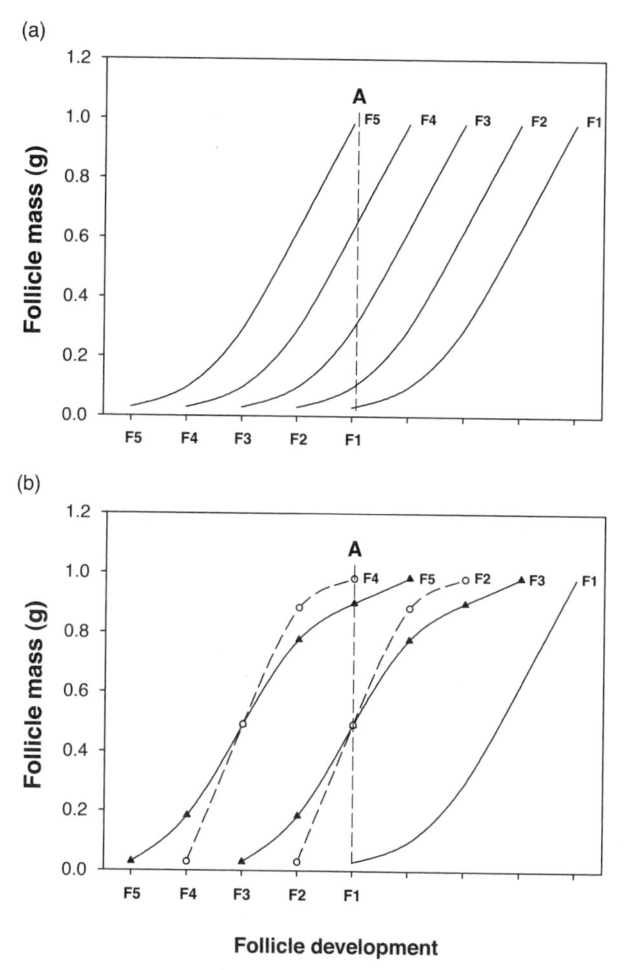

Fig. 2.6. Different models for patterns of rapid yolk development in successive follicles following selection and recruitment to the vitellogenic growth phase: (a) the traditional hierarchical model; (b) a more flexible model proposed by Badyaev et al. (2005).

time that the first follicle ovulates (point A in fig. 2.6a), the female's ovary should contain a graded series of follicles that are approximately 100% (F1), 64% (F2), 30% (F3), 10% (F4), and 3% (F5) of final F1 size (based on the exponential growth curve shown in fig. 2.6). Conversely, at the same time point, the Badyaev model predicts two large follicles similar in size (F4, F5), two medium follicles identical in size (F2, F3), and one very small follicle (F1). Measured follicle sizes for European starlings and zebra finches match the prediction of the traditional model very well: the mean follicle masses for successive F1 to F5 follicles as a proportion of the final mass are 1.00, 0.64, 0.30, 0.10, and 0.03 g, respectively, with 95% confidence intervals less than ± 5% (T. Williams, unpublished data). These results suggest that each follicle follows the constant exponential growth curve predicted by the traditional model and that there is very little flexibility in growth rates of individual follicles, at least in these two species. Other studies have reported variation in the estimated duration of RYD among individuals and between years within populations (Hatchwell and Pellatt 1990; Meathrel 1991), so it is clear that there can be plasticity in follicle development, but further studies are needed to confirm the intra-individual variability in follicle growth reported by Badyaev et al. (2005) in the house finch.

2.1.3. Vitellogenesis and Lipoprotein Metabolism

Vitellogenesis and oocyte growth in birds and other oviparous species are characterized by estrogen-dependent hepatic synthesis of two main yolk precursors: VTG and yolk-targeted very low-density lipoprotein (VLDLy), which are the primary sources of yolk protein and lipid, respectively, providing all of the nutrients and energy required by the developing embryo (Burley and Vadehra 1989; Walzem 1996; T. Williams 1998). Vitellogenin and VLDLy are synthesized by the liver, secreted into and transported in the blood stream, and taken up by developing follicles via receptor-mediated endocytosis (Barber et al. 1991; Wallace 1985). Around 90% of yolk dry weight is derived from VLDLy and VTG, with about 10% of yolk solids comprising other plasma proteins (albumen, α2-glycoprotein, imunoglobulins, binding proteins).

The onset of vitellogenesis involves major changes in the physiology, and specifically lipoprotein metabolism, of breeding females. Vitellogenin production alone comprises approximately 50% of the daily hepatic protein synthesis of the laying chicken and may triple the amount of protein secreted into the blood (Gruber 1972). Plasma lipid concentrations increase from about 3·mg·neutral·lipid/ml plasma in non-laying turkeys (*Meleagris gallopavo*) to 21·mg/ml in laying turkeys (Bacon et al. 1974). Similarly, plasma triacylglyceride concentrations increase from 0.5–1.5 μmol/ml·plasma in non-laying chickens to 20–50 μmol/ml plasma in laying chickens (Griffin and Hermier 1988). Studies on seabirds (Vanderkist et al. 2000), waterfowl (Gorman et al. 2009), and passerines (Challenger et al. 2001; Christians and Williams 1999; Vézina and Williams 2003) have shown that similar changes occur in free-living birds. Moreover, these studies have confirmed that changes in plasma yolk-precursor levels are very closely coupled to the demands of yolk formation and the rate of follicle growth (fig. 2.7).

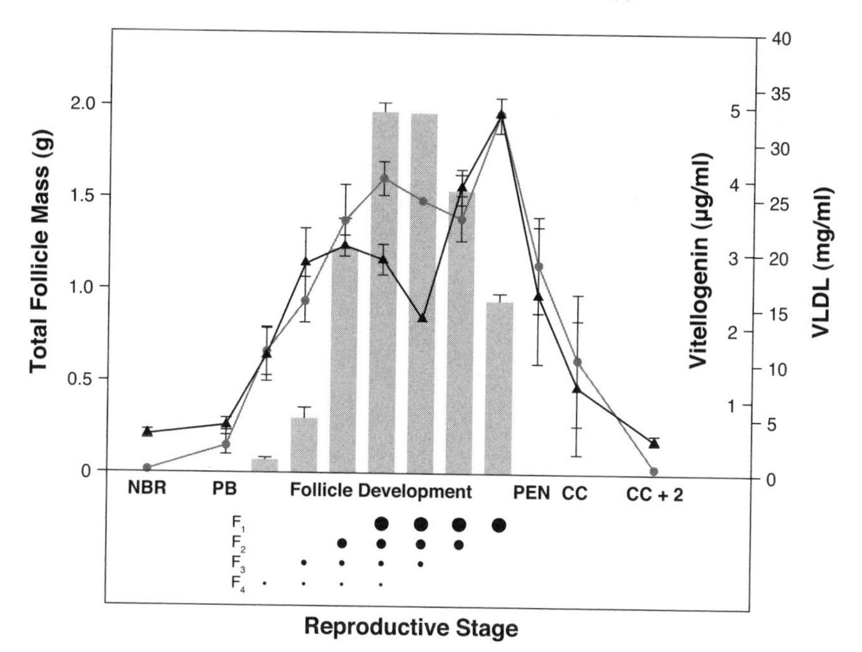

Fig. 2.7. Variation in plasma concentrations of vitellogenin (circles) and yolk-targeted very low density lipoprotein (triangles) in relation to total yolky follicle mass during the laying cycle (gray bars) and pattern of follicle development (bottom panel) in female European starlings. NBR, non-breeders; PB, pre-breeders; PEN, day of penultimate egg; CC, clutch completion; CC + 2, 2 days after clutch completion. (Reproduced from Challenger et al. 2001.)

Three forms of vitellogenin have been identified in the chicken: VTG-I, VTG-II, and VTG-III (S. Wang et al. 1983). VTG-II is the most abundant form, and the relative levels of VTG-I, VTG-II, and VTG-III RNA in the liver of the laying chicken are approximately 4:10:1, similar to the ratio of these VTG proteins synthesized in adult estrogenized roosters (K. Evans et al. 1988; S. Wang et al. 1983). These different vitellogenins have been shown to have different estrogen responsiveness in the developing chicken embryo (K. Evans et al. 1988) and different hepatic synthesis rates in estrogenized roosters (S. Wang et al. 1983). However, it appears that nothing is known about either the occurrence, differential pattern of expression, or functional significance of these different individual vitellogenins during yolk formation in free-living species.

In addition to changes in the *total* amount of lipoprotein produced and transported during egg production, there is also an estrogen-dependent shift in VLDL synthesis from the production of larger, "generic" VLDL particles to that of smaller yolk-targeted VLDL particles (fig. 2.8), along with associated changes in apolipoprotein composition. Generic VLDL particles range in size from 30 nm to more than 200·nm, have at least six associated apolipoproteins (including apoA-I, apoB, and apoC), and are synthesized constitutively in non-breeding and breeding females and males. They function in the transport of triacylglycerides throughout the body for storage in adipose tissue or for metabolism, primarily in muscle tissue. Generic VLDL is therefore a fundamental lipid-transport vehicle that has apolipoprotein B-100 (apoB100) as its integral structural protein and receptor ligand and is secreted from the liver at all stages of the life cycle (Walzem 1996).

VLDLy particles are smaller than generic VLDL, ranging in size from 15 to 55 nm in domestic fowl, and they have only two associated apolipoproteins: apoB and apoVLDL-II. VLDLy comprises 12% protein and 88% lipid (70%–75% triglyceride, 20%–25% phospholipids, and 4% cholesterol), and its specific function is therefore to deliver lipid-rich triacylglycerides to the developing oocyte (Walzem 1996; Walzem et al. 1999). ApoVLDL-II inhibits lipoprotein lipase activity, probably by limiting access to the water needed for triacylglycerol hydrolysis, and this resistance of VLDLy particles to hydrolysis by extra-ovarian tissues preserves the triacylglycerol-rich VLDLy for uptake by the developing ovarian follicles (Boyle-Roden and Walzem 2005; Schneider et al. 1990; Walzem 1996). Cross-injection studies on turkeys and chickens using radiolabeled generic VLDL or VLDLy have confirmed that immature and laying birds utilize generic VLDL and VLDLy differently: a greater proportion of generic VLDL is deposited into non-reproductive tissues, whereas more VLDLy is incorporated into ovarian follicles (Bacon et al. 1978). ApoVLDL-II also appears to be responsible for the reduction in the diameter of the VLDLy particles that is thought to be critical for enabling the VLDLy particles to pass through the pores in the granulosa basal lamina of the ovary (see fig. 2.2; Schneider et al. 1990), providing access to the receptors on the oocyte plasma membrane (Griffin and Perry 1985). The granulosa basal lamina has been described as a "molecular sieve" that excludes apoB-based lipoproteins larger than a certain size (>44 nm in the chicken; Perry and Gilbert 1979). The "sieving" property of the granulosa basal lamina precludes the possibility of dietary lipids being transferred *directly* to the yolk since these are transported from the intestine as much larger diameter (150 nm) "portomicrons" (Bensadoun and Rothfeld 1972).

Almost all work on yolk-precursor synthesis and uptake has been done in the domestic chicken—and commonly in estrogenized roosters! In the chicken there is a near complete shift in hepatic lipoprotein metabolism from the synthesis of "generic" VLDL to VLDLy, a change that perhaps evolved as a correlated response to selection for maximum egg

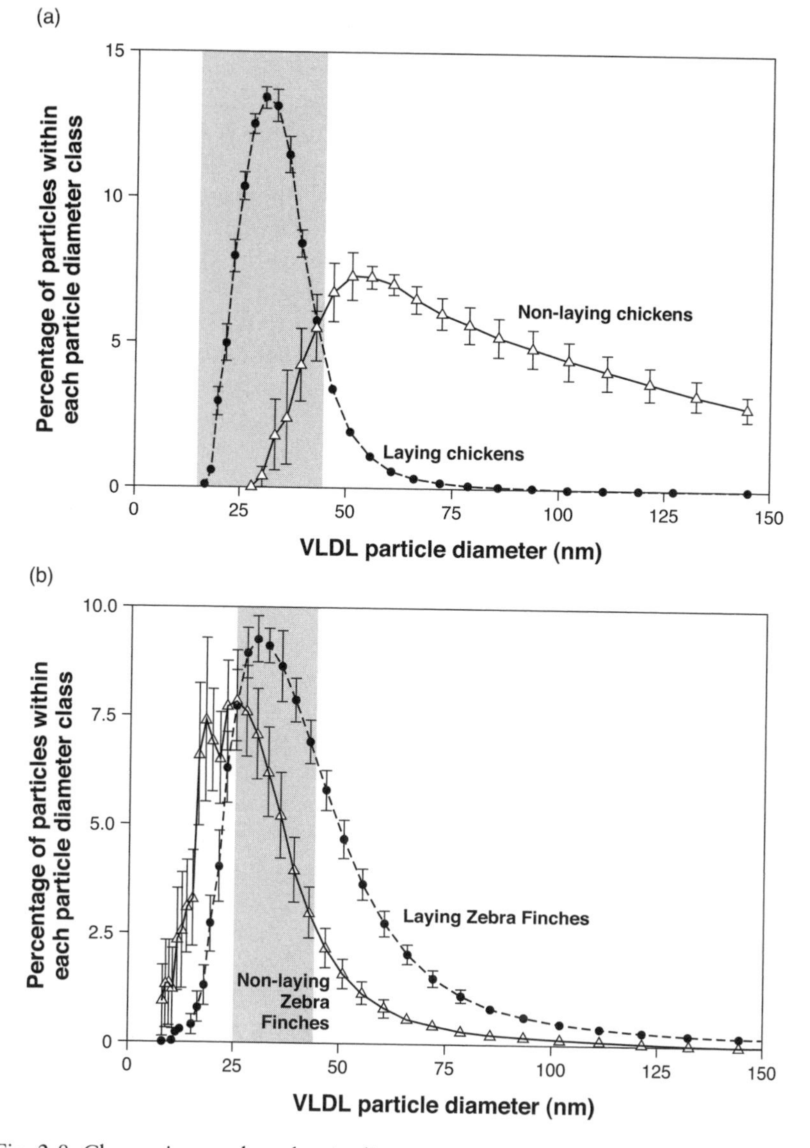

Fig. 2.8. Change in very-low density lipoprotein (VLDL) particle diameter distributions in non-laying (triangles) and laying (circles) females of (a) the domestic hen and (b) the zebra finch. Gray shading indicates the estimated range of particle diameters for yolk-targeted VLDL. (See text for details; reproduced from Salvante et al. 2007a.)

production in this domesticated species (Walzem 1996). Even in laying Japanese quail, another domesticated species but one not so strongly selected for egg production, the distribution of VLDLy particle sizes is more heterogeneous than in the chicken (Chen et al. 1999). In vivo studies in laying chickens suggest that only about 10% of the lipid associated with VLDLy can be hydrolyzed by lipoprotein lipase (Griffin and Hermier 1988). However, this partial hydrolysis, coupled with use of small amounts of generic VLDL that are synthesized by the kidney (Tarugi et al. 1988; Walzem et al. 1999), might be sufficient to meet the hen's own energy requirements given the optimal conditions under which domesticated egg production occurs: light-controlled facilities, a diet regime tailored for egg production, vaccinations against disease, and husbandry practices that eliminate parasites (Etches 1996). Salvante et al. (2007a) suggested that such a radical, near complete shift to VLDLy production would be unlikely to occur in free-living birds, where egg production can occur under far less favorable environmental conditions, such as low ambient temperatures, particularly early in the breeding season. Changes in lipoprotein metabolism during egg production provide a potential mechanism underlying parent-offspring conflict and/or "costs of reproduction" (Bernardo 1996) in that females must balance the costs and benefits of allocation to eggs or offspring (greater smaller-particle VLDLy production) while meeting their own energy needs (maintaining larger generic VLDL particles). This mechanism fits the criteria for a resource-allocation-based trade-off where the currency is the differential allocation of energy-rich lipids (see chapter 7). A more complete shift in hepatic synthesis toward VLDLy, which is less easily metabolized by the female, might compromise the condition of laying females during periods of high energy demand, which in turn might negatively impact subsequent breeding performance (Salvante et al. 2007b).

In captive-breeding zebra finches, the VLDL particle size distributions in males and fed and fasted non-breeding females are very different from those of non-laying chickens. Non-breeding female zebra finches have a much higher proportion of small-diameter VLDL particles (57% are less than 30 nm, compared with only 0.5% in the chicken), and the median VLDL particle size is 22.9 nm (SD 1.5) compared with 68.0 nm (SD 5.1) in the non-laying chicken, which might be related to the higher metabolic rate of small bodied passerines (Salvante et al. 2007a). During egg production there is then a significant shift in VLDL particle size distribution toward *larger* VLDL particle sizes (fig. 2.8). This results in a significantly greater proportion of circulating VLDL particles in the range of VLDLy particle sizes (25–44 nm) that are predicted to be available for yolk uptake based on Griffen and Perry's sieving hypothesis (1985) for the basal lamina (Salvante et al. 2007a). In the only study to date on VLDL metabolism in a free-living bird, VLDL particle sizes in the greater scaup (*Aythya marila*) during egg production were also within the range predicted for VLDLy (25–44 nm), but there was only limited evidence for major shifts in VLDL particle size distribution during the egg production and laying cycle (Gorman et al. 2009). These results suggest that changes in lipoprotein metabolism, specifically an increase in the proportion of VLDLy particles of specific diameters suitable for yolk uptake, might be an essential component of egg production in both domesticated and non-domesticated species. However, egg-laying females in non-domesticated species appear to exhibit less complete shifts in lipid metabolism, perhaps due to constraints on maintaining a sufficient pool of generic VLDL to meet the female's own maintenance requirements.

2.1.4. Mechanisms of Receptor-Mediated Yolk Uptake

Uptake of large amounts of lipoprotein into developing follicles involves the binding of yolk precursors to receptors on the oocyte plasma membrane and their transport across the cell membrane via receptor-mediated endocytosis (see fig. 2.2). A single, 95-kDa receptor protein (VTG/VLDL-R, termed the LR8 receptor in chicken) binds both VLDLy and VTG with high affinity and has been well characterized in the chicken (George et al. 1987; Stifani et al. 1990; Stifani et al. 1988) and quail (Elkin et al. 1995). VTG/VLDL-R is a member of the LDL-receptor gene family and a homologue of the mammalian VLDL receptor (Schneider 2007, 2009). The gene for chicken VTG/VLDL-R has a high degree of sequence similarity to the mammalian LDLR gene (84%), and the zebra finch VTG-R gene shows high sequence identity with the chicken receptor gene (57% identity, Han et al. 2009), with both mapping to the Z chromosome. This receptor is distinct from the LDL receptor that functions to provide lipoprotein-derived cholesterol for steroid hormone synthesis in ovarian follicle cells (Hummel et al. 2003; Schneider 2007). A second chicken-oocyte-specific receptor, the 380-kDa LRP380, also binds VTG and might function with VTG/VLDL-R in regulating oocyte growth (Stifani et al. 1991). It is becoming clear that both the LR8 and LRP380 receptors also bind and transport various non-lipoproteins to the yolk, including α2-macroglobulin, riboflavin-binding protein, and clusterin (MacLachlan et al. 1994; Mahon et al. 1999). Many of these proteins and other factors appear to be transported "piggy-back" fashion in a complex with VTG (MacLachlan et al. 1994), and this arrangement might be a key mechanism for the uptake of numerous factors essential for embryo development (although none of these have been studied as "maternal effects" to date). In birds the VTG/VLDL-R is essential for oocyte growth: a chicken strain, the restricted ovulator (or R/O strain), which carries a single mutation of the VLDL-R locus (a cysteine has been replaced with a serine), fails to produce eggs (Bujo et al. 1995). As a consequence of the failure to deposit yolk precursors, R/O females are hyperlipidemic, having plasma levels of VLDLy and VTG that are four- to five-fold higher than those of normal females.

The regulation and dynamics of the VTG/VLDL receptor in relation to vitellogenesis and follicle development have been characterized in a range of taxa, and these studies have confirmed that VTG/VLDL-R function is a key component of yolk-precursor uptake by developing follicles and may play a role in regulating variation in oocyte growth (Hiramatsu et al. 2004; Nuñez Rodriguez et al. 1996; Perazzolo et al. 1999; Schonbaum et al. 2000). In birds both the receptor protein and mRNA transcript are expressed, and are most abundant, in pre-vitellogenic follicles of pre-laying females (George et al. 1987), and expression of VTG/VLDL-R mRNA is low in the ovary of non-breeding females (Han et al. 2009). In the pre-vitellogenic follicles of chicken, receptors are located centrally, not cortically, which may explain why there is no yolk-precursor uptake at this point, and Shen et al. (1993) proposed that at the onset of rapid yolk development, this pre-existing pool of receptors is redistributed to the periphery of the oocyte, with negligible de novo synthesis of receptors during the final stage of oocyte growth. Recent data from the zebra finch support this hypothesis (Han et al. 2009): the expression of VTG/VLDL-R mRNA was highest in the ovary and smallest pre-F3 follicles, and it decreased linearly from the F3 to F1 hierarchical follicles (fig. 2.9).

Shen et al. (1993) suggested that this pattern of receptor expression was due to the fact that the nuclear and biosynthetic machinery necessary for transcriptional, translational, and post-translational events would be compromised by mechanical distortion during the later stages of rapid oocyte growth, when the space not occupied by yolk is less than 0.1% of the oocyte's volume in a layer 2 μm thick! This redistribution of receptors during RYD might therefore be a key component of VTG/VLDL-R function that influences oocyte growth. Another potential source of variation in yolk uptake is in the recycling of receptors between the cytosol and the oocyte plasma membrane following internalization during endocytosis. The recycling of VTG/VLDL receptors is likely to be important in supporting high rates of yolk uptake in the absence of de novo receptor synthesis in later stages of follicle development (Perazzolo et al. 1999). However, almost nothing is known about individual variation in, or hormonal regulation of, VTG/VLDL-R expression and/or recycling in birds (see section 4.4.1).

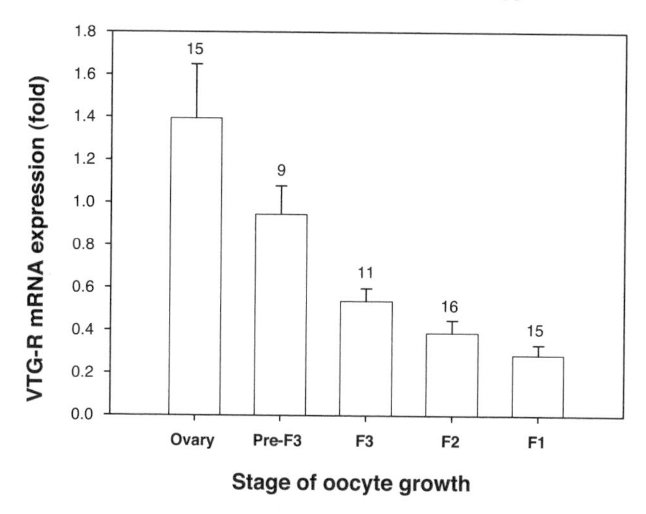

Fig. 2.9. Variation in expression of VTG/VLDL-R mRNA (normalized to β-actin mRNA) in the zebra finch in relation to follicle development for the ovary (including white, pre-vitellogenic follicles) and vitellogenic follicles of different stages (from the smallest pre-F3 to the largest F1 follicle). (Reproduced from Han et al. 2009.)

2.2. Oviduct Structure and Function

The avian oviduct, which extends from the ovary to the cloaca, is a highly differentiated organ comprising five morphologically and functionally distinct regions: (a) the infundibulum, (b) magnum, (c) isthmus, (d) shell gland or uterus, and (e) vagina (A. Gilbert 1979; A. King and McLelland 1984; Solomon 1983; see fig. 2.1). The oviduct comprises a tube of connective tissue, in which the major blood vessels are located; circular and longitudinal muscle layers; and a mucosal layer, which is extensively folded into ridges. The mucosal layer consists of an outer secretory layer and an inner epithelial layer with tubular gland cells, which form 90% of the oviduct cells and synthesize and secrete albumin proteins; ciliated cells; and avidin-secreting, non-ciliated goblet cells (Burley and Vadehra 1989; Etches 1996). Egg transport within the oviduct is effected primarily by peristaltic contractions of the longitudinal and circular muscles (which function as stretch receptors responding to mechanical stretch from the ovum), perhaps aided by ciliary action, with hormonal modulation of muscle contraction by prostaglandins E and F and arginine vasotocin.

The top part of the oviduct, the infundibulum, is not connected physically to the ovary, but it engulfs the ovulated yolk and oocyte following ovulation, perhaps facilitated by dense populations of ciliated cells. Although the specific mechanism by which the ovum is transferred is not known, it is not dependent on ovulation per se since inanimate objects placed in the body cavity can be transferred to the oviduct (A. Gilbert 1979). The ovum spends 15–30 minutes in the infundibulum, where the first layers of albumen are deposited in the lower tubular portion. Fertilization of the oocyte therefore has to occur very shortly after ovulation in the upper funnel-shaped part of the infundibulum before albumen deposition prevents the sperm from reaching the oocyte. The fertilized ovum then passes into the largest part of the oviduct, the magnum (comprising 65% of total oviduct dry mass in laying female zebra finches), where the remainder of the albumen is laid down over a 2–3-hour period (all transit times are for chicken). The main albumen proteins include ovalbumin (54% total protein), ovotransferrin (12%), ovomucoid (11%), and lysozyme (3%) (Burley and Vadehra 1989), although albumen is actually a complex mixture of 40 different proteins, each of which has specific functions in the embryo (more than enough to fuel the "maternal effects" field for several more decades!). A clearly distinguishable translucent band, which lacks tubular gland cells, separates the magnum from the isthmus. Although the isthmus is the shortest region of the oviduct, the ovum remains here for 1–2 hours, while the inner and outer shell membranes are added. The shell gland, or uterus, has thick walls, giving it a bulbous appearance, and it is distinguished from the distal end of the isthmus by the "red region" (A. Gilbert 1979). The egg remains here for 18–20 hours to complete the formation of the shell. During the early stages of shell formation, prior to calcification, the tubular glands in the distal part of the uterus secrete water and ions, which are transferred to the egg to give it its final, mature size (a process known as "plumping"). Calcification involves high rates of calcium carbonate secretion, and the shell gland sequesters and concentrates ionized Ca^{2+} by means of a calcium-binding protein located in the tubular gland cells, a process influenced by vitamin D and parathyroid hormone (Etches 1996). Finally, shell pigments and the cuticle are deposited by ciliated and non-ciliated epithelial cells 3–0.5 hours prior to oviposition (A. Johnson 2000). The main egg pigments are porphyrins, which are associated with red and brown egg colors, and biliverdin, which gives the eggs blue and green coloration (Kennedy and Vevers 1975). The mechanism(s) determining patterning and egg coloration is still not known, although it has been suggested that rotation of the egg in the shell gland might contribute to streaking.

In poultry, an egg takes approximately 25·hours to pass down the entire length of the oviduct, and egg albumen is synthesized de novo in the oviduct during this period. Albumen secretion appears to proceed continuously, although protein synthesis increases when an egg is in the magnum in the chicken, and the magnum does not become depleted of albumen, containing sufficient protein for the deposition of two eggs just before an egg enters (Etches 1996; Yu and Marquardt 1973). Ovarian growth and the synthesis of albumen proteins are both dependent on the actions of gonadal steroids, but deposition of albumen is thought to be directly due to mechanical stimulation or distortion caused by the ovum moving down the oviduct (A. Gilbert 1979). Lavelin et al. (2001) confirmed that mechanical stress regulates gene expression in the uterus, but only in laying birds, thus suggesting a requirement for "priming" the mechanical sensitivity, perhaps involving estrogens or progesterone (Lavelin et al. 2002). This mechanism suggests that the *amount* of albumen that gets deposited in each egg will be correlated directly with yolk size, with larger yolks providing a greater mechanical stimulus for albumen deposition. In other words, yolk size and egg albumen content are not regulated independently; rather, the former determines the latter, with yolk size therefore playing a key role in determining the size and composition of the whole egg (T. Williams, Hill, and Walzem 2001).

The size of the oviduct varies markedly with sexual activity and is tightly coupled to the reproductive stage: the maximum oviduct size is attained only 24 hours before the first follicle is ovulated, and the oviduct regresses as soon as the last follicle is ovulated (fig. 2.10). In the zebra finch the mass of the oviduct increases from less than 50 mg to 650 mg over 5 days at the onset of egg production and has a highly regulated size-function relationship (T. Williams and Ames 2004). Specifically, at the end of egg-laying, this linear organ regresses rapidly from the top down as soon as the more proximal regions have completed their function but while the distal regions are still functional. In the approximately 24-hour period after the last follicle is ovulated, but before the last egg is laid, the proximal infundibulum/magnum and the isthmus regions shrink by 56% and 38%, respectively. During the same period there is no change in the combined mass of the shell gland and vagina, but these sections will regress by 34% over the course of about 24·hours after the last oviposition. This maintenance of the functional capacity of the oviduct until the last oviposition is supported by the fact that there is no decline in the absolute or relative mass of oviduct-dependent egg components (shell and albumen) for later-laid eggs. Yu and Marquardt (1973) showed that the rate of growth of the magnum during oviduct development is much greater than that of more distal sections of the oviduct; that is, the pattern of growth at the start of the laying cycle also closely reflects functional demands. T. Williams and Ames (2004) suggested that this arrangement would minimize the time that the different components of the oviduct are maintained in a functional state and thus reduce the energy cost of maintaining the complete oviduct. Although there appears to be little known

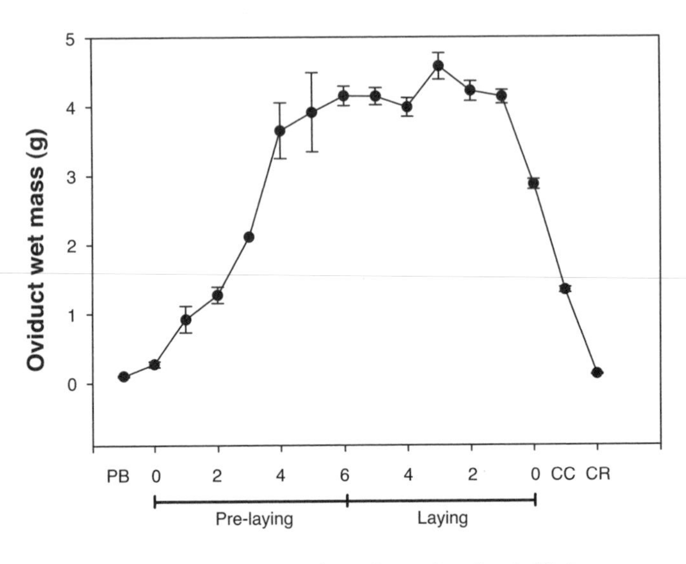

Fig. 2.10. Variation in oviduct mass during the laying cycle in European starlings. Numbers on x-axis indicate the number of yolky follicles; PB, pre-breeding; CC, clutch completion; CR, chick-rearing. (Based on data in Vézina and Williams 2003.)

about the specific mechanisms involved in oviduct regression, that study suggests that these mechanisms must be specific to each region of the oviduct (e.g., differential timing of receptor expression or apoptosis) rather than involving a more generic, humoral signal such as down-regulation of plasma estrogen or progesterone levels (Burley and Vadehra, 1989).

2.3. Regulation of the Timing of Egg-Laying (Oviposition)

In many species, follicle recruitment into the pre-ovulatory, vitellogenic follicle hierarchy occurs with circadian rhythmicity every 24–26 hours, ovulation of the largest follicle occurs every 24–26 hours, and the yolk takes approximately 24–26 hours to pass down the entire length of the oviduct. Egg-laying, or oviposition, therefore occurs in many species with daily and circadian rhythmicity. Information on the daily timing of egg-laying is available for many free-living species, and it also reveals a high level of between-species variability in the timing and frequency of oviposition (Astheimer 1985). However, the mechanism regulating timing of oviposition has been studied in detail only in the domestic chicken, and even here, many specifics are not fully understood. Detailed studies on hens have provided the basis for a model of timing of ovulation and oviposition whereby a circadian rhythm of neuroendocrine events (e.g., luteinizing hormone [LH] release) generates a limited eight-hour window, or "open period," during which a pre-ovulatory LH and progesterone surge can occur to initiate ovulation. For example, in chickens on a 14L:10D cycle, with lights on at 0600, ovulations (follicle rupture) are restricted to the 8-hour period from 0600 to 1400. Follicles that attain fully mature size in synchrony with this circadian open period are ovulated and laid approximately 24 hours later (fig. 2.11). However, for follicles that mature outside of the open period, ovulation is delayed until the start of the next open period, such that the interval to the next oviposition is about 40 hours, and there is a "laying skip," or "laying gap," with no egg laid on the intervening day. A second key component of this model in chickens is that the duration of the ovulatory cycle in a laying sequence, between one ovulation and the next, is 24–28 hours, with only a small proportion of laying females approximating a 24-hour ovulatory cycle. This means that successive follicles mature later and later in the open period, until the point is reached where an F1 follicle matures after the end of the open period (fig. 2.11). Since this follicle is then ovulated at the start of the next open period, synchrony between follicle development and the open period is effectively "reset" with another sequence of eggs laid on successive days before the next laying skip. The neuroendocrine and circadian underpinnings of this model remain poorly understood but might involve a gradual decrease in progesterone-synthesizing activity in response to LH as the laying sequence progresses, along with a corresponding increase in the time taken to reach some "threshold" required for the acute, pre-ovulatory release of LH (Kamiyoshi and Tanaka 1983).

This oviposition model makes two important predictions if it applies to birds in general: (a) egg-laying should be restricted to a limited period each day, equivalent to an "open period," and (b) laying gaps should occur, but these should be more common with later-laid eggs, assuming that first eggs are ovulated early in the open period and that the duration of the ovulatory cycle is greater than 24 hours. There is some support for both of these predictions in free-living birds (see below and section 2.3.1); however, available data suggest that there are also significant species differences, at least in the timing of the restricted period of oviposition, and potentially in the duration of the ovulatory cycle. Even in other domesticated species, although successive ovipositions occur during a restricted period, the timing of laying differs from that in the chicken: turkey, 1030 –1800, lights on 0600; Japanese quail, 1530–1800, lights on 0600; Chukar partridge (*Alectoris chukar*), 0700–1500, lights on 05:00

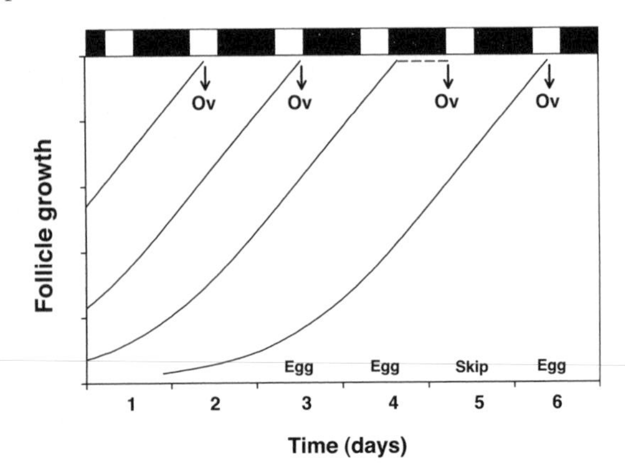

Fig. 2.11. The open period model for regulation of oviposition and laying skips in the domestic hen. White bars indicate the open period within each 24-hour cycle; Ov, timing of ovulation. (Based on Etches 1996.)

(Etches 1996). This difference suggests more variable circadian timing and duration of the open period, from 3.5 hours in quail up to 8 hours in the chicken. In contrast, in captive-breeding female zebra finches on 14L:10D (lights on at 0700), 95% of eggs (n = 103) were laid between 0700 and 0900, and the mean daily interval between eggs was 24.0 ± 0.1 hours (range 23.0–24.8; Christians and Williams 2001).

Many free-living passerines lay one egg per day, and in some species, the daily timing of egg-laying is highly synchronous and occurs very soon after sunrise (S. Gill 2003; Oppenheimer et al. 1996). For example, female yellow warblers (*Dendroica petechia*) return to the nest within 10 minutes of sunrise, and oviposition is completed within 30 minutes, with little variation among individuals (fig. 2.12). In these species, therefore, the interval between successive ovipositions, and presumably the duration of the ovulatory cycle, must be very close to 24 hours. In contrast, other passerines lay less synchronously and egg-laying occurs later in the day. European starlings lay between 0600 and 1200, with most eggs laid between 0800 and 1000 (Feare et al. 1982), and in the American robin (*Turdus migratorius*), the mean time of laying in one study was 1132, with eggs laid between 0730 and 1800 (Weatherhead et al. 1991). Some large-bodied non-passerines, including most ducks, such as the common eider (*Somateria mollisima*; Watson et al. 1993), also have laying intervals of 1 day, but in the eider the mean estimated time of laying was 1349 (1230–1500). Most other non-passerines and some sub-oscine Passeriforms (e.g., Tyrannidae, Pipridae, Cotingidae) have laying intervals that are typically 2–4 days (Astheimer 1985), and the Ancient murrelet (*Synthliboramphus antiquus*) lays a two-egg clutch with a 7-day laying

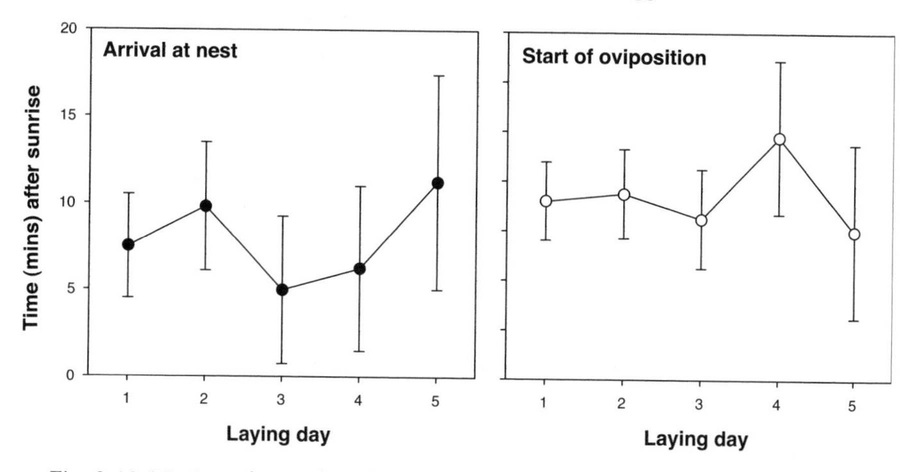

Fig. 2.12. Timing of arrival at the nest (left) and timing of oviposition (right) in relation to sunrise in the yellow warbler. (Based on data in McMaster et al. 1999.)

interval. In some species it is clear that laying intervals greater than 24 hours are associated with longer intervals between the *initiation* of yolk development (follicle recruitment) rather than with *differences* in the duration of RYD for successive yolks. For example, in the Adelie penguin (*Pygoscelis adeliae*), the second and third follicles initiate RYD 3 days apart and this timing dictates a 3-day laying interval in this species (Astheimer and Grau 1985). Laying intervals that are multiples of 24 hours could therefore still be explained by a circadian-based mechanism (albeit somewhat modified from that for the chicken).

The adaptive significance, if any, of variation in timing of oviposition is poorly understood. Schifferli (1979) provided a general explanation for birds laying early in the day, suggesting that females can then avoid having an egg in the shell gland during the day, which might be susceptible to damage or might constrain normal diurnal activity. This hypothesis is clearly not consistent with data for the many species that actually lay later in the day. Astheimer (1985) suggested that long laying intervals allowed for, or might be caused by, extension of the total duration of yolk formation and albumen synthesis, which would decrease overlap in the development of successive follicles and possibly lead to a lower daily cumulative demand for egg production and potentially lower daily energy expenditure (DEE) (see fig. 2 in Astheimer 1985). Clearly, given the extensive, and largely unexplained, interspecific variation in the timing and frequency of oviposition, a robust phylogenetic analysis of these comparative data would be worthwhile and informative. Similarly, from a mechanistic perspective, intraspecific variation in timing of oviposition is virtually unstudied. While in some species there appears to be little phenotypic variation in timing (McMaster et al. 1999), in other species both the timing of oviposition and therefore potentially also the duration of the ovulatory cycle appear to vary considerably among individuals.

2.3.1. Follicle Atresia and Laying Skips

The open-period model of oviposition provides one possible explanation for why birds would skip a day when laying eggs, but an alternative explanation is that one or more pre-ovulatory follicles fails to develop and gets reabsorbed, leaving a gap in the sequence of developing follicles. This question is important since follicular atresia and reabsorption have been proposed as mechanisms for manipulation of offspring sex ratio, with "selective abortion" of follicles of a particular offspring sex (Pike and Petrie 2003). The apparent widespread occurrence and variability of "facultative" manipulation of the sex ratio requires that the timing of follicle atresia can be highly variable and that, under some circumstances, the frequency of atresia can be high. Is this the case? Gaps in the laying of eggs has been well documented in a range of species. In captive-breeding female zebra finches, among 349 clutches of 4–7 eggs, 268 (76.8%) were completed without females skipping a day, in 76 clutches (21.8%) females had one 24-hr interval in which no egg was laid, and in 5 clutches (1.4%) females skipped two consecutive days before laying the next egg. Females were more likely to skip laying an egg later in the laying sequence, which is consistent with the open-period model. Laying skips occurred in 20%–70% of clutches in other species laying large clutches of 10–12 eggs (Kennamer et al. 1997; Nilsson and Svensson 1993). In contrast, skips occurred in 8%–15% of clutches (n = 78 and 65) in the European starling, which lays only a 5-egg clutch on average, and in only 4 of 32 clutches (12.5%) in house martins (*Delichon urbica*), which lay 3–4 eggs (Bryant 1975). In some species, laying gaps are more frequent later in the laying sequence (Lessells et al. 2002), while in other closely related species the reverse is true (Nilsson and Svensson 1993). Few data are available on the frequency of follicle atresia itself, and it should not be assumed that all laying skips involve atresia. Atresia appears to occur only at low frequency in pre-ovulatory, vitellogenic follicles (A. Gilbert et al. 1983; A. Johnson and Woods 2007); for example, in European starlings the frequency of atresia was about 40% for F4 follicles and 10% for F3 follicles but zero for F1 and F2 follicles (Challenger et al. 2001). These observations suggest that atresia is not sufficiently frequent or variable to provide a widespread mechanism of sex-ratio adjustment; for example, "selective abortion" could allow sex-ratio adjustments late in the laying cycle, through atresia of F3 or F4 follicles, but not early in the laying cycle (F1 or F2 follicles).

Many studies have assumed that laying skips reflect environmental stress, with the female temporarily suspending egg-laying as a response to low ambient temperatures (Ta) or low food availability. In house martins, laying skips occurred in all clutches ($n = 4$) initiated during a period of low relative abundance of aerial insects, but in no clutches ($n = 28$) initiated when insect abundance was high (Bryant 1975). In experimental studies, laying skips were less frequent when nest-box temperatures were increased (Yom-Tov and Wright 1993) and when laying females were provided with supplemental food (Nilsson and Svensson 1993). However, Lessells et al. (2002) showed that laying skips occurred throughout the laying period in blue tits (*Cyanistes caeruleus*) and were not clustered around specific dates or low temperature events, as would be predicted if they reflected periods of environmental stress. In addition, the occurrence of laying skips seen even in captive birds with ad lib food is not consistent with nutritional stress. It seems more parsimonious to suggest that rather than reflecting constraint due to low resource availability, laying skips are an inevitable consequence of the circadian-based regulatory system, with a limited open period and an ovulatory cycle longer than 24 hours.

2.4. Hormonal Control of Ovarian and Oviduct Function

It is widely assumed that ovarian function is primarily regulated by "upstream" events occurring at the level of the hypothalamus and pituitary (the HPG axis) that, in turn, are regulated by higher neural centers in the brain (central nervous system) that receive and integrate information from a wide range of physiological and environmental cues to time breeding events (see section 2.6 and chapter 3). This idea of "top-down" control is embedded in a century of endocrinology and is so widespread that rarely are questions asked about "downstream" events at the level of the ovary, or possible intra-ovarian mechanisms. It seems at least plausible that downstream mechanisms play a critical role in timing of the final stages of female reproductive development and egg formation. Variation in the hormonal regulation of female reproduction at the level of the ovary, oviduct, or liver might also explain individual, or phenotypic, variation in reproductive traits values (e.g., egg or clutch size). Similar ideas about the "autonomy" of ovarian function have been proposed in relation to circadian rhythmicity (Nakao et al. 2007) and ageing (Hsin and Kenyon 1999); there are no shortage of systemic and intra-ovarian hormones that might play such a role (Onagbesan et al. 2009); and newly identified reproductive hormones have been localized to the ovary (Bentley et al. 2008). For these reasons I will focus here on "direct" regulation of ovary and oviduct function by peripheral and intra-ovarian hormones.

Then I will briefly describe central regulation by the central nervous system, hypothalamus, and pituitary, concentrating on studies of female birds and ovarian function. Finally, I will describe how environmental information is hormonally integrated with the reproductive axis, focusing again on ovarian function.

2.4.1. The Ovary

The expression of steroidogenic enzymes and production of ovarian steroid hormones, which are essential for initiation of vitellogenesis by the liver, has been well characterized in birds, although almost entirely from work on the domestic fowl (A. Johnson and Woods 2007). Vitellogenesis, or yolk-precursor production, is estrogen dependent, and the treatment of chicks and immature and non-breeding females with exogenous estradiol rapidly increases plasma yolk-precursor levels (Mitchell and Carlisle 1991; Wallace 1985; T. Williams and Martyniuk 2000). Earlier studies used pharmacological doses of estradiol (up to 10–50 mg/kg), but treatment of non-breeding female zebra finches with 1 mg/g estradiol-17β (E2) increases plasma vitellogenin and VLDL to the physiological levels seen in egg-laying females (T. Williams and Martyniuk 2000). Interestingly, male birds retain the ability to synthesize and secrete yolk precursors when treated with estrogens (Bergink et al. 1974), and, ironically, estrogenized roosters have been the model of choice for much of the detailed molecular and biochemical characterization of lipoproteins and receptors described above. Estrogens initiate gene transcription and synthesis of VTG and apo-VLDL-II proteins and also reprogram hepatic lipid and protein synthesis to support the transition from generic VLDL to VLDLy synthesis, for example, estrogen stabilizes apoVLDL-II mRNA, increases intracellular phospholipid transfer proteins that support VLDLy assembly, and alters phospholipid and fatty acid synthesis and composition to provide a supply of lipids for VLDLy assembly (Walzem 1996 and references therein).

Vitellogenesis in response to exogenous estrogens can involve a so-called memory effect (Bergink et al. 1974), in which a second treatment with estradiol (secondary stimulation) causes a more rapid and larger increase in plasma vitellogenin than that seen with the initial exposure to estradiol (primary response). Early work on estrogenized roosters showed that there are large differences among individuals in the timing and magnitude of the responses to both primary and secondary stimulation, as well as differences in the time-dependent response of the different vitellogenins (S. Wang and Williams 1983). In wild-caught European starlings, females exposed to artificially long photoperiods (18L:6D) showed increased oviduct growth, and plasma VTG levels, in response to secondary stimulation with exogenous E2 if they had previously been exposed to E2 on short days (T. Williams, unpublished data), results that suggest that this effect also occurs in non-poultry species (see also Sockman et al. 2004). This memory effect appears to involve changes in estrogen-receptor expression and chromatin remodeling and estrogen-dependent changes in the promoter regions of the VTG and apo-VLDLII genes (Edinger et al. 1997). The relevance of these observations to individual variation in the onset, timing, and magnitude of vitellogenesis, or potential differences between reproductively naive females and experienced females, in relation to fitness has not, to my knowledge, been investigated.

In free-living European starlings, levels of plasma E2 increase rapidly at the onset of yolk development in parallel with the increase in plasma VTG levels, reaching a maximum in females with a complete follicle hierarchy (≥4 yolky follicles; fig. 2.13). Plasma E2 levels then decrease linearly throughout the later stages of follicle development, returning to pre-breeding values *before* the final yolky follicle is ovulated. In contrast, plasma VTG levels remain elevated for a further 48 hours and only decrease to baseline at clutch completion (T. Williams, Kitaysky, and Vézina 2004) . A similar pattern was seen for fecal E2 levels in canaries (*Serinus canaria*; Sockman and Schwabl 1999). T. Williams, Kitaysky, and Vézina (2004) suggested that maintenance of high circulating levels of yolk precursors until after the last yolky follicle has ovulated in the absence of elevated plasma E2 levels was due to the relatively long half-life of VTG (1–2 days) coupled with the high circulating yolk-precursor levels earlier in RYD (Redshaw and Follett 1976; Schultz et al. 2001).

In contrast to *initiation* of vitellogenesis, hormonal control of VTG/VLDL receptor function and yolk uptake remains poorly understood in any avian species. Early studies in the amphibian *Xenopus laevis* suggested that insulin, gonadotropins (hCG), and VTG itself increase the rate of VTG uptake and involve hormone-induced modulation of the VTG/VLDL-R internalization rate rather than changes in receptor number (Opresko and Wiley 1987). Tyler et al. (1990) suggested that higher plasma levels of VTG itself might induce up-regulation of VTG receptor numbers and/or an increase in the rate of turnover of VTG-R in the rainbow trout (*Oncorhynchus mykiss*). Conversely, Follett and Redshaw (1974) reported that estradiol alone *decreased* the accumulation of radioactive vitellogenin by the ovaries of *Xenopus*, whereas estradiol together with FSH substantially elevated the rate of yolk-precursor uptake. They suggested that exogenous estradiol may have lowered endogenous secretion of FSH by the pituitary, thereby reducing the endocrine (FSH) stimulation of oocyte growth. In support of this hypothesis, S. Palmer and Bahr (1992) found that exogenous FSH increased yolk deposition in the

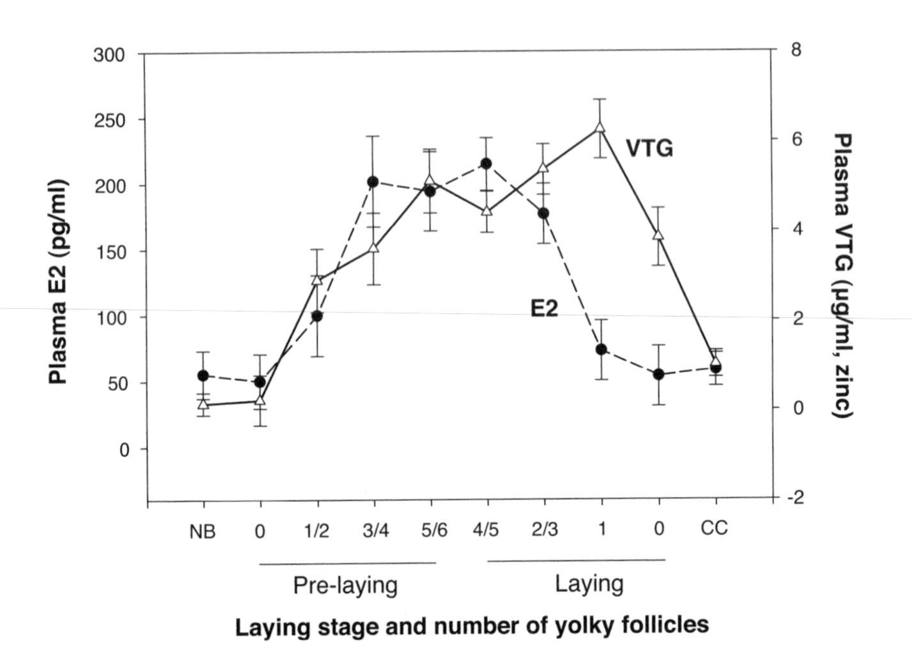

Fig. 2.13. Variation in plasma levels of 17β-estradiol (E2, circles) and vitellogenin (VTG, triangles) during the laying cycle in the female European starling. Numbers on x-axis indicate the number of yolky follicles; NB, non-breeder; CC, clutch completion. (Reproduced from T. Williams, Kitaysky, and Vézina 2004.)

yolky follicles in chickens, particularly at low doses. However, Christians and Williams (2002) found that treatment of female zebra finches with porcine FSH *decreased* both egg size (by 10%) and clutch size, perhaps via a reduction in the number of developing follicles. Follicle-stimulating hormone is assumed to be a key hormone in the regulation of follicle development, and it has been implicated in the regulation of egg size, clutch size, and in the trade-off between these traits in lizards (Sinervo 1999; Sinervo and Licht 1991). However, we still know surprisingly little about how FSH regulates folliculogenesis and follicle or yolk *size* in birds (Bentley et al. 2007).

Many other hormones, growth factors, and/or receptors are known to be expressed in the ovary, but the functions of these remains largely unknown, especially in relation to control of the VTG/VLDL-R (A. Johnson 2000; Onagbesan et al. 2009). The ovary is the main site for the synthesis of inhibins and activins , which are under the control of FSH, but the function of these hormones in follicular development is poorly defined (Knight et al. 2005; Onagbesan et al. 2004). Vasoactive intestinal peptide (VIP), the releasing factor for prolactin, is localized in the nerve terminals of the thecal layer and may function in steroidogenesis or differentiation of pre-hierarchical follicles, but it has not been implicated in regulation of rapid follicle growth per se (Flaws et al. 1995). Other growth factors and their receptors are expressed in ovarian tissue, suggesting that these might be involved in regulating ovarian function via intra-ovarian paracrine effects. These include members of the insulin-like growth factor (IGF), EGF, transforming growth factor-β, fibroblast growth factors, and tumor necrosis factor-α families (Onagbesan et al. 2009). The primary functions of many of these growth factors appears to be in regulating granulosa cell differentiation and steroidogenesis or differentiation and apoptosis in pre-hierarchical follicles, rather than RYD or VTG/VLDL receptor function (Onagbesan et al. 2009; Woods et al. 2007). However, IGF-I and IGF-II might be involved in regulating reproductive performance in the laying chicken (Onagbesan et al. 1999). The expression of several growth factors, such as IGF-I (Onagbesan et al. 1999) and inhibin/activin (P. Johnson et al. 2005), varies systematically during follicle development from F4 to F1 follicles, and Nagaraja et al. (2000) showed that variation in IGF-I gene expression was related to that in egg weight, but not laying rate, in the chicken. Conversely, Kim et al. (2004) reported significant differences in the egg-laying productivity of different IGF genotypes in Korean native chickens. Without doubt, the list of "candidate" regulatory factors will rapidly increase once genomic analysis of the ovary becomes possible in birds. A comprehensive survey of the genes involved in maturation and development of the rainbow trout ovary showed that approximately 240 genes are developmentally regulated during this process (von Schalburg et al. 2005).

Glucocorticoid hormones, such as corticosterone, play central roles in maintaining a daily homeostatic energy balance at baseline levels as well as in mediating physiological and behavioral responses to stress. Both functions could represent important factors that laying females might integrate with ovarian function. Early studies on chickens investigated the effects of corticosterone on ovary function, often using pharmacological doses (Etches 1996). More recent studies of the corticosterone-mediated stress response have focused on parental care (see chapter 6), and the specific role of corticosterone in regulation of ovarian function remains largely unknown. In seasonally breeding avian species, the predominant pattern is for baseline corticosterone and, in some species, stress-induced corticosterone levels, to be highest during breeding compared with pre- or post-breeding stages (fig. 2.14; though on average, stress-induced levels are similar in breeding and post-breeding birds, Romero 2002). Several studies have reported seasonal peaks in baseline plasma corticosterone coincident with periods of egg production (Hegner and

Wingfield 1986; Silverin and Wingfield 1982; Wingfield 1994; Wingfield and Farner 1978), but since these have been correlative studies, it is not clear if elevated plasma corticosterone is associated with "positive" physiological changes associated with egg production, for example, resource mobilization, or if they reflect a "negative" response to stress during egg production. It is routinely asserted that glucocorticoids can suppress the gonadal axis (Sapolsky et al. 2000), but such claims typically refer to effects of *stress-induced hormone levels*, and the assertion is based largely on studies of non-avian vertebrates. The physiological functions of changes in baseline corticosterone (e.g., in response to "normal" changes in metabolic demand) and of stress-induced corticosterone are very different, but this distinction has not always been fully appreciated in experimental studies of egg-laying females (Breuner et al. 2008; Sapolsky et al. 2000). Deeley et al. (1993) suggested that glucocorticoids and estrogens might up-regulate the VTG II gene, whereas glucocorticoids inhibit apoVLDL-II gene expression (based on in vitro studies). Walzem (1996) rationalized Deeley et al.'s results by suggesting that suppression of apoVLDL-II might redirect triglycerides for the chicken's own use as part of a generalized stress response, whereas continued expression of VTG might facilitate the reestablishment of yolk deposition once the stressor is removed. However, Petitte and Etches (1991) reported that any elevation of plasma corticosterone above the "normal physiological range" inhibited rapid follicle growth, causing ovarian regression and decreases in plasma estrogens and progesterone in hens. In female zebra finches exogenous corticosterone treatment, using silastic implants, elevated plasma corticosterone to high physiological (stress-induced) levels and was associated with a smaller proportion of females who initiated laying and with a delayed onset of egg-laying (Salvante and Williams 2003). Furthermore, corticosterone-treated females had higher plasma triglyceride levels but lower plasma VTG levels 7 days post-treatment, a finding consistent with Walzem's suggestion (1996) that corticosterone inhibits yolk-precursor production and shifts lipid metabolism away from production of VLDLy and toward production of generic VLDL. However, even though corticosterone-treated female zebra finches had significantly higher plasma corticosterone levels during laying (45.9 ± 9.0 ng/ml) compared with controls (7.9 ± 6.8 ng/ml), there was no difference in circulating yolk precursors at this stage or in mean egg mass, clutch size, or egg composition (Salvante and Williams 2003). These studies suggest that high stress-induced corticosterone levels can inhibit, or perhaps delay, vitellogenesis and egg-laying, although clearly some females can lay normally even with high circulating corticosterone levels. However, these studies do not provide information on the regulatory effects of changes in *baseline* corticosterone on ovary function. Love et al. (2005)

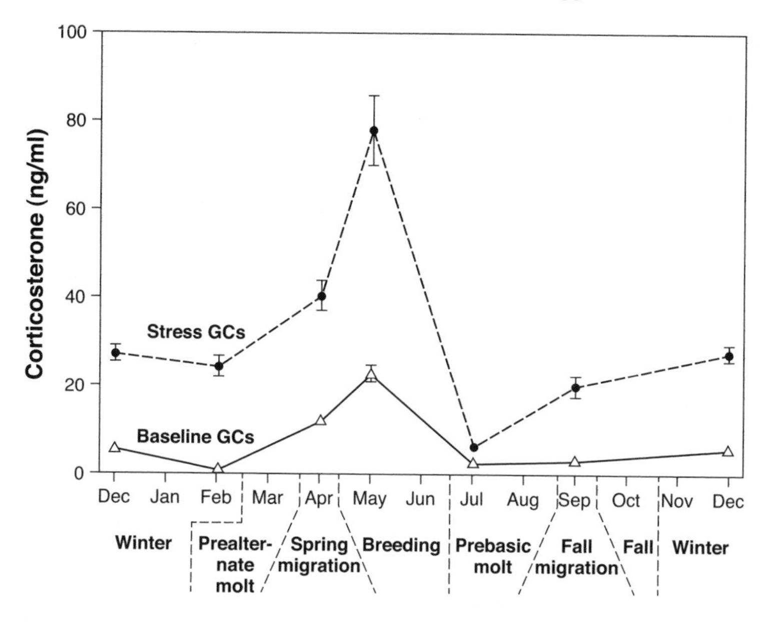

Fig. 2.14. Seasonal variation in baseline and stress-induced plasma glucocorticoid (corticosterone) levels in white-crowned sparrows. (Reproduced from Romero 2002, with permission of Elsevier.)

have conducted the only study in which females were treated with corticosterone *before* laying and in which plasma corticosterone was elevated within *baseline* levels. In this study, with free-living European starlings, corticosterone treatment had no effect on the timing of (re)laying, the probability of (re)laying, egg size, or clutch size.

2.4.2. The Oviduct

Estrogens, specifically E2, and progesterone are considered to be the primary hormones involved in growth, differentiation, and function of the oviduct. Oviduct mass increases in a very sensitive, dose-dependent manner in response to exogenous estrogen treatment in non-breeding adult females (T. Williams and Martyniuk 2000) and chicks (Millam et al. 2002), and the chick oviduct has been an important model system in understanding the mechanisms of action of steroid hormones (Dougherty and Sanders 2005). However, in females in captivity, maximum oviduct growth is typically far less than that seen in free-living, laying females. Many studies of the hormonal regulation of oviduct function using the immature chick oviduct involve only the initial stages of oviduct growth—up to approximately 30 mg wet weight, which is less than 0.1% of the oviduct mass of the laying chicken (50 g, Etches 1996). In female European starlings exposed to stimulatory long days (18L:6D) *combined with* exogenous estradiol treatment (1 μg/g), the oviduct mass averaged only 0.45 g, which is approximately 11% of mean oviduct mass (4 g) in free-living birds. This result suggests that, as with the final critical stages of ovarian development, other factors that are required for full oviduct development are lacking in captive birds. However, in general these experimental studies in captive or immature females tend to assume that continued growth and development of the fully mature oviduct in laying females is also estrogen- and progesterone-dependent. In the immature chick oviduct the appearance of ovalbumin-secreting cells and release of protein into the lumen occur after 4 and 6 days, respectively, of exogenous E2 stimulation (Kohler et al. 1969). This observation agrees with the 4–5-day interval between the initial increase in plasma estradiol and functioning of the oviduct after the first ovulation in free-living birds (compare figs 2.10 and 2.13). As with onset of vitellogenesis, there is a clear "memory effect" for E2 stimulation of oviduct function, with a much more rapid onset of ovalbumin secretion during the second exposure to estrogens (Schimke et al. 1975). Estrogen stimulates cytodifferentiation of the tubular gland cells, ciliated cells, and goblet cells, gene transcription, and synthesis and glycosylation of albumin proteins, along with hyperplasia and, to a lesser extent, cell hypertrophy (Yu and Marquardt 1973). Estrogen also "primes" the oviduct tissue for the subsequent effects of progesterone via the up-regulation of progesterone receptors. The effects of progesterone on the oviduct appear to require synergy with estrogens because progesterone alone stimulates only very little growth in the chick oviduct. Indeed, chronic treatment with exogenous progesterone causes a decrease in the oviduct mass, as well as reduced ovarian mass, a smaller number of hierarchical follicles, reduced F1 follicle size, and lower egg-production rates in laying hens, turkey and, Japanese quail (Bacon and Liu 2004; Liu and Bacon 2004, 2005).

2.5. Hypothalamic and Pituitary Regulation of Gonadal Function

Ovarian and oviduct function ultimately rely on neuroendocrine information that is, most probably, relayed to the reproductive system from the pituitary, hypothalamus, and higher neural centers in the brain: the HPG axis (fig. 2.15). The HPG axis integrates sensory information, both environmental and physiological, and transduces it into humoral hormonal signals, which then modulate gonadal function. The HPG axis can also be affected by inputs from other major endocrine axes, such as the hypothalamic-pituitary-thyroid axis and the hypothalamic-pituitary-adrenal axis. The structure and function of the hypothalamus and pituitary and the role of the HPG axis in regulating seasonal breeding in birds have been extensively reviewed elsewhere (Bentley et al. 2007; A. Dawson 2008; A. Dawson et al. 2001), so only a brief overview will be provided here. Gonadal maturation is ultimately determined by the secretion rate of chicken gonadotropin-releasing hormone-I (cGnRH-I), and possibly gonadotropin-inhibiting hormone (GnIH; see below), from the hypothalamus in response to the integration of environmental stimuli. Two forms of GnRH have been identified in poultry and several free-living species: cGnRH-I and cGnRH-II, both of which have potent effects on LH release (Sharp and Ciccone 2005). A third GnRH (ir-lamprey GnRH-III), which is also gonadotrophic, has recently been identified in the hypothalamus of songbirds (Bentley et al. 2004). cGnRH-I is synthesized in the cell bodies of specialized neurosecretory cells (which form loosely defined "nuclei" in the hypothalamus), released at the median eminence, and transported to the anterior pituitary gland via the portal blood system. cGnRH-I neuron function is controlled via integration of (a) a very large number of neural inputs from higher brain centers encoding photoperiodic information received from extra-retinal photoreceptors and (b) other neural inputs encoding information from non-photoperiodic cues (Halford et al. 2009; Nakanea et al. 2010). Recent studies also suggest a role for thyroid hormones in the photoperiodic control of GnRH synthesis, specifically the enzymatic conversion of thyroxin (T4) to triiodothyronine (T3) (Watanabe et al. 2007; Yoshimura et al. 2003). Integration of environmental cues also requires a time-measurement system based on both endogenous circadian and circannual rhythms, that is, a "clock" and "calendar" that interact with photoperiod, perhaps in a taxa-specific manner (see Bradshaw and Holzapfel 2007; A. Dawson et al. 2001; Gwinner 2003).

GnRH and GnIH control synthesis and release of the gonadotrophic hormones, LH, and FSH from the anterior pituitary. The synthesis and release of LH is regulated by GnRH, , but the mechanisms regulating FSH secretion remain largely unknown, although they might involve inhibin and activin (Sharp and Ciccone 2005). LH is thought to primarily regulate gonadal steroidogenesis and ovulation, and FSH is thought to regulate spermatogenesis and follicle development (although little work has been conducted specifically using *avian* FSH). A three-cell model of ovarian steroidogenesis has been proposed for pre-ovulatory follicles: granulosa cells, in particular those distal to the germinal disc, are the predominant source of progesterone, which serves as the precursor for synthesis of androstendione and testosterone by cells of the theca interna, with subsequent aromatization of testosterone to estradiol in the theca externa cells (A. Johnson and Woods 2007). As pre-ovulatory follicles

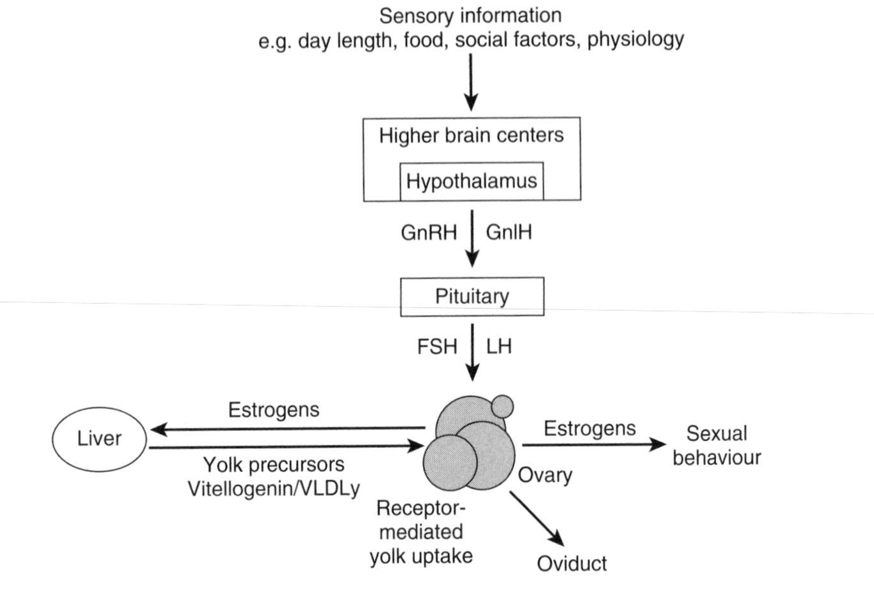

Fig. 2.15. Diagrammatic representation of the organization of the hypothalamo-pituitary-gonadal axis, which integrates environmental information and regulates ovary, oviduct, and liver function for vitellogenesis, yolk, and egg formation in female birds.

mature from F5 to F1, the predominant steroid hormone produced changes from estradiol in F5 follicles to progesterone in F1 follicles (Bahr et al. 1983) This process might be associated with a requirement for maximizing progesterone secretion at the time of the pre-ovulatory LH surge for ovulation of the F1 follicle (A. Johnson and Woods 2007). In addition to regulating vitellogenesis and yolk uptake, gonadal steroids have many roles that affect other aspects of physiology and also behaviors such as aggression, territoriality, nest building, copulation, and the development of secondary sexual characteristics, plumage (in some birds), and bill color (Adkins-Regan 2005).

For almost 40 years, GnRH, LH, and FSH have been considered to be the main regulatory components of the HPG axis, but recently two other regulatory neuropeptides have been identified that might prove to be important in control of seasonal reproduction: GnIH and kisspeptins (Bentley et al. 2008; de Tassigny and Colledge 2010; Greives et al. 2008; Tsutsui 2009). The cell bodies of GnIH neurons are localized to the paraventricular nucleus of the hypothalamus, and their axons form a dense network extending to many other brain regions, including the pre-optic area, median eminence, and brain stem. GnIH neurons interact directly with GnRH neurons (Ubuka et al. 2008), and the GnIH receptor is also expressed in the quail pituitary (Yin et al. 2005). GnIH inhibits the release of both LH and FSH in a dose-dependent manner in vitro (Ciccone et al. 2004). GnIH treatment decreases plasma testosterone levels and spermatogenic activity in male quail (Ubuka et al. 2006), and it decreases plasma LH and copulation solicitation in female quail (Bentley et al. 2006). Taken together these data suggest that GnIH functions to inhibit reproductive function via inhibition of gonadotropin release at both the hypothalamic and pituitary level. GnIH expression is photoperiodically regulated and is mediated by melatonin, observations that support the idea that GnIH plays a role in seasonal reproduction (Tsutsui et al. 2007). Bentley et al. (2008) detected GnIH peptide and the GnRH receptor in thecal and granulosa cells of the ovary in quail and European starling and suggested the potential for an autocrine/paracrine role in the regulation of steroidogenesis and perhaps oocyte development. Kisspeptins comprise a family of amidated peptides encoded by the *Kiss-1* gene. They have been shown in mammals to be involved in the initiation of puberty and regulation of seasonal breeding and might play similar roles in birds (de Tassigny and Colledge 2010; Navarro et al. 2007; Revel et al. 2007).

2.6. Hormonal Integration of Environmental Information

In almost all non-domesticated birds, breeding is restricted to a particular time of year that coincides with a seasonal improvement in the environmental conditions required for successful rearing of offspring, for example, greater food availability, higher temperatures, or increased rainfall (Murton and Westwood 1977; chapter 3). Annual cycles of reproduction are most common, with birds in temperate and higher latitudes breeding in spring and early summer. In some tropical species, egg-laying can occur in any month of the year, but individual birds do not lay continuously and many still show approximate annual cycles. Other tropical species do have seasonal breeding cycles, albeit ones that are not clearly associated with changes in day length but perhaps with cycles of rainfall or food availability, and males fully regress their testes between breeding attempts for at least some months of the year (Beebe et al. 2005; Wikelski et al. 2003). Even temporal "opportunistic breeders" show regular, seasonal declines in reproductive activity, with regressed gonads and low circulating levels of reproductive hormones between breeding attempts (Deviche and Sharp 2001; Hahn 1998; Hau et al. 2004; Perfito et al. 2007). These species appear to have regular annual cycles of "reactivation" of the reproductive system but breed only when conditions are suitable, so they are very similar in this regard to temperate-zone seasonal breeders.

The timing of breeding is dictated by two sets of factors, which were first distinguished by J. Baker (1938). In general, individuals that time breeding so that they have chicks in the nest when food is most available will produce more offspring and therefore have higher fitness. The availability of food required for offspring growth thus sets a selective, or evolutionary, value on the timing of breeding (Lack 1968) and is the "ultimate" factor explaining *why* birds have evolved to breed when they do. However, before the young hatch, their parents must undergo extensive physiological and behavioral preparation for breeding, which involves maturation of the reproductive system, finding a nest site and mate, and laying and incubating eggs. These events can take 6–8 weeks to complete and often must be initiated well in advance of any seasonal increase in food availability (see fig. 3.2). It follows that birds must be able to *predict* the onset of the breeding season, and they do so using information from environmental cues, or "proximate" factors. Proximate factors therefore explain *how* birds time their breeding season. The main proximate factor used by birds to time their breeding season outside of the tropics is the annual cycle of day length or photoperiod: since day length varies absolutely predictably each year, the increasing day length in spring therefore provides long-term information, or *initial predictive information*, for the timing of breeding (A. Dawson et al. 2001). However, since there is usually small-scale, local variation in environmental conditions from year to year, other non-photic factors, such as temperature, food availability, and social interactions, are thought to provide *essential supplemental information* that allows birds to fine-tune the onset of egg-laying itself. In addition to controlling the initiation of reproductive development at the start of the breeding season, day length permits most birds to predict in advance the time for the termination of breeding, a process called *photorefractoriness* (A. Dawson 2002; A. Dawson and Sharp 2007; Nicholls et al. 1988). This ability is important since if laying were to continue until food availability began to decline, the young would be in the nest at a time when there is insufficient food to rear them successfully. Chapter 3 will consider the environmental regulation of timing of breeding in more detail in an ecological and evolutionary context; here I will focus primarily on relevant information from experimental studies of captive birds that highlight the hormonal responses of the HPG axis to photic- and non-photic stimuli between captive- and free-living birds and between male and female birds.

2.6.1. Photoperiodic Control of Gonadal Function

In a wide range of species, male birds in captivity will undergo complete testicular growth, maturation, and spermatogenesis when transferred from short, winterlike, artificial photoperiods (8L:16D) to long, spring- or summerlike photoperiods (e.g., 18L:6D; Murton and Westwood 1977). Reproductive maturation in captive males on artificial photoperiods is very similar or identical to changes seen in free-living birds in response to the gradually increasing day length in spring (A. Dawson and Goldsmith 1982; 1983). In captivity, the transfer of male birds from 8L:16D to day lengths longer than 13L causes a rapid increase in the levels of plasma LH and FSH, which is presumably associated with greater GnRH secretion from the hypothalamus. GnRH secretion has not been measured directly, but in some species, photostimulation involves a transient increase, then a gradual decrease, in hypothalamic GnRH *content*, which would be consistent with its increased secretion followed by decreased GnRH-1 synthesis, which ultimately "switches off" of the HPG axis with onset of absolute photorefractoriness (A. Dawson et al. 2001). In the absence of other non-photic cues, photostimulation of captive males is sufficient to elevate their plasma LH and FSH levels to those typical of free-living birds (A. Dawson and Goldsmith 1982, 1983; see also Wingfield and Farner 1993 and references therein). Testis growth rates are proportional to the length of the artificial stimulatory photoperiod (A. Dawson and Goldsmith 1983), and maximum testis size in photostimulated captive birds can be equivalent to that of free-living birds; for example, the peak testis mass was 408 ± 14 mg in white-crowned sparrows breeding in Alaska and 356 ± 42 mg in captive males on 20L:4D (J. King et al. 1966; Wingfield et al. 1980). Treatment of captive males with exogenous GnRH also stimulated increases in plasma LH and testosterone—to levels as high as those in breeding birds (Jawor et al. 2006; Wingfield and Farner 1993)—and exogenous LH or testosterone treatment stimulated Leydig cell maturation, the appearance of primary spermatocytes, and increased testis mass, even in non-photostimulated males (Murton and Westwood 1977). Few photoperiodic-manipulation studies have measured plasma testosterone levels, but it is clear that in males, long-day photostimulation also initiates steroidogenesis via LH release, although this response is often delayed compared with the increase in plasma LH, and peak testosterone levels are lower (Wingfield and Farner 1993). These data support the idea that the annual change in day length is the most important environmental cue used for synchronizing and initiating gonadal development in *male* birds, which is consistent with a top-down model of regulation in which photostimulation of GnRH neurons in the hypothalamus is sufficient to stimulate gonadotropin release and, at least to some extent, initiate spermatogenesis and steroidogenesis in the testes.

In contrast to male birds, females of most non-domesticated species held in standard captive conditions (small cages) will *not* undergo complete ovarian development: the ovary can develop to the pre-vitellogenic phase, but the onset of vitellogenesis, yolk formation, and egg production is very unusual in captivity (J. King et al. 1966). In part for this reason, male birds have been used much more widely in experimental studies, and far fewer data are available for females. Farner et al. (1966) stated that "the accumulation of data on the ovarian cycle, both in the field and the laboratory, has been necessarily at a much slower rate [than for testis function]." Forty years later our knowledge of regulatory mechanisms controlling the final stages of ovarian development in females remains rudimentary (G. Ball and Ketterson 2008). Female birds show some hypothalamo-pituitary response to long-day photostimulation in captivity; for example, female European starlings show a slight, but not significant, increase in hypothalamic GnRH content during the first 6 weeks after transfer from 8L:16D to 16L:8D, after which GnRH content declines rapidly to very low levels as the birds became refractory (A. Dawson et al. 1985). Pituitary FSH content also increases rapidly in photostimulated females within 2 weeks of transfer from short days to 18L:6D (A. Dawson et al. 1985). Female starlings transferred from short days to 13L or 18L showed a rapid increase in plasma LH and FSH, with peak levels that are much *higher* than those seen in males (A. Dawson and Goldsmith 1983; fig. 2.16). Plasma LH concentrations in photostimulated female starlings are also somewhat higher than those in free-living birds during nest-building or laying (1.5–3.0 ng/ml), but FSH levels in captive females are fourfold higher (A. Dawson and Goldsmith 1982). Female European starlings held in outdoor aviaries on natural day-length cycles showed annual cycles of plasma LH (and prolactin) very similar to those of free-living females (A. Dawson and Goldsmith 1984). Some data from older studies suggest that there can be marked individual variation in the changes in plasma LH and FSH for a given photoperiod (Lewis 1975; Wingfield et al. 1980) and that this variability might represent an interesting line of inquiry. However, although these studies support the idea that females *can* respond to photoperiod alone at the level of the hypothalamus and pituitary, are there any effects of photostimulation on "downstream" events such as ovarian steroidogenesis, vitellogenesis, or follicle development?

Stetson et al. (1973) showed that a combined exogenous LH and FSH treatment increased both ovary and oviduct mass in captive white-crowned sparrows, compared with birds that were photostimulated but not hormone treated, but only after 40 days of long-day photostimulation (16L:8D). Ovary and oviduct mass reached 40%–50% of maximum with photostimulation and gonadotropin treatment combined (based on data for free-living females in J. King et al. 1966). However, the maximum follicle diameter in LH/FSH-treated females was only 3.2 mm, which is considerably less than the largest follicle diameter in free-living birds (5 mm; J. King et al. 1966). Furthermore, in a second group of females that

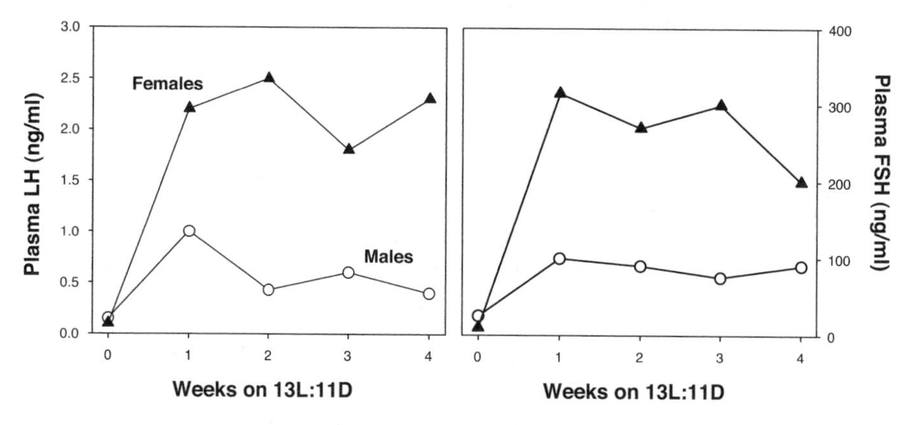

Fig. 2.16. Comparison of changes in plasma luteinizing hormone (LH) and follicle-stimulating hormone (FSH) in male (circles) and female (triangles) European starlings after transfer from short days (8L) to a stimulatory 13L photoperiod. (Based on data in Dawson and Goldsmith 1983.)

were photostimulated for 16 days on 20L:4D, although exogenous gonadotropin treatment had a *significant* stimulatory effect on ovary and oviduct growth, these organs only reached 1%–3% of maximum size. These results are in contrast with those in the photostimulated female starlings described above, where high endogenous plasma LH and FSH levels were not associated with ovarian development. However, since plasma gonadotropins were not measured in Stetson et al.'s study (1973), it is not clear if these effects were physiological or pharmacological. Their study does suggest that exogenous LH and FSH stimulated ovarian steroid production, which, in turn, stimulated oviduct development. Few photoperiod-manipulation experiments with wild birds have measured plasma steroid or yolk-precursor levels directly. Female European starlings with *previous* experience of long days do have significantly elevated plasma VTG levels after a second 8-week photostimulation on 18L:6D, although these are only about 10% of peak values in free-living females during egg formation (Sockman et al. 2004).

As described above, pituitary FSH and plasma LH and FSH all rose rapidly to peak levels in *female* European starlings only 2 weeks after transfer from 8L:16D to 18L:6D (A. Dawson and Goldsmith 1983). However, 12 days after transfer from 8L to 18L, female starlings did not have plasma estradiol or yolk-precursor (VTG, VLDLy) levels above baseline, and no birds showed vitellogenic follicle, ovary, or oviduct growth (T. Williams, unpublished data). Females did respond readily to exogenous estradiol treatment on 18L, but only in terms of elevated plasma VTG levels and increased oviduct mass, with no effect on VLDLy levels measured either as plasma triglyceride levels or apoVLDL-II expression. These data are consistent with a *female-specific* mechanism whereby initiation of vitellogenesis and follicle development is partly regulated by "downstream" mechanisms operating at the level of the ovary and/or liver (Caro et al. 2009). The lack of response in VLDLy might be particularly significant: VTG production involves induction of genes and synthesis of protein products (VTGs) that are not normally expressed, and do not function, in the non-breeding female. In contrast, VLDLy production involves "reprogramming" of existing hepatic lipoprotein synthesis and a transition from generic VLDL production. The latter pathway plays a very important function in non-breeding females: to meet the female's own energy demands of self-maintenance. Therefore, the "decision" to switch to almost exclusive production of VLDLy at the onset of egg production might be particularly costly to females, leading to the evolution of a more "conservative," more highly regulated mechanism, downstream from the hypothalamus and pituitary, for initiation of vitellogenesis and VLDLy synthesis.

2.6.2. Supplemental, Non-photoperiodic Cues and Ovarian Function

It is widely assumed that females do not undergo full ovarian development and egg-laying in captivity because they lack the additional environmental cues that are required for completion of these processes and that are hard to replicate in the laboratory (G. Ball and Ketterson 2008; Farner et al. 1966). Wild-caught females can complete egg production and egg-laying when maintained in large aviaries (2 × 2 × 2.5 m) at low bird density, even with artificial light and temperature (Visser et al. 2009). However only a handful of, albeit well-cited, studies have coupled photoperiod manipulation with manipulation of other "supplementary" factors (e.g., food, temperature, social factors) in *female* birds (see chapter 3 for more studies of temperature effects). Visual and audiovocal cues, such as the presence of a nest or "cooing" associated with courtship, have been well studied in the ring dove (*Streptopelia risoria*). Increases in follicle diameter were greatest in females exposed to male coos (fig. 2.17), although follicle growth also occurred in response to the female's own coos and those of other females (Cheng et al. 1988). These results provide a clear example of extra-hypothalamic input controlling HPG function in both sexes, but the underlying physiological mechanisms remain poorly understood (Cheng 2005). Only recently have changes in GnRH synthesis and release been measured in relation to courtship interactions, and then only in male birds (Mantei et al. 2008), and similar detailed work has not been conducted on other species. T. Stevenson et al. (2008) reported modulation of the GnRH system in relation to social context in house sparrows

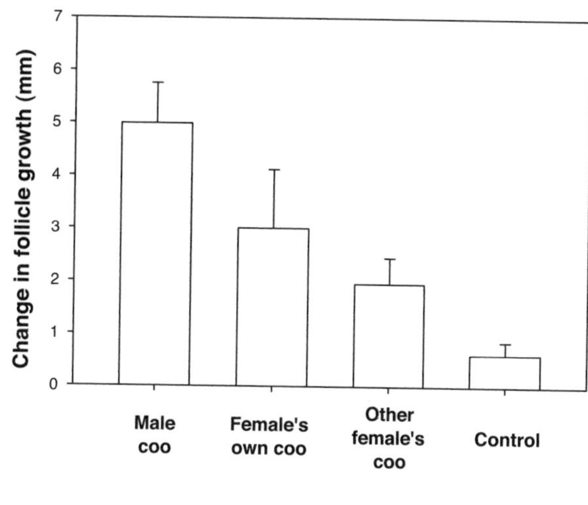

Fig. 2.17. Change in follicle growth (mm, diameter) in female ring doves in response to nest-coo vocalizations. (Based on data in Cheng et al. 1988.)

(*Passer domesticus*), although there was no direct effect of male presence on plasma LH, estradiol, or ovary development in females. Silverin and Westin (1995) transferred willow tits (*Parus montanus*) from natural winter day length to 20L:4D in either single-sex groups or mixed-sex pairs. The initial increase in largest follicle diameter was greater in paired than single females, although the final maximum follicle diameter did not differ by treatment, increasing to only approximately 14% of that in free-living birds. M. Morton et al. (1985) reported that male song augmented pre-vitellogenic ovarian growth rates in female white-crowned sparrows (*Z. l. gambelii*), but this effect was significant only at intermediately long day lengths (12.5 L, 14L), not at shorter or longer day lengths. Finally, Meijer (1991) exposed female starlings to a period of food restriction on a stimulatory 12L:12L photoperiod, which caused a 21% decrease in body mass, but food-restricted females did not differ in beak color (indicative of circulating gonadal steroids) or follicle growth compared with controls (although the maximum follicle diameter was <2 mm compared with 12 mm at the one-egg stage in free-living starlings).

Several experiments suggest a role for "experience" in the regulation of ovarian development (Sockman et al. 2004). Baptista and Petrinovich (1986) hand-raised 36 female white-crowned sparrow nestlings under natural photoperiods, and all 36 (100%) subsequently laid eggs between February and August of the following year when housed singly in small cages with visual and auditory stimuli from males in neighboring cages.

However, egg-laying occurred on an atypical schedule: 9 females (25%) laid only one egg each, and other females laid single egg "clutches" irregularly rather than on consecutive days, as occurs in free-living birds. Furthermore, mean clutch sizes were generally smaller (less than three eggs) compared with free-living birds (three to seven eggs). Baptista and Petrinovich (1986) suggested that these birds "may have imprinted on the characteristics of their holding cages" during hand-rearing "and accepted these as substitutes for stimuli from natural habitats that wild birds require in order to complete egg formation." However, hand-reared European starlings exposed to stimulatory photoperiods mimicking natural spring day lengths showed no ovarian or oviduct development (T. Williams, unpublished data). Given the common assumption that cues other than photoperiod play a key role in regulating gonadal function in females, it is therefore surprising how few experimental studies have investigated these alternative cues, and it is even more surprising that existing and recent studies have actually been conducted on male birds (O'Brien and Hau 2005; Perfito et al. 2008). I will return to this issue in chapter 3, where I consider regulation of timing of egg-laying in free-living birds.

2.7. Future Research Questions

- What role does vascularization of the ovary or individual follicles play in determining yolk uptake rates and intra-clutch and inter-individual variation in final follicle or egg size?
- How is the molecular/cellular process of follicle selection integrated with, and determined by, environmental cues to generate daily and seasonal patterns of follicle development, clutch size, and timing of breeding (onset of egg development)?
- Does variation in expression of intra-ovarian growth hormones, or other regulatory factors at the level of the ovary, explain individual variation in reproductive "phenotype," for example, follicle or egg size?
- Is there a discrete, time-dependent phase of pre-vitellogenic growth in free-living birds that must be completed before the onset of steroidogenesis, vitellogenesis, and yolk development?
- Do shifts in lipoprotein metabolism from generic VLDL to lipoprotein lipase–resistant VLDLy represent a widespread mechanism for trade-offs in egg-laying females?
- Does the open-period model of oviposition derived from studies of the domestic hen provide a satisfactory explanation for the variable laying intervals and timing of laying seen in free-living species?

- Does a "memory effect" of primary versus secondary exposure to high levels of estrogens significantly affect vitellogenesis, yolk-precursor production, and yolk formation in reproductively naive, first-time breeding females compared with experienced females?
- What role does avian FSH play in ovarian function, in particular, in relation to egg size, number of eggs, and trade-off between these traits (as has been demonstrated in other oviparous vertebrates)?
- Are there "female-specific" regulatory mechanisms for the onset of vitellogenesis and for VLDLy synthesis downstream of hypothalamic-pituitary control at the level of the ovary and/or liver that are critical for regulation of individual variation in timing of breeding?

CHAPTER 3

Timing of Breeding

As DESCRIBED in chapter 2, a long-standing view has been that the timing and duration of breeding are *ultimately* determined by the food available to support offspring growth during the chick-rearing phase of reproduction (Lack 1968). In almost all free-living birds, breeding is therefore restricted to a particular, often limited, time of year coincident with a seasonal improvement in environmental conditions associated with the greater availability of food that is required to support successful rearing of offspring, for example, longer day lengths, higher Ta, or increased rainfall. There is abundant evidence in a wide range of avian species with widely varying ecologies and life histories that, in general, at the population level birds do successfully time their breeding to "match" the seasonal variation in food availability in most years (fig. 3.1; although recently in some populations, but apparently not others, the timing of breeding has become increasingly "mismatched" with prey availability due to climate change [Dunn et al. 2011; Visser 2008]; see chapter 1). However, there is also good evidence that at the *individual* level, many females in any given population will actually breed later than the "optimum" time for having chicks in the nest coincident with peak food availability. This observation led to the suggestion of some form of constraint on timing of breeding (Perrins 1970) such that there is a trade-off between costs of breeding too early and the fitness benefits of matching the offspring's needs to variation in food availability.

Given that on average birds do time their breeding seasons appropriately, the "decision" to initiate egg-laying can be highly synchronous within populations in any given year, even though annual mean date of initiation of egg-laying can vary markedly among years. In some colonial or semi-colonial nesting species, all females can initiate laying over a 5–10-day period. In European starlings in British Columbia, Canada, all first clutches (n = 77) were laid over 8 days in 1 year (2008), with 79% laid over 4 days. Similarly, for starlings in the United Kingdom, although the first-egg date varied annually from 9 April to 19 April over a 7-year period (Feare 1984), most first clutches were started within a 10-day period in any one year, and in Belgium all first clutches were laid within 6.7 days (range 4–10 in different years, Verheyen 1980 in Feare 1984). This high level of synchrony within years holds true for larger, colonially

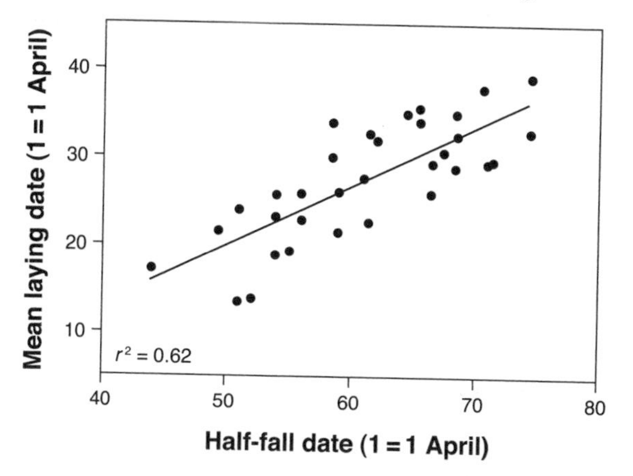

Fig. 3.1. Relationship between mean laying date of great tits and availability of winter moth caterpillars, based on half-fall date of frass. (Reproduced from Charmantier et al. 2008, with permission of the American Association for the Advancement of Science.)

nesting species: in the lesser snow goose (*Chen caerulescens caerulescens*), the laying of first eggs occurs over a 10-day period, with 90% of the nests initiated over a 4-day period in all years (Cooke et al. 1995), and in macaroni penguins (*Eudyptes chrysolophus*), 95% of eggs were laid over 8, 13, and 13 days in 3 years (T. Williams 1995). In contrast, in other species initiation of egg-laying can occur over a longer period, from several weeks up to 40–50 days (T. Arnold 1992; Camfield et al. 2010; Daan, Dijkstra, and Tinbergen 1990; Travers et al. 2010), although with advancing laying dates it can be difficult to be certain that later-laid clutches are true first clutches rather than replacement clutches. As van Noordwijk et al. (1995) pointed out, even if there is positive selection for earlier laying (see below), in many populations the average individual female is never more than a few days away from the optimum timing of breeding, so within-year variation is small compared with between-year variation in laying dates of the same individual females. Nevertheless, as described below, differences in laying dates of just a few days can have significant effects on seasonal trends in other reproductive parameters, such as clutch size, and on fitness.

Different species vary markedly not just in breeding synchrony but also in the number of breeding attempts they make per year. Some populations of blue and great tits and collared and pied flycatchers (*Ficedula hypoleuca*), all of which have been well studied, appear to have a much more restricted period of seasonally increased food availability than many other species, perhaps lasting only 2–3 weeks, and are typically single-brooded not multiple-brooded (fig. 3.2). In such species selection will act on the timing of a single event: the initiation of the first, and

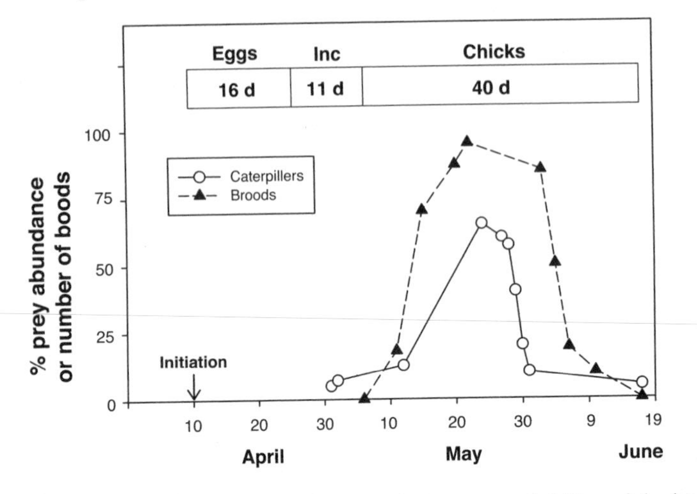

Fig. 3.2. Timing of breeding in blue tits in relation to availability of the bird's caterpillar prey showing that egg-laying is initiated well in advance of the seasonal increase in food availability. (Based on data in Gibb 1950.)

only, breeding attempt. In other species, such as the European starling (fig. 3.3) and the European coot (*Fulica atra;* Brinkhof 1997), the period of increased food availability can be much longer, during which multiple broods might be reared. Here selection should act on the timing of breeding to maximize overall fitness from multiple breeding attempts, although timing of initiation of the first breeding attempt is still probably of importance in determining the timing of subsequent events. Even in species such as the oceanic roseate tern (*Sterna dougallii*), which breeds in the tropics, where seasonal increases in food availability can be much more prolonged and less marked, egg-laying is still initiated synchronously at the start of a phase of increasing food availability (Monticelli et al. 2007). The main argument pursued in this chapter is that the timing of the *initiation* of a single breeding event, or the *initiation* of the first of multiple breeding events within the same breeding season, is completely dependent on the *female-specific* reproductive process (sensu chapter 2) of timing of egg production and egg-laying.

3.1. Early-Season Events Are Critical in Determining Timing of Breeding

Egg production and egg-laying, which initiate a breeding attempt, are temporally separated from the chick-rearing phase by several weeks or even months due to the time required for the extensive physiological and

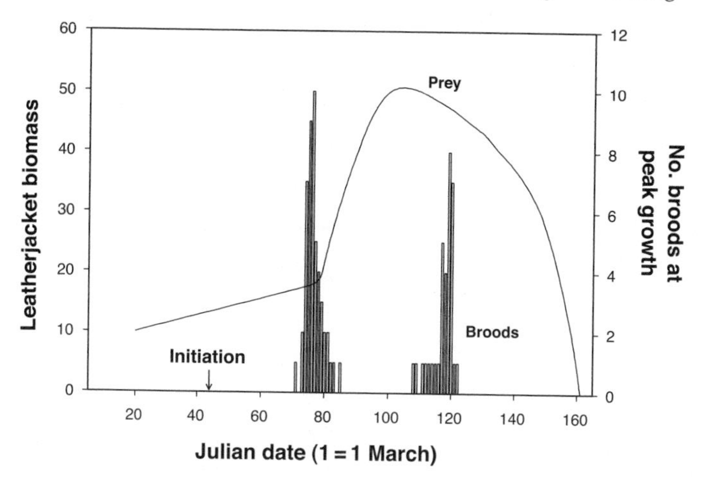

Fig. 3.3. Timing of breeding in European starlings in relation to seasonal variation in availability of leatherjacket (Tipulidae) prey. (Based on data in Dunnett 1955.)

behavioral "preparation" for breeding: maturation of the reproductive system, finding a nest site and mate, and laying and incubating eggs. In the European starling, rapid yolk development for the first developing follicle must be initiated about 22 days before the chicks hatch, including 12 days for egg formation and laying and 10 days for incubation. Similarly, in blue and great tits, egg formation must be initiated about 27 days before hatching, and therefore about 3 weeks before any seasonal increase in the abundance of the main caterpillar prey (fig. 3.2). In long-lived seabirds this pre-hatch phase of breeding can extend for 10–13 weeks, including up to 20–30 days for yolk development and laying of the single-egg clutch and 10 weeks for incubation. This temporal separation of egg production and chick-rearing highlights why birds, and many other animals, use environmental cues for timing of laying, particularly photoperiod, that can reliably *predict* when food will be plentiful much later in the season (see chapter 2). The timing of laying should therefore be determined by cues from the "environment at the time of decision making" early in the breeding season, whereas the fitness benefits of the timing decision are determined by the "environment at time of selection" during the nestling phase (Visser et al. 2004). I will argue below that this timing of "decision making," that is, initiating reproductive maturation in anticipation of breeding, might not be the same for both sexes, in which case the "environment" at the time of decision making and the environmental cues used in this process might be very different for females and males (see section 3.5). If the early-season environment, when egg

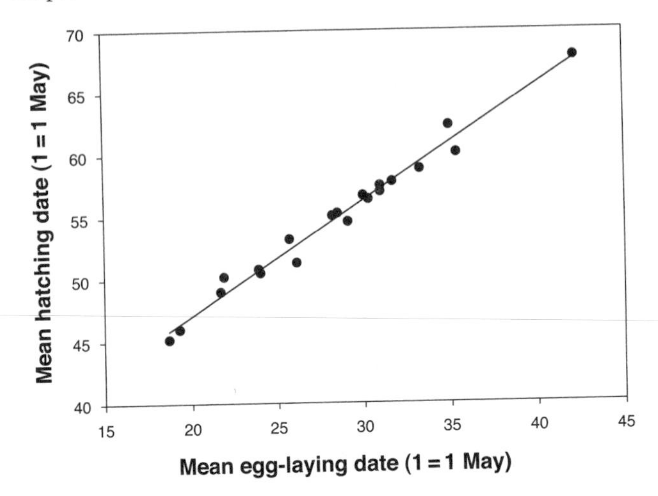

Fig. 3.4. Relationship between laying date and hatching date in the lesser snow goose, showing that later season events are highly correlated with, and predicted by, date of *initiation* of laying. (Based on data in Cooke et al. 1995.)

production occurs, is highly correlated with the late-season environment, when chicks are being reared, then the use of predictive cues will allow for the accurate timing of breeding even many weeks in advance. In many populations the mean timing of egg-laying and mean timing of hatching are indeed highly correlated (fig. 3.4). Thus, the initiation of egg production *is* the critical event in the timing of breeding, such that synchronization with food availability is primarily achieved by varying the first-egg date, and the duration of stages of reproduction subsequent to the initiation of laying (completion of laying, incubation, and hatching) are relatively constant (Cresswell and McCleery 2003; but see Matthysen et al. 2010). The pivotal nature of the initiation of egg production confirms the importance of focusing on physiological mechanisms in female birds; that is, environmental cues and physiological mechanisms controlling initiation of *egg production* are the key to understanding the timing of seasonal breeding.

3.2. Fitness Consequences of Timing Decisions

Reproductive success generally, though not always, declines with later egg-laying dates during the breeding season (Perrins 1970): on average, individuals that breed relatively early produce larger clutches (see chapter 5) and more fledglings with higher growth rates and higher survival, and they recruit more young to the breeding population than do individuals that breed relatively late (figs. 3.5 and 3.7). These seasonal trends in reproductive success have been confirmed in population studies in a wide range of species, for example, passerines (Verboven and Visser 1998), ducks and geese (R. Dawson and Clark 2000; Lepage et al. 2000), seabirds (Spear and Nur 1994), and raptors (Daan, Dijkstra, and Tinbergen 1990). In multiple-brooded species, early-breeding individuals also have a higher probability of rearing a second brood in the same breeding season. Brinkhof et al. (2002) manipulated the timing of reproduction in European coots, exchanging first clutches of equal size but different laying date between nests, to create "delayed" and "advanced" pairs. The probability of starting a second brood in the same breeding season and the number of second-brood young (i.e., future fecundity) declined following an experimental delay of the hatching date of the first brood, whereas an experimental advance of hatching date raised the probability of second broods being reared. Individuals breeding relatively early might also derive other benefits in that they can initiate post-breeding molt earlier or molt more slowly over a more protracted period, which can increase feather quality (A. Dawson et al. 2000). Earlier breeding individuals are also less likely to show molt-breeding overlap (see chapter 7). Finally, early-breeding individuals (and their offspring) might benefit from a longer post-breeding period to recover, and/or undergo hyperphagia before migration or onset of winter, which might enhance their survival until the next breeding period (Nilsson and Svensson 1996). Thus, the *female-dependent* early-season decision to initiate *egg-laying* has important consequences for many later stages of breeding and subsequent life-history stages.

Seasonal declines in some reproductive parameters with later laying date can be marked; for example, significant decreases in clutch size can occur over just a few days, highlighting the importance of even small-scale variation in egg-laying dates. In the greater snow goose (*Chen caerulescens atlanticus*), clutch size decreases from about five eggs in early clutches to three in late ones at a rate of –0.20 eggs per day (Lepage et al. 2000). In the marsh tit, clutch size decreases by 0.1–0.2 eggs per day in different years (H. Smith 1993), and in great tits the number of surviving young decreases by up to 10% for each day's delay in breeding date (Perrins and McCleery 1989). Even in highly synchronous species, such as the European starling, clutch size can decrease by 0.3 eggs per day (see fig. 5.3), and recruitment success of fledged young also declines sharply with season (H. Smith 2004). Lifetime fecundity or recruitment can decrease rapidly with advancing laying date (Brommer and Rattiste 2008; Garant, Kruuk, et al. 2007). In great tits breeding in the Netherlands, lifetime fitness decreases by about 50% over a 10-day range of laying dates, and 75% over 20 days (M. Visser, pers. comm.). In great tits breeding

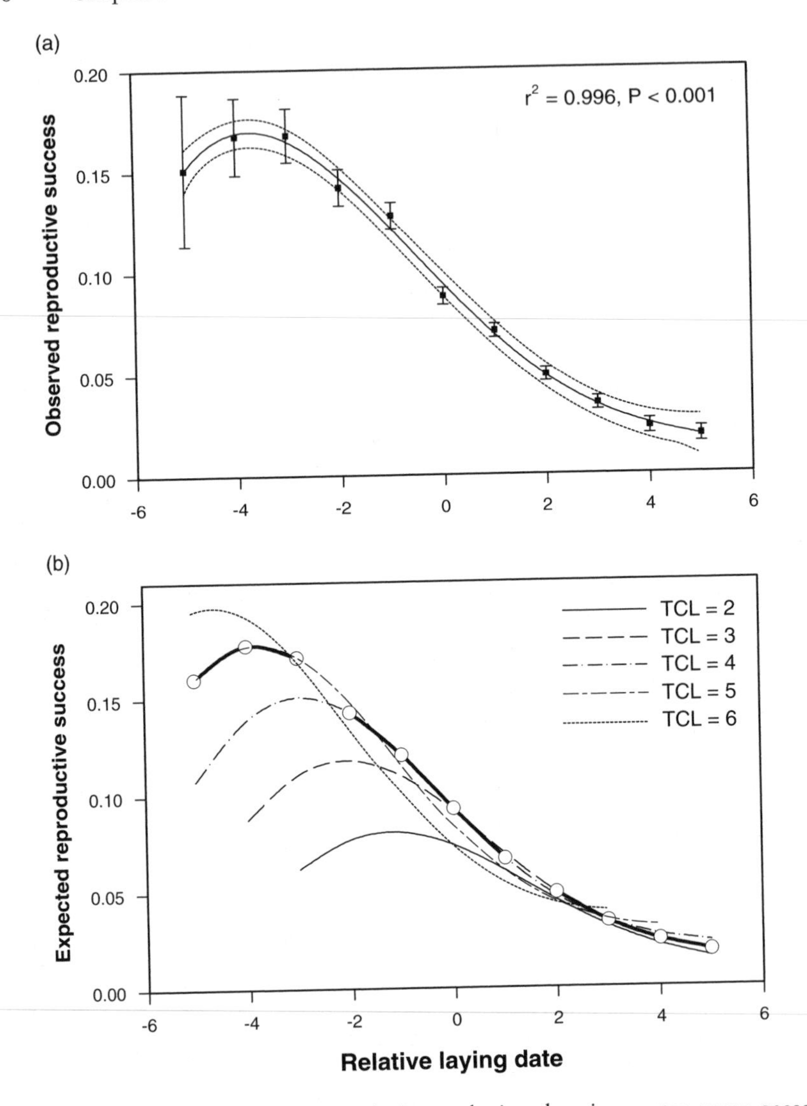

Fig. 3.5. Reproductive success in relation to laying date in greater snow geese: (a) composite measure of reproductive success (mean number of young surviving to the first winter) in relation to laying date; (b) expected reproductive success in relation to laying date over the range of possible clutch sizes. (Reproduced from Lepage et al. 2000, with permission of John Wiley and Sons.)

in the United Kingdom, the reduction in lifetime fitness with advancing laying date is more marked in birds laying larger clutches (see fig. 3.7), presumably because mistiming of breeding relative to any peak in food availability has more severe consequences for individuals trying to rear more chicks. Therefore, even though egg-laying can be highly synchronous in many populations, small-scale differences in timing of only a few days can have significant fitness consequences *and physiological response mechanisms underpinning timing of breeding must be capable of regulating these relatively small differences in timing.*

Why is there a seasonal, laying-date-dependent decline in reproductive performance? Two main explanations have predominated: first, seasonal declines might be due to time-dependent decreases in habitat quality or food availability (the *date hypothesis*). This idea predicts that the reproductive performance of *all* individuals would be affected by date in the same way. Alternatively, seasonal declines might reflect quality differences between individuals, where lower-quality birds have lower reproductive output and breed later than higher-quality birds (the *quality hypothesis*). These two hypotheses generate clear predictions about the effects of experimentally manipulating egg-laying or hatching date (Verhulst et al. 1995), but the causes of these seasonal patterns are still not well understood, in part because all potential methods for experimental manipulation of timing have some flaws or confound direct manipulation of timing with indirect manipulation through effects on other traits such as body condition (Verhulst and Nilsson 2008). Verhulst and Nilsson (2008) reviewed 33 studies that used either egg removal and re-laying (which can only delay timing) or cross-fostering to manipulate the timing of breeding. Only 7 studies measured local recruitment as the fitness endpoint, and 6 of these supported the date hypothesis (all studies were on blue or great tits), with 1 study supporting both hypotheses. For studies involving a broader range of reproductive endpoints, such as growth rate, immunity, fledging mass, and fledging success, 12 studies supported a date effect, 9 studies supported a quality effect, and 2 found support for both.

In terms of physiological mechanisms, the date hypothesis largely argues that seasonal declines in reproductive performance are imposed on all individuals by external, ecological factors, for example, decreased food availability later in the season. This mechanism appears to be relevant for some species (perhaps Paridae, European coot) but not necessarily for others where seasonal declines in clutch size occur even though feeding conditions improve during laying (Dunn et al. 2011; Lepage et al. 2000; Monticelli et al. 2007). In other species, seasonal increases in nest parasitism (C. Brown and Brown 1999) or reduced availability of suitable habitat (Dzus and Clark 1998) might be important. In contrast, the quality

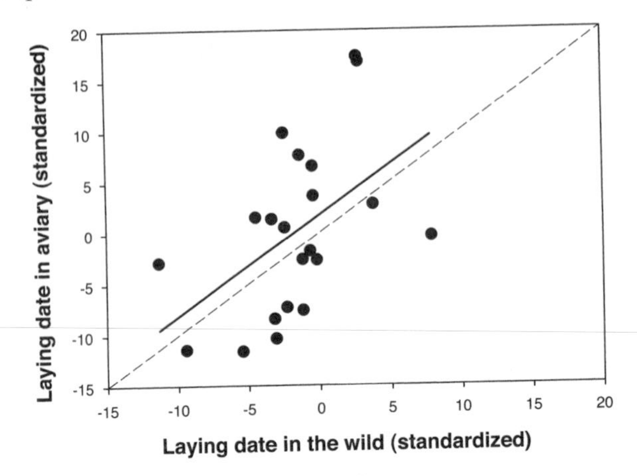

Fig. 3.6. Repeatability of laying date in individual female great tits when laying date was obtained both in the wild and subsequently in aviaries in captivity. (Reproduced from Visser et al. 2009, with permission of the Royal Society.)

hypothesis implies that there are intrinsic differences among individuals that determine variation in reproductive performance; these differences must have a physiological basis, but they remain poorly understood.

3.3. Selection on Timing of Breeding

The fitness benefits of earlier breeding should generate directional selection for earlier breeding, and if there is significant additive genetic variance for this trait then we might expect to see an evolutionary response to selection. Timing of breeding is a repeatable trait: individual variation in timing persists when wild birds are brought into captivity (Visser et al. 2009), and individual variability is maintained in captive birds exposed to artificial temperature and photoperiod and with ad lib food (fig. 3.6). Earlier studies estimated the heritability of laying date to be as high as 0.3–0.4 using single-generation mother-daughter regression (Gustafsson 1986; Lundberg and Alatalo 1992; van Noordwijk et al. 1981). More recent estimates from "animal models" of breeding values are generally lower: 0.11–0.19 (Brommer et al. 2008; Sheldon et al. 2003; but see Caro et al. 2009, where h^2 = 0.4), which suggests that heritability estimates might sometimes be inflated by common environmental effects in closely related individuals (van der Jeugd and McCleery 2002). However, despite putative positive selection and heritable genetic variation, there was, until recently, little evidence for evolutionary change in laying date, and laying date in birds has been considered one of the classic examples of a trait showing unexplained evolutionary stasis (Merilä et al. 2001; Price et al. 1988).

Several recent studies have utilized long-term data sets (often with more than 5,000 breeding records) for individuals with known genetic relatedness, an approach that allows the use of animal models to estimate selection differentials and selection gradients for phenotypic and additive genetic variation (A. Wilson et al. 2010). Selection differentials measure both direct selection acting on laying date and indirect selection acting via correlated traits such as egg or clutch size. Selection gradients provide a measure of the direct selection on a trait, holding the effects of other traits constant. These studies all indicate that there is strong "fecundity selection" for timing of laying; that is, individuals that lay earlier than the average date recruit more offspring to the breeding population than those individuals that lay later than the average date (Brommer and Rattiste 2008; Garant, Kruuk, et al. 2007; Sheldon et al. 2003). For example, using data from more than 4,000 females over a 20-year study of collared flycatchers (*Ficedula albicollis*) in Gotland, Sweden, Sheldon et al. (2003) estimated that negative fecundity selection gradients on laying date acted on both phenotypic and additive genetic variation, with selection varying little among years. Similarly, using more than 5,000 breeding records from the long-term great tit data set from Wytham Woods, Oxford, Garant, Kruuk, et al. (2007) estimated significant, and similar, linear fecundity selection differentials and fecundity selection gradients for laying date. Thus, on average, earlier laying females *do* recruit more offspring to the breeding population from each breeding attempt, and there should be consistent directional selection for earlier laying (fig. 3.7).

If there is significant directional selection for earlier laying, why do some individuals within a population "miss" the optimum window of timing and lay relatively late? The general answer to this question is that there are costs to breeding early (or other benefits to breeding late) or constraints that prevent some females from breeding early (Perrins 1970; van Noordwijk et al. 1995). Perrins (1970) recognized that in order to have chicks in the nest when food was most abundant, females must lay their eggs earlier in the season "at a time when food is not so plentiful" (see figs. 3.2 and 3.3), and he proposed that egg-laying cannot be initiated until the female is able to find food enough food to form eggs "without risk to herself." This relatively simple idea, which has been interpreted by some as suggesting that food availability is a *proximate* factor determining laying date (Drent 2006), has been very influential and has important implications for potential *physiological mechanisms* (see section 3.8). Other potential examples of "constraints" are situations where low temperatures or snow cover prevent earlier laying in Arctic-nesting species (Dickey et al. 2008) or where socially mediated breeding synchrony

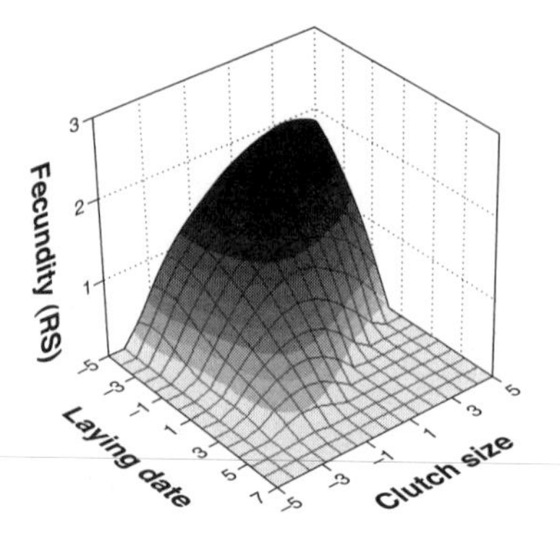

Fig. 3.7. Relationship between fecundity (number of recruits per breeding attempt) and variation in standardized laying date and clutch size for great tits in Wytham Woods, Oxford. (Reproduced from Garant, Kruuk, et al. 2007, with permission of John Wiley and Sons.)

means that earlier laying, "asynchronous" individuals have lower breeding success (T. Reed et al. 2006). Some studies have suggested that, under certain circumstances, individuals might gain an advantage by delaying their laying date. For example, in migratory species individuals that arrive at the breeding ground in poor condition following migration might delay laying to improve body condition through feeding and hence increase their clutch size (the *cost-of-delay hypothesis*, Drent and Daan 1980). This hypothesis assumes that there are optimal combinations of clutch size and timing of breeding in relation to initial condition of individuals and the onset of condition gain determined by their arrival date at the breeding grounds as predicted by condition-dependent individual optimization models (Rowe et al. 1994). Some evidence in support of this model exists in greater snow geese (Bety et al. 2003), but the generality of these models needs to be explored in other species.

Fecundity selection for earlier laying would be countered if there are survival costs to laying early (i.e., viability selection). In collared flycatchers there is evidence of weak negative viability selection on the egg-laying date, but there was a *positive* selection gradient on the additive genetic variance underlying breeding time; females with values for earlier timing of breeding had lower survival. Thus, the positive relationship between timing and survival reduced the overall force of selection for laying date by about 82% (Sheldon et al. 2003). Similarly, in cliff swallows, adult survival to the following year was positively correlated with laying date in females, but not males, which might prevent directional selection for laying date (C. Brown and Brown 1999). In contrast, in great tits there was a negative viability selection differential and selection gradient for laying date; earlier-laying females have greater reproductive success *and* greater adult survival to the next year (Garant, Kruuk, et al. 2007). In studies where the laying date has been *experimentally* advanced, limited data suggest that earlier-breeding parents suffer higher mortality rates than controls both when supplemental food is provided (Nilsson 1994) and when there is no provision of extra food (Brinkhof et al. 2002). These results indicate that in some species higher mortality can be a cost of early breeding and that parents trade off this survival cost against current reproductive success, even though the latter increases when breeding time is advanced (Verhulst and Nilsson 2008).

A few studies have reported fluctuating selection for laying date; that is, selection varies in direction between years (van Noordwijk et al. 1981; Svensson 1997; Visser et al. 1998; but see Sheldon et al. 2003). In cliff swallows local recruitment of offspring was highest for early-nesting females in 5 of 7 years but was higher for intermediate-date breeders in the remaining years (C. Brown and Brown 1999). Similarly, in greater snow geese the overall relationship between laying date and nesting success was curvilinear (and this was the case in 3 of 7 individual years); early and late nests had a higher failure rate than those initiated near the median laying date (Lepage et al. 2000). However, for first-year survival there was a nearly constant decline of reproductive success during the season in all years (fig. 3.5), with lower success only in the very earliest breeding birds (Lepage et al. 2000). Charmantier et al. (2008) showed in great tits that when peak food availability occurred soon after laying (e.g., due to relatively warm post-laying temperatures), the very earliest laying birds had the highest fitness, but when this interval was longer (in relatively cool years), fitness was lowest for early- and late-laying birds. Garant, Kruuk, et al. (2007) also found some evidence for significant non-linear selection on laying date, indicating that convex or stabilizing selection was acting on this trait.

Finally, selection for earlier laying in females might be constrained by "hidden" selection due to indirect genetic effects: genes for laying date might have pleiotropic, sexually antagonistic effects, where alleles that are associated with higher fitness in females are associated with lower fitness in males. Brommer and Rattiste (2008) showed in common gulls (*Larus canus*) that there was an indirect, male genetic effect (4.8%) contributing to the variances in laying date, which was significantly negatively correlated with a direct female effect. This result suggests that genes for early laying in females are associated with genes for a delaying male effect on his partner's laying date. However, other studies have found little or no evidence for a male effect on laying date in passerines (McCleery et al. 2004; Sheldon et al. 2003) or shorebirds (van de Pol, Heg, et al. 2006).

3.4. Constraint, Individual Optimization, and the Search for Mechanism

Perrins' suggestion (1970) that egg-laying is initiated only once a female is able to find enough food to form eggs implies that food availability can act as a simple *constraint* preventing some birds, for example, poor foragers or birds in low-quality habitats, from laying earlier. In terms of underlying physiological mechanisms, this places food availability or *resources* front and center, so we need to identify mechanisms by which food availability per se can be integrated with, and modulate the function of, the reproductive (HPG) axis. However the second idea, that a female forms eggs at the time when she can do so "without risk to herself" (Perrins 1970), also captures the idea that *if* individual females do start laying too early, before they have sufficient food, they risk potential negative consequences of this inappropriate timing decision in the longer term. Life-history theory formalizes this second point and predicts that variation in the timing of reproduction is an "individually based compromise" involving trade-offs between different fitness components, including reproductive success *and* survival, which are optimized at the level of the individual female (Drent 2006). Thus, individuals that breed relatively early might recruit more offspring per breeding attempt but will have lower adult survival and potentially a shorter reproductive life span, whereas individuals that breed relatively late recruit fewer offspring per breeding attempt but might have higher survival rates. This *individual optimization hypothesis* argues that although a later-laying female might have lower fitness compared with an earlier-laying female, she herself would have lower fitness if she laid earlier (or later) since she would be moving away from the "optimum" laying date that maximizes her own fitness. This life-history approach therefore also requires that we identify *physiological mechanisms* that underpin the longer-term negative consequences or costs of reproduction associated with the "risk" of making inappropriate timing decisions (Harshman and Zera 2007).

There is considerable empirical support for the individual optimization hypothesis from experimental studies for laying date and clutch size (e.g., Pettifor et al. 1988, 2001; Brinkhof et al. 2002; but see Tinbergen and Both 1999; Tinbergen and Sanz 2004). However, an important point, though one that is often ignored, is that although each female might breed on her "optimum" date, thus maximizing her inclusive fitness, a large proportion of individual females in any given population effectively have zero fitness; that is, they fail to recruit any offspring into the breeding population over their lifetime (Newton 1989a; see fig. 3.8). For example, in small, short-lived, hole-nesting birds, such as the blue tit, only 14% of fledglings survive to reproduce, and 35% of these do not produce

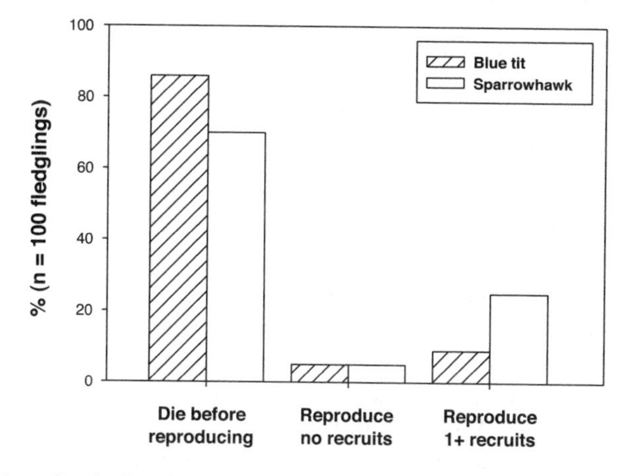

Fig. 3.8. Mortality before breeding, unsuccessful breeding, and successful breeding (%) among 100 fledglings in blue tits and sparrowhawks. (Based on data in Dhondt 1989 and Newton 1989a.)

recruits themselves, so that only 9% of fledglings produce all recruits in the next generation (Dhondt 1989). Similarly, in the longer-lived sparrowhawk (*Accipter nisus*), 70% of fledglings die before reproduction, and 17% of birds surviving to reproduce produce no recruits, despite attempting to breed for up to 4 years (Newton 1989b). Among individuals that *do* reproduce successfully, there is huge variance in the number of recruits they themselves produce, from 1 to 62 in the blue tit and from 1 to 24 in the sparrowhawk. As a consequence, a relatively small number of high-quality individuals produce the majority of the offspring in the next generation, and the majority of individuals that fledge successfully (a commonly used metric of breeding success) die without producing young themselves (Clutton-Brock 1988; Newton 1989a).

Relatively simple ideas about constraints, such as the *food constraint hypothesis* of Perrins (1970) and the more comprehensive life-history perspective, have important implications for the types of *physiological* mechanisms that will be involved in determining timing of breeding. There must be "immediate" or short-term mechanisms that allow for timing of the current breeding attempt, primarily involving the integration of information from environmental cues with the HPG axis. Then there must also be longer-term mechanisms that determine the cost or benefit an individual obtains for any particular timing decision (for example, reduced future fecundity or survival). Both of these must operate against a strong background of "individual quality, "which determines relative fitness and which itself must have a mechanistic basis. Furthermore, as discussed in the next section, there is no a priori reason why these mechanisms should be the same in males and females. Thus, we need to identify several distinct, potentially interacting and not mutually exclusive, and potentially sex-specific physiological mechanisms underlying timing of breeding:

1. Mechanisms that allow birds to use environmental information weeks or months in advance of egg-laying as "initial predictive" cues to time breeding or to set a reproductive window during which egg-laying can occur if other factors are in place
2. Mechanisms that fine-tune the actual timing of egg-laying, that is, the decision to initiate rapid yolk development (It should be noted that a priori the mechanisms for (1) and (2) could be the same and constitute what Visser [2008] calls the "response mechanism.")
3. Mechanisms that allow small-scale adjustment of timing of subsequent breeding events (e.g., hatching) *after* RYD or egg-laying have been initiated
4. Mechanisms that generate longer term, future "costs of reproduction" of the particular timing decisions in individuals that breed too early or too late relative to their "optimum," for example, higher mortality or lower future fecundity (These will be considered further in chapter 7 since they are a recurring theme in this book applying to multiple life-history traits and multiple life-history stages.)

3.5. Sex-Specific Response Mechanisms for Timing of Breeding

I will now consider the question of whether the same environmental cues and physiological mechanism(s) are used to time breeding in males and females. Onset of breeding clearly requires both males and females to have functionally mature reproductive systems at approximately the same time; not only must the eggs be developed and laid by the female, but they also must be fertilized by the male. Therefore, at the population level, the date at which the testes and ovaries reach their predicted size of functional maturity is expected to be highly synchronous (Partecke et al. 2005), although this timing can differ among populations (fig. 3.9). However, J. King et al. (1966) showed that the peak in testis size in male white-crowned sparrows occurs about two weeks before the peak in the ovarian cycle in females, and even more extreme sex differences in the timing of reproductive maturation are seen in other populations. Several blue tit populations that breed at similar latitudes in the Mediterranean region differ in their timing of breeding by up to one month (Blondel et al. 1999). However, males show similar timing in the onset of pituitary and gonadal recrudescence among these different populations (Caro et al. 2006), whereas female reproductive development is asynchronous by up to one month between populations and closely matches local differences in the timing of breeding (Caro et al. 2009). Local adaptation of breeding in these populations therefore involves marked asynchrony in the final stages of gonadal maturation between males and females of a single population. Somewhat earlier functional maturity in males than females would be consistent with evolutionary theory: (a) sperm production is less costly than egg production so the costs to males of mistimed reproductive development will be less than that for females (A. Hayward and Gillooly 2011), and (b) early reproductive maturation in males can be beneficial in terms of access to higher quality females and, potentially, multiple matings (e.g., extra-pair copulations).

Regardless of timing of attainment of functional maturity, is the time taken from the *initiation* of gonadal development to functional maturity the same in males and females, as is widely assumed to be the case? If males and females initiate reproductive development at the same time, and if the preparatory phase of reproductive development before fertilization and egg-laying is the same in both sexes, then it would be parsimonious to assume that both sexes rely on the same cues and cue values (e.g., a critical day length or *warmth sum* [see section 3.7.1]) to time breeding. In fact the timing and rate of gonadal development prior to egg-laying is clearly very different in female birds compared with male birds (fig. 3.10). In field studies the time between initiation of rapid growth and maximum testis size is about 6 weeks in passerines (Partecke et al. 2005; Wingfield and Farner 1978). Similarly, in experimental studies maximum testis size is also reached after about 6 weeks on 13L:11D (A. Dawson and Goldsmith 1983; Silverin et al. 2008), a day length that approximates that prevailing in March and April, when rapid testis growth occurs in free-living birds (even though testis development does occur at a very low rate in males even on short days, perhaps starting as soon as the birds break refractoriness (A. Dawson and Goldsmith 1983; A. Dawson and Sharp 2007s). In contrast, gonadal development leading to functional reproductive maturity in *females* occurs more rapidly: the time from the initiation of vitellogenesis to ovulation of the first egg in female European starlings is only 6–7 days (Challenger et al. 2001; Vézina and Williams 2003), and plasma estradiol levels remain low prior to this period of rapid ovarian development (see fig. 2.13). This idea of timing of reproductive maturation in females does not include pre-vitellogenic ovary growth but, as discussed in section 2.1.1, the functional significance of this phase of ovary development, in the context of constraints on timing, is not known. Thus, I would argue that males initiate rapid gonadal development on average 5 weeks before females, and this timing suggests that the environmental cues, or the

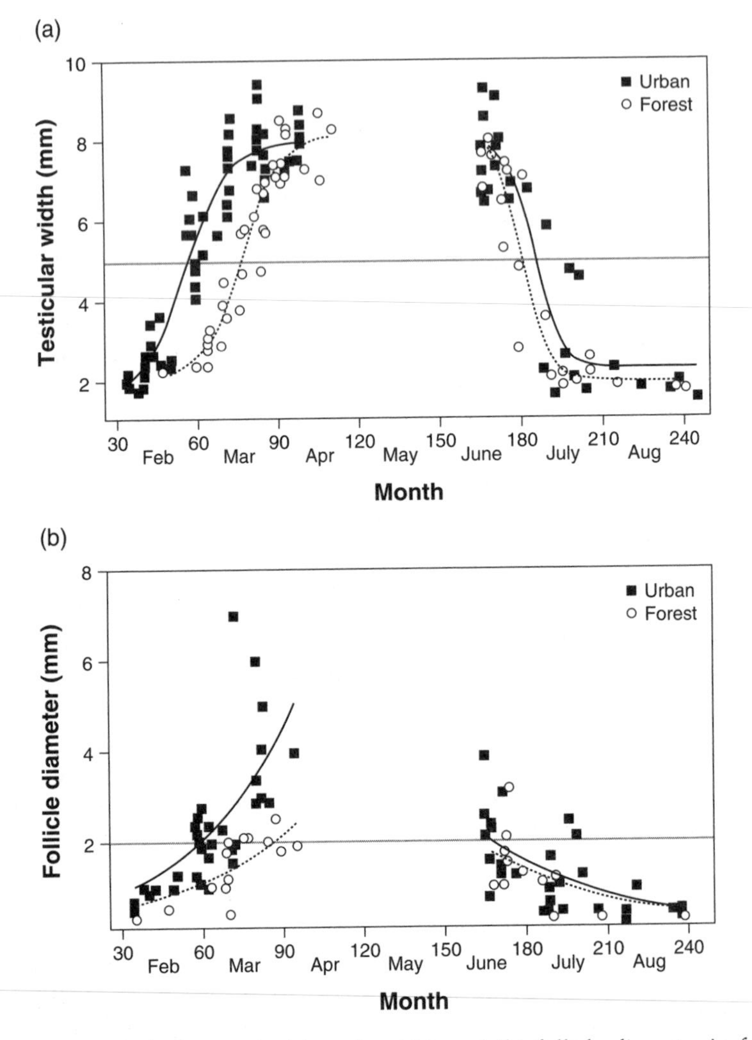

Fig. 3.9. Seasonal changes in (a) testis width and (b) follicle diameter in free-living urban (squares) and forest (circles) European blackbirds. Horizontal black line indicates minimum gonadal development above which egg-laying is possible within a few days. (Reproduced from Partecke et al. 2005, with permission of John Wiley and Sons.)

value of these cues, used to time breeding will be very different between the sexes. For example, for the male- and female-specific dates of rapid gonadal development for European starlings shown in figure 3.10, the prevailing day lengths are approximately 12.3L (5 March) and 15.2L (15 April), respectively.

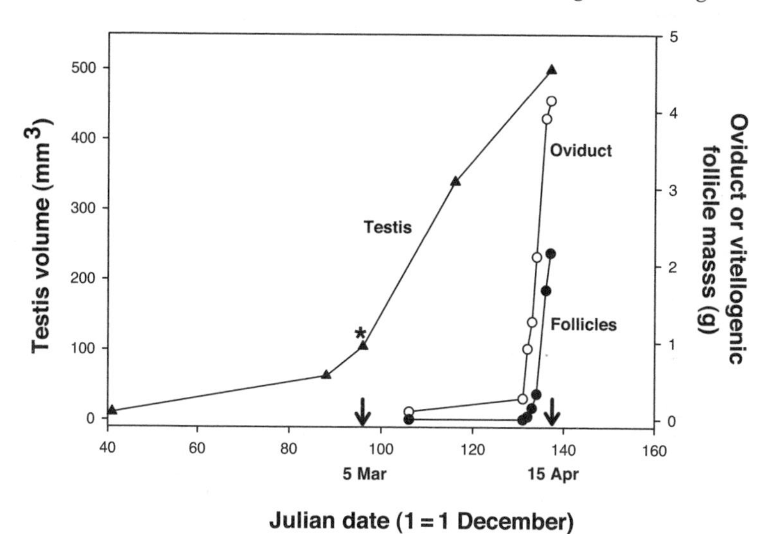

Fig. 3.10. Sex-specific differences in timing and rate of seasonal gonadal maturation in male (triangles) and female (circles) European starlings. (Based on data in Dawson 2003 and on unpublished data from T. Williams.)

A second key difference between males and females is the *length* of time the gonads are maintained in a fully developed state in any one breeding cycle. Although frequent repeated sampling of free-living males is rarely possible, data from field populations suggests that individual males maintain full-sized, functionally mature testes for several weeks or even months, probably again associated with the low cost of sperm production (relative to ovarian development, A. Hayward and Gillooly 2011) and the potential benefits of extra-pair copulations in males. In white-crowned sparrows, maximum testis mass is maintained for up to 60 days, from spring arrival through to late incubation, and in the multi-brooded Puget Sound white-crowned sparrows (*Z. l. pugetensis*), testis mass decreases slightly during the feeding of nestlings from the first brood but remains large (>400 mg) throughout the period from spring arrival to incubation of the second clutch (Wingfield and Farner 1978). In contrast, maximum (functional) ovary and oviduct size in females is very tightly coupled to the period of egg formation, which lasts at most 10–15 days in small passerines (depending on clutch size; see figs. 2.10 and 2.13). Following ovulation of the last egg, the female reproductive system rapidly regresses to a non-breeding state (Vézina and Williams 2003; T. Williams and Ames 2004) and must be fully redeveloped during subsequent breeding attempts. Changes in ovarian size and function, and changes in plasma steroid and gonadotropin levels in

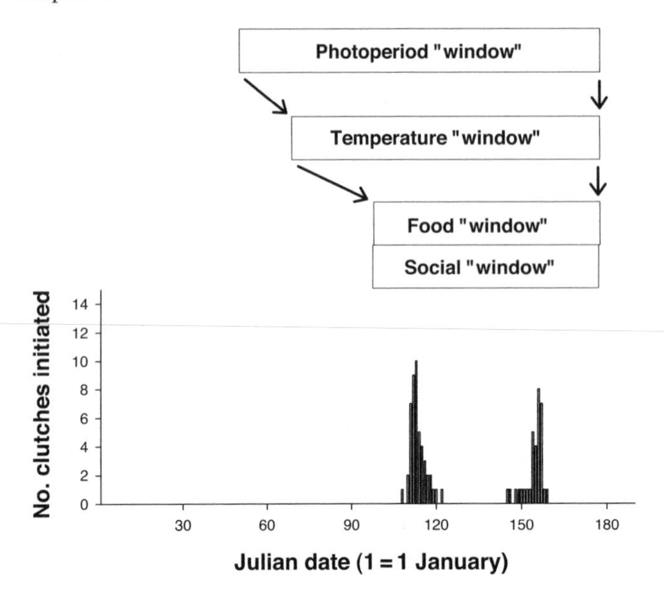

Fig. 3.11. Conceptual model for the integration of initial predictive information (photoperiod) and supplementary factors in relation to timing of breeding; clutch size data are for European starling. (Based on unpublished data from T. Williams.)

females, are therefore much more cyclical in natural populations than changes in testis size and function in males (for example, see figs. 3 and 4 in Wingfield and Farner 1978). The mechanistic significance of this point lies in the fact that rapid testicular regression is considered a hallmark of photorefractoriness, the mechanism that is considered to terminate reproduction at the end of the breeding season (A. Dawson et al. 2001; Nicholls et al. 1988). Photorefractoriness ensures that birds do not initiate breeding attempts late in the season when environmental conditions, such as declining food availability, would not support successful chick-rearing. Therefore, in males refractoriness and testicular (gonadal) regression typically *do* occur at the end of the breeding season and mark the end of reproductive activity. However, in females ovarian regression and cessation of vitellogenesis occur early in each breeding attempt (at clutch completion), and these events *do not* mark the shutdown of the HPG axis or the end of reproductive activity: females still incubate and rear chicks and can lay subsequent clutches. Furthermore, this process can occur repeatedly throughout the breeding season in females that produce multiple clutches such that the association between gonadal regression, refractoriness, and termination of reproductive activity is very different for males and females, and the mechanism underlying testis regression is unlikely to explain ovarian regression.

As I described in chapter 2, it is widely accepted that day length is the main proximate environmental factor regulating the timing of reproduction, perhaps determining a "photoperiodic window" during which breeding can occur. Then, within this broader window other factors, such as temperature, food availability, and social factors, fine-tune the actual timing of egg-laying (fig. 3.11). In the next section I will consider three environmental factors— photoperiod, temperature and food availability—in turn focusing on (a) how these factors might regulate individual variation in timing of egg-laying, (b) how they might regulate sex-specific differences in timing of gonadal development, and (c) how information from these cues is integrated into HPG-axis function to specifically regulate ovarian and oviduct development, vitellogenesis, and yolk uptake (for a comprehensive consideration of social factors see Adkins-Regan 2005).

3.6. Physiological Mechanisms Associated with Photoperiod (Day Length) as a Proximate Factor

Photoperiod (day length) is a key proximate factor used by birds to time breeding for the simple reason that day length varies predictably each year. Increasing day length in spring can therefore provide accurate long-term, predictive information for general timing of breeding each year (A. Dawson et al. 2001). However, is day length *sufficient* as well as necessary for timing of breeding such that it alone determines (or constrains?) individual variation in timing, as has been suggested for some populations (Lambrechts et al. 1997)? Or does day length simply set a very broad "reproductive window" during which reproduction can occur if other proximate factors allow (fig. 3.11)? Visser et al. (2004) suggested that since the change in photoperiod with date is the same every year, it cannot play a role in how birds adapt to year-to-year variation in optimal breeding time. Thus, additional, supplemental cues should play a more important role in determining *individual* variability. However, this conclusion assumes that there are relatively fixed, species-specific, photoperiodic responses to particular day lengths, which seems unlikely (A. Dawson 2008). If there is individual, sex-specific or population-level variation in the day length required to initiate gonadal development, or in the *rate* of response to a given day length, then variation in, and selection for, the photoperiodic response could provide a mechanism underlying individual variation in timing of breeding both within and between years.

3.6.1. Individual Variation in Photoperiodic Response

Few studies have documented individual variation in rates of gonadal development in response to fixed stimulatory photoperiods (but see section 2.6.1) even though such experiments would be very straightforward using an individual variation paradigm (T. Williams 2008). In fact, few studies have reported species-specific (A. Dawson 2008; Sharp and Moss 1981) or population-level (Lambrechts et al. 1996; Silverin et al. 1993) variation in photoperiodic responses. In free-living birds some species have broad latitudinal breeding ranges, the implication being that different populations undergo seasonal gonadal recrudescence under very different day lengths. For example, great tits breed between 35°N and 70°N (Sanz 1998) and egg-laying occurs about one month later at 70°N (c. 25 May) compared with 40°N (c. 24 April). Assuming males initiate rapid testis development on average 6 weeks before egg-laying in both populations, the prevailing day lengths at this time would be approximately 12L (13 March) and 16L (14 April) at 40°N and 70°N, respectively. In a common-garden experiment on gonadal growth on male great tits, photoperiod-induced rapid testicular growth was initiated when day length exceeded 11L in birds from northern Italy at 45°N and when it exceeded 12L hrs for birds from Norway at 69°N (Silverin et al. 1993). If these represent population-specific day lengths required for the initiation of testis development, then they are exceeded on 28 February and 18 March at 45°N and 69°N, respectively. This in turn would predict that the earliest possible dates for onset of egg-laying based on *male functional maturity* are 11 and 29 April, approximately 2 weeks and 4 weeks before the earliest recorded laying dates at these latitudes (Sanz 1998). Similar calculations for male European starlings breeding at 49°N, where 11L is exceeded on 1 March, predict attainment of functional reproductive maturity in *males* around 10 April, the mean first-egg date at this latitude (but later than the earliest laying date recorded: 29 March in 2010). Silverin et al. (1993) concluded that selection has led to the evolution of higher photoperiodic thresholds at more northern latitudes, and Coppack and Pulido (2004) also concluded that "intraspecific variation in egg-laying date is attributable to among-population differences in photoresponsiveness" in species that have a "rigid photoperiodic control system." However, it is clear that in some populations at least, and especially at northern latitudes, population differences in the day length at initiation of rapid testis development, as revealed by experiments on *captive birds*, are very different from prevailing day lengths at these different latitudes when *free-living* males actually undergo testis development. Thus, day length alone would predict a slightly earlier onset of egg-laying than that observed based on the timing of male gonadal development.

The calculations above assume that a relatively fixed period, of approximately 6 weeks, is required for rapid testis development in males under the prevailing day lengths experienced by free-living birds. In captive birds maximum testis size occurs 6 weeks after onset of photostimulation on 13L:11D, but after only 2 weeks on 18L:6D (A. Dawson and Goldsmith 1983). This observation suggests that there *can* be marked variation in the rate of testis development (although free-living birds experience day lengths as long as 18L only at high latitudes). This variation might ensure, for example, that testis maturation and other male traits (e.g., singing behavior) are synchronous with later female development and the egg-laying period. However, there is no evidence that differences in the *rate* of male development explain local differences in the timing of breeding. For example, in males of two populations of Mediterranean blue tits breeding in different habitats with a 1-month difference in the onset of egg-laying, the initiation of seasonal HPG development (GnRH-I cell number and immunoreactivity, testis volume, song activity) occurred at approximately the same time; in late winter, in both populations (Caro et al. 2005; Caro et al. 2006). The timing of the later stages of seasonal testicular growth and singing activity *did* differ between populations, occurring more rapidly in the earlier-breeding Muro population. However, the difference in male timing (10–15 days) was much less than the 1-month difference in egg-laying dates between populations. Caro et al. (2006) concluded that the difference in the photoperiodic response between males from the two habitats, if any, is very limited. Rather, "strong selection pressures on these two populations to adapt the timing of their breeding seasons to their local environment may have acted mostly on the *female egg-laying dates*, and not so much on the initiation and rate of seasonal recrudescence of the hypothalamo-hypophysial-testicular activity in males."

Consideration of the actual day lengths experienced by free-living birds during seasonal gonadal recrudescence (in the absence of real data for female birds!) highlights why it is so important to consider sex-specific differences in the timing of initiation and rate of gonadal development (fig. 3.10). For *female* great tits at 45°N and 69°N, if vitellogenesis and follicle and oviduct development are initiated only 1 week before egg-laying, prevailing day lengths at this time would be 13.6L and 24L, respectively. Thus, the photoperiods during gonadal development are very different in the two sexes, especially at higher latitudes, and the prevailing day lengths for females are markedly different from the "critical" day lengths determined for initiation of testis development from experimental studies of captive males. In European starlings, the prevailing day length is approximately 12.3L and 15.2L during the period of rapid gonadal development in males and females respectively (based on dates in fig. 3.10). There are a number of possible explanations for this difference, although we do not currently have the data to determine the correct one. First, compared with males, females might have a higher "threshold" day length (greater photostimulation) required to initiate reproductive development. Second, there might be a longer time- and photoperiod-dependent phase of pre-vitellogenic follicle or ovary development in females that must be completed before the phase of final, rapid gonadal development can be completed (as discussed in section 2.1.1). If this were the case, then prevailing day lengths for the *initiation* of development might be similar in males and females. There is only a single example in which a simple photoperiodic response mechanism has been proposed as the basis of intraspecific, adaptive variation in laying date based on analyses of female data (Lambrechts et al. 1997). Even in this example there appear to be population-level differences in sensitivity and response to non-photoperiodic factors that might contribute to timing decisions (Caro et al. 2007). The alternative, therefore, is that the "reproductive window" for breeding might be identical in both sexes, with the same day-length requirement, but that initiation of ovarian development is delayed relative to testis development because the former requires other non-photic, environmental cues (G. Ball and Ketterson 2008; Farner et al. 1966). This hypothesis predicts that non-photic cues are more important in females than males and, specifically, that non-photic cues should have significant, direct effects on the initiation and rate of vitellogenesis and follicle and oviduct development. I will explore the evidence for this conclusion in subsequent sections of this chapter.

Although few data are currently available on sex-specific differences or heritable, individual variation in photoperiodic responses, several lines of evidence suggest that such variation must exist. Some relevant information can be obtained from studies of populations or species at the same latitude, which are therefore under the same photoperiod regime, but with different local timing of breeding (Lambrechts et al. 1996; Partecke et al. 2005). In blue tits breeding in the Mediterranean region, the local differences in the timing of breeding of 1 month persist in pairs maintained under controlled, captive conditions in common-garden experiments with photoperiods that mimic the natural day length cycle (A in fig. 3.12) and with accelerated but increasing day lengths (B in fig. 3.12; Lambrechts et al. 1997; Lambrechts et al. 1996). Although extreme photoperiod regimes (C in fig. 3.12) can override these population differences, the ecological and evolutionary relevance of that is not clear (individuals are never exposed to day lengths >15.3L at these latitudes). These data are consistent with local adaptation: later-laying populations have a higher photoperiodic threshold such that breeding differentiation between populations is *proximately* caused by variations in the response

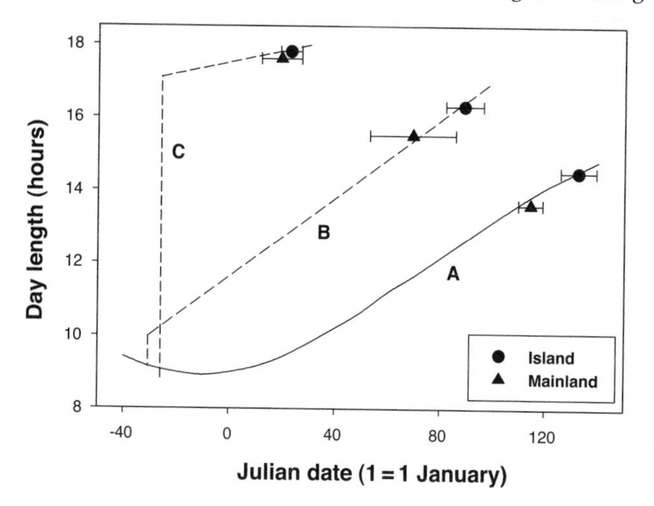

Fig. 3.12. Population-level variation in laying dates in response to different artificial photoperiod regimes in captive-breeding mainland (triangles) and island (circles) blue tits. (Redrawn from Lambrechts et al. 1996, with permission of the Royal Society.)

to photoperiod. This observation suggests that the photoperiodic response has a genetic basis with heritable individual variation that can respond to selection. Under natural day lengths, red grouse (*Lagopus lagopus scoticus*) begin to lay eggs 3 weeks earlier than willow ptarmigan (*L. l. lagopus*), with egg-laying occurring when day lengths reach 15 and 19 L, respectively, in the two species. However, willow ptarmigan × red grouse hybrids have intermediate photo-responsiveness, as would be predicted for a trait with a simple genetic basis (Sharp and Moss 1981). Other evidence indirectly supports the idea that the photoperiodic response mechanism has a genetic basis that can respond to selection. Domesticated species have a much more constant relationship between testis growth and duration of the daily photoperiod, probably as a response to selection for egg-laying, for example, gonadal growth is proportional to photoperiod only between 11.5L and 13L in quail (Etches 1996) but between 11L and 18L in the European starling (A. Dawson and Goldsmith 1983). A recent whole-genome analysis in chickens identified the locus for the thyroid-stimulating hormone receptor as a potential target of selection associated with the loss of strong photoperiodic or seasonal control of reproduction during domestication (Rubin et al. 2010). Finally, Helm and Visser (2010) reported heritable variation of circadian period length, a component of the photoperiodic response mechanism, in great tits. However, much more work on individual variation in the different components of the photoperiodic response mechanism is clearly warranted.

3.6.2. Where Would Sex-Specific Photoperiodic Response Mechanisms Reside?

If there is sex-specific (or individual) variation in the photoperiodic response mechanism, current thinking would suggest that it would be controlled at the level of the central nervous system (e.g., by photoreceptors themselves, or in the timing mechanisms in the suprachiasmatic nucleus) or the hypothalamus or pituitary (see section 2.5). In other words, lack of reproductive maturation or ovarian function in females is caused by insufficient "upstream" hypothalamic-pituitary stimulation of the gonads. This idea remains plausible given the lack of data specifically characterizing photoperiodic responses of female birds and given the lack of data characterizing these responses at the level of the ovary or in relation to vitellogenesis. Recent studies in free-living species have analyzed genotypic variation among individuals for components of the molecular clock that generates the circadian rhythms underpinning timing of breeding. Liedvogel et al. (2009) analyzed allelic variation in the *Clock* gene in a single population of blue tits. In females, but not in males, individuals with fewer polyglutamine (poly-Q) repeats bred earlier in the season in one of two years, and these individuals produced a higher number of fledged offspring. Liedvogel et al. (2009) therefore suggested that the among–population allelic variation previously documented at the *Clock* locus (Johnsen et al. 2007) could be linked to within-population variation in the timing of breeding. However, in a subsequent study, Liedvogel and Sheldon (2010) found only low variability in the poly-Q locus of the *Clock* gene in great tits, and genotypic variation was not associated with reproductive phenotype (timing of egg-laying) in females or with reproductive success.

As described in chapter 2, some experimental studies suggest that there appear to be *no* differences in the photoperiodic response in males and females at the hypothalamic-pituitary level, downstream from photoreceptors and the central clock. For example, captive male and female European starlings show similar, rapid increases in plasma LH and FSH when transferred from short days (8L:16D) to either 13L or 18L (A. Dawson and Goldsmith 1983). A. Dawson et al. (1985) also showed that in female starlings transferred from 8L:16D to 18L:6D, hypothalamic GnRH content rose after 2 weeks on long days and elevated levels were maintained for 6 weeks before decreasing, a pattern identical to that seen in male birds. Similarly, in European blackbirds (*Turdus merula*) there were no differences in the timing or pattern of secretion of pituitary gonadotropins (LH) or gonadal steroids (testosterone, estradiol) in either sex between urban and forest populations, even though urban birds advanced the timing of final gonadal maturation by 20 days in males and by 28 days in females compared with forest birds (Partecke et al. 2005). Partecke et al. (2005) speculated that the HPG axis might not be the only hormonal cascade that translates and integrates environmental information to regulate gonadal function. In red grouse and willow ptarmigan Sharp and Moss (1981) showed that despite a 3-week difference in mean laying dates, the critical day length for initial increase in plasma LH was not different between these two species (<12 L), even though egg-laying occurred when day lengths reached 15 and 19L respectively. They also suggested that differences in the onset of the timing of breeding "may be caused by modifications in neural or endocrine pathways *downstream* of the biological clock."

These few studies are consistent with an alternative *female-specific* hypothesis in which the initiation of vitellogenesis, follicle development, and the timing of egg-laying is regulated by "downstream" mechanisms operating at the level of the ovary and/or the liver. These "downstream" mechanisms might be directly regulated by photoperiod or might involve integration of other environmental information (e.g., temperature, food) at the level of the ovary. Few data are available to test this idea, since so few studies have measured ovarian function or vitellogenesis directly. However, the idea of autonomous or semi-autonomous control of ovarian function has been proposed elsewhere in relation to photoperiodic and circadian regulation of ovulation (Nakao et al. 2007). As described in chapter 2, the ovulatory cycle is thought to be regulated by a circadian-clock-driven diurnal rhythm of neuroendocrine responsiveness to the positive feedback effect of progesterone on LH release, which in turn has traditionally been thought to be *centrally* regulated by circadian oscillators residing in the suprachiasmatic nucleus. However, based on the expression of clock genes in the ovary and on the circadian patterns of expression of genes in pre-ovulatory follicles involved in progesterone synthesis, Nakao et al. (2007) suggested that the circadian clock regulating ovulation might lie in the ovary itself. As G. Ball (2007) said, "The ovary knows more than you think!" To further address this issue, more data on rates of ovarian development and on up-regulation of vitellogenesis in relation to variation in daily photoperiod would be invaluable, although the data might only be obtainable in semi-natural aviary breeding systems where females will complete egg production. Such studies need to be comprehensive and include "downstream" components of the female reproductive system (e.g., vitellogenin synthesis and uptake), not just measurement of the hypothalamus-pituitary system. Molecular studies measuring mRNA or protein expression should also include reproductive tissues, not just hypothalamic tissue, as evidenced by the recent finding of expression of GnIH peptide and GnRH receptor in thecal and granulosa cells of the ovary (Bentley et al. 2006).

3.7. Physiological Mechanisms Associated with Temperature as a Proximate Factor

In temperate- and high-latitude-breeding birds, the timing of egg-laying is highly correlated with annual variation in spring temperatures: in relatively warm springs females breed earlier than they do after a cold spring. This population-level response is mainly the result of individual birds breeding earlier in warmer years (Nussey et al. 2007). Dunn (2004) summarized a large number of studies and showed that 79% (45 of 57) of species showed a significant negative relationship between the date of laying and spring air temperature. This response is considered to be adaptive because it fine-tunes or synchronizes the birds' phenology with the temperature-dependent development and growth rates of the insect prey that are used to feed nestlings (Visser et al. 2006). However, this fine-tuning requires that temperature affects the timing and rate of female gonadal development and the timing of egg-laying since, as we have seen, initiation of egg-laying is what determines the timing of subsequent breeding stages such as chick-rearing. If birds are responding to temperature itself, then it could operate in two ways and involve two different mechanisms: (a) temperature might act indirectly as a relatively long-term "information" cue that predicts seasonal increases in food availability and determines laying dates within the broader reproductive window set by photoperiod, or (b) temperature might act directly through short-term effects on the rate of ovarian development or female energy expenditure, that is, on temperature-dependent physiological or metabolic processes. It is also possible that birds do not respond to temperature itself but rather to some other cue that is correlated with temperature, for example, the rate of development or availability of insect prey (Bryant 1975) or vegetation (Perrins and McCleery 1989). These processes are not mutually exclusive (Dunn 2004), but as we will see, we currently know very little about the *physiological mechanisms* that might mediate and regulate any of these possible temperature-dependent effects.

3.7.1. Temperature as a Long-Term "Information" Cue

Most ecological studies have calculated some form of warmth sum to investigate the relationship between Ta as a long-term predictive cue and

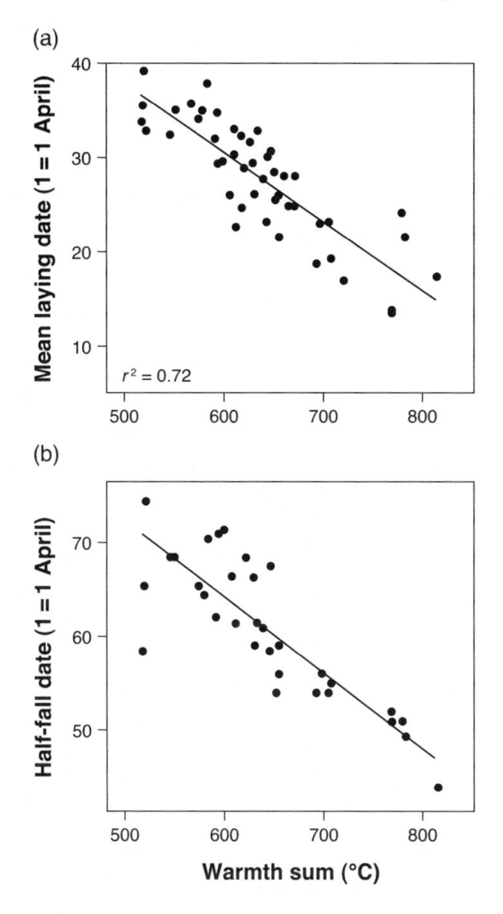

Fig. 3.13. Relationship between (a) temperature (spring warmth sum for 1 March–25 April) and mean laying date of great tits and (b) temperature and half-fall date of winter moth caterpillars. (Reproduced from Charmantier et al. 2008, with permission of the American Association for the Advancement of Science.)

laying date in correlational studies. In their analysis of the long-term Wytham Wood's tit data, Perrins and McCleery (1989) calculated the warmth sum as the sum of the daily maximum temperatures during the period 1 March–25 April, and this value was highly negatively correlated with laying date. A problem with this approach, from a mechanistic perspective, is that mean laying date in this population varies from 10 April to 10 May. So in some years egg-laying occurs on average 10 days *before* the end of the calculated warmth-sum period, whereas in other years it occurs 20 days *after* the end of the warmth-sum period. In a re-analysis of these data, Charmantier et al. (2008) also calculated a warmth sum for the period 1 March–25 April, and this period included the mean egg-laying date in about one-third of the years analyzed. This study confirmed that spring temperatures were highly correlated both with laying date and with food availability (fig. 3.13, where the half-fall date is a standard measure of the timing of peak larval biomass of the winter moth [*Operophtera brumata*]). These analyses therefore confound the potential long-term predictive effects of temperature with "direct" short-term effects, at least in some years (Gienapp et al. 2005).

If egg-laying dates were correlated with March Ta only, before the onset of vitellogenesis, temperature might then act as a long-term predictor of future conditions, whereas correlations with late April Ta, after the onset of ovarian development, would be more consistent with temperature having direct effects. Van Balen (1973) calculated warmth sums over different time periods and showed that laying date was most highly correlated, statistically, with temperatures between 1 March and 20 April ($r = 0.782$). Laying dates in this population ranged from 15 April to 4 May, so again this warmth-sum period captured potential direct effects of temperature in at least some years. A warmth sum calculated only for the period 1–20 April was almost as good a predictor of laying dates ($r = 0.718$), suggesting that earlier, March temperatures are relatively unimportant. Conversely a warmth sum calculated for the period 11–20 April, which would mainly capture "direct" effects of temperature, did not predict the timing of laying ($r = 0.41$). Gienapp et al. (2005) highlighted the fact that these studies do not account for within-year variation in phenology among individuals, which, as we have seen, is a key determinant of fitness. What is required is analysis of individual observations, not population means. Gienapp et al. (2005) used a proportional hazards model to analyze temperatures in relation to individual laying dates and concluded that current temperatures (i.e., those closest to first egg dates) play a less important role than the long-term temperature trend in determining laying date (a finding consistent with van Balen's analysis [1973]). However, Gienapp et al.'s analysis (2005) suggested that "short-term temperature signals" were still visible in the data. Brommer et al. (2008) used a sliding-window approach to identify the period that provided the best correlation between temperature and laying date in the common gull; this was 31 March–26 April in a population with earliest laying dates of 1 May, again more consistent with a longer-term effect of temperature. Husby et al. (2010) similarly showed that temperatures relatively far in advance of laying best predicted the laying date, although the period that explained most of the variation in the onset of laying was 15 February–25 April in the Wytham Woods great tit population and 13 March–20 April in the Hoge Veluwe, Netherlands population. Finally, Visser et al. (2009)

experimentally manipulated temperature in great tits breeding in aviaries and showed that temperatures over a 3-week period just before mean egg-laying date was the best predictor of variation in timing of breeding (but see Visser et al. 2011). These studies all support the view that temperature mainly acts as a longer-term, predictive "information" cue, with short-term direct effects of temperature being less important.

Several recent studies have analyzed individual variation in the laying date–temperature relationship, or reaction norm, to investigate the fitness consequences of *plasticity* in laying date (slope) and of laying earlier or later in an average environment (elevation; see fig. 1 in Brommer et al. 2005). These studies provide evidence for heritable, individual variation in slope and/or elevation and for selection on these traits (Brommer et al. 2005; Brommer et al. 2008; Charmantier et al. 2008; Husby et al. 2010; Nussey et al. 2005). Female great tits in the Netherlands that lay earlier in the average environment (low elevation) and respond more strongly to temperature (more negative slope) had significantly more of their offspring recruited into the population as breeding adults. Nussey et al. (2005) suggested that these earlier/more plastic birds would be expected to match their reproductive timing better with the peak in caterpillar biomass, especially as spring temperatures become increasingly warm. T. Reed et al. (2009) documented among-individual differences in nonlinear (i.e., degree of curvature), as well as linear, components of the reaction norm of laying date in relation to a composite variable of environmental conditions (including sea-surface temperature, an upwelling index, and a large-scale atmospheric phenomenon, the Northern Oscillation Index) in common murres (*Uria aalge*) in California. In particular, individual reaction norms were distinct from each other in years with the poorest environmental conditions (higher variance in laying date among individuals) but converged as environmental conditions became more favorable. This relationship would be consistent with the idea that when conditions were challenging, many females were constrained to lay later (T. Reed et al. 2009). Any physiological response mechanism(s) for temperature cues must therefore be able to accommodate this adaptive phenotypic plasticity at both the population and individual level.

Studies on individual variation in laying date–temperature relationships are useful in another way—in identifying the sensitivity, or "tolerances," required of any underlying physiological response mechanism(s). Female collared flycatchers laid on average 2.1 days/°C earlier with warmer local temperatures, and individual slopes varied from −2.5 to +2.2 days/°C around the average slope, so that a variation in mean temperature of only 4°C–5°C over several months captured the full range of laying date responses (Brommer et al. 2005). In blue tits breeding in

Corsica, the laying date varies by approximately 4 days for each 1°C change in mean March temperature (Bourgault et al. 2010). Similarly, the average difference in temperature between a "warm" and "cold" spring for great tits breeding in the Netherlands is only 4°C (Visser et al. 2009). So whatever mechanism underlies the response to temperature in birds, it must be very sensitive, in that a difference of only a few degrees changes the timing of egg-laying by several days, a shift that is sufficient to generate significant fitness consequences.

3.7.2. Correlates of Temperature as Information Cues

Although the studies described above support the idea of temperature as a long-term, predictive information cue, they do not show that birds are responding directly to temperature itself. Rather, birds might be responding to indirect (visual) cues that are themselves directly correlated with temperature. It is generally assumed that visual cues from prey are unlikely to be informative given that prey availability often does not increase until *after* egg-laying, which is coincident with young in the nest (although it is possible that the presence and/or abundance of earlier developmental stages, such as larvae, might provide phenological information [Visser and Lambrechts 1999]). An alternative visual cue might be one provided by developing vegetation, such as bud burst. The timing of egg-laying can be highly correlated with bud burst for tree species that are the main hosts for the bird's specific insect prey or with other measures of vegetation phenology, for example, flowering (Møller 2008; Slagsvold 1976; Visser and Lambrechts 1999). Bourgault et al. (2010) suggested that laying date is causally related to vegetation phenology, although they also pointed out that these correlational analyses do not show that females necessarily use bud burst per se as the proximate cue for timing of breeding (see also D. Thomas et al. 2010). In Bourgault et al.'s study (2010), bud burst preceded the date of peak caterpillar abundance by about 25 days, and these events were highly correlated, so bud burst could be a long-term cue. Nevertheless, leaf phenology is not an absolute cue for the timing of breeding since local populations of the same species can initiate laying at different times relative to leafing, and local populations can initiate laying earlier relative to leafing during late springs (Nilsson and Källander 2006). Furthermore, great tits started laying at the normal "average" date in a year when oak leafing was much delayed because of events in the preceding year (Visser et al. 2002). The only experimental work to investigate the effect of leaf phenology does not support the idea that this cue plays a direct role in the timing of gonad development or egg-laying, at least in *Parus* species. Visser et al. (2002) maintained great tits on naturally increasing day lengths and provided birds with branches of oak trees (*Quercus* spp.) that differed only in the stage of development of buds on the branches, ranging from zero (winter rest) to three (fully unfolded leaves). Egg-laying started about a week after the first introduction of the branches into the aviaries, but there was no significant difference in laying date between treatments, indicating that there were no effects of either visual or chemical cues from buds.

3.7.3. Possible "Direct" Effects of Temperature on Timing

Is there evidence that temperature has direct, short-term effects on the rate of gonadal growth or female energy expenditure during egg production such that higher ambient temperatures "lift a constraint" on earlier onset of reproduction (Visser et al. 2009)? An important issue here is, again, the potential sex-specific difference in the timing of ovary/oviduct versus testis development (see fig. 3.10). If the rapid phase of testis development takes approximately 6 weeks in males, then temperature could affect individuals directly over this relatively long time period, but if most ovarian development occurs only 1 week prior to laying, then the period during which females would be responsive to the direct effects of temperature might be more limited. In contrast, if temperature has a direct, metabolic effect, it should also be more important in females because the resource cost of egg production is much greater than the resource cost of testis function and sperm production (A. Hayward and Gillooly 2011).

Experimental manipulation of ambient temperature (5°C, 20°C, or 30°C) had no effect on ovarian follicle development in female white-crowned sparrows (Wingfield et al. 1996). Furthermore, temperature had no effect on photoperiodic induction of plasma LH or FSH in either sex at 5°C and 20°C, although plasma LH increased significantly earlier in females at 30°C. In a follow-up study with Mountain white-crowned sparrows (*Z. l. oriantha)* transferred from 9L:15D to 15L:9D at either 5°C, 20°C, or 30°C, temperature had no effect on photoinduced LH release in females, and although exposure to 30°C increased the largest follicle size, there was no effect of temperature on ovary development at 5°C or 20°C (Wingfield et al. 2003; fig. 3.14). In male birds temperature has no effect on the timing of initiation, or rate, of testis development, despite large differences in temperature treatment (25°C vs. 5°C, and 18°C vs. 8°C in 2 years), although lower temperatures delay development of photorefractoriness and gonadal regression at the end of the breeding season (A. Dawson 2005; Perfito et al. 2005). Silverin et al. (2008) showed that there was no effect of temperature (4°C vs. 20°C) on the initiation and rate of testis growth or on increases in plasma

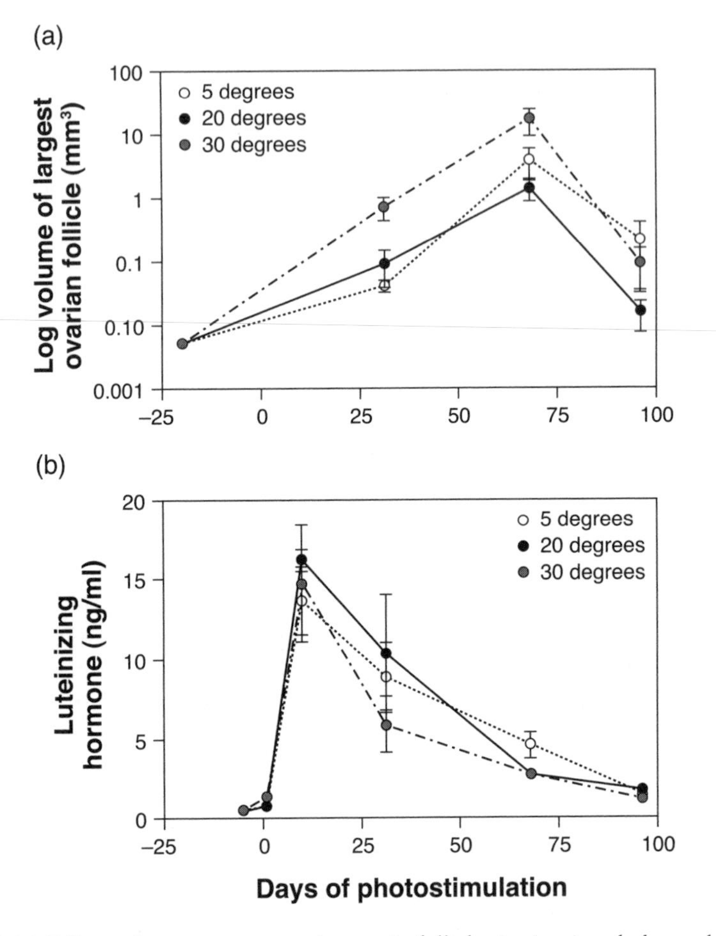

Fig. 3.14. Effects of temperature on changes in follicle size (top) and plasma levels of luteinizing hormone (bottom) in female white-crowned sparrows transferred from 8L:16D to 20L:4D at day 0. (Redrawn from Wingfield et al. 2003, with permission from Elsevier.)

LH in northern populations of male great tits (Sweden, northern Norway). However, in southern populations (Italy), testis growth and the photoperiod-dependent increase in plasma LH were delayed by up to 1 month at the lower temperatures. This result seems counterintuitive given the prediction that supplementary factors should be more important at higher latitudes, where there is a greater interval between experimentally determined "critical" day lengths and actual laying dates (see section 3.6.1).

Several important observations need to be made on the above studies. First, the range of temperatures used in experimental studies to date, between 5°C and 30°C, is much greater than the 4°C range of temperatures free-living birds are exposed to between extreme "warm" and "cool" years (Visser et al. 2009). Estimated warmth sums for experimental studies range from 280°C to 1120°C, which are much greater than the 500°C–800°C observed in field studies (see fig. 3.13). It could be argued that if a large temperature difference has no effect on the timing of maturation, then a smaller one is unlikely to have any effect, but it is possible that birds simply do not perceive the very large differences in Ta used in experiments as an appropriate cue. Second, most experiments have investigated the modulatory effects of temperature on the rate of photostimulated gonadal growth after transfer of birds from short to long day lengths and have used very long, stimulatory day lengths (20L). These experimental photoperiods are often much greater than the prevailing day lengths that free-living birds are exposed to during seasonal gonadal recrudescence, and, therefore, these experimental conditions are not ecologically realistic. A recent study in mammals suggests that extremely long day lengths can inhibit the effects of supplementary cues and that the use of intermediate day lengths can "unmask" reproductive responses to non-photic environmental cues (Paul et al. 2009). A third problem with the studies conducted to date is that captive females did not show normal or complete ovarian development, for example, in Wingfield et al.'s study (2003), "ovarian development" extended over 50 days, a time-frame very different from the period of rapid ovarian development seen in free-living birds. Thus, any future experimental studies should preferably use photoperiods and temperatures within ecologically relevant ranges.

Relatively few studies have looked at the effects that Ta prior to laying has on female energy expenditure, especially in free-living birds. I. Stevenson and Bryant (2000) showed that DEE was negatively correlated with Ta during laying in great tits (fig. 3.15). They suggested that this correlation reflected a "thermal constraint" on egg production due to greater thermoregulatory requirements or temperature-dependent food availability that would oppose selection for earlier laying. However, only one experimental study has attempted to manipulate female energy expenditure prior to laying in free-living birds. Nager and van Noordwijk (1992) manipulated temperatures 2–4 weeks before laying by heating and cooling nest boxes, such that mean nightly temperature was 1.4°C higher (warmed) or 2.9°C lower (cooled) than control boxes (mean 5.6°C). They found no effect of temperature on laying date, although a limitation of this study was that females were exposed to different temperatures only for the periods of the day when they were roosting in the boxes. In a study of captive European starlings with ad lib food,

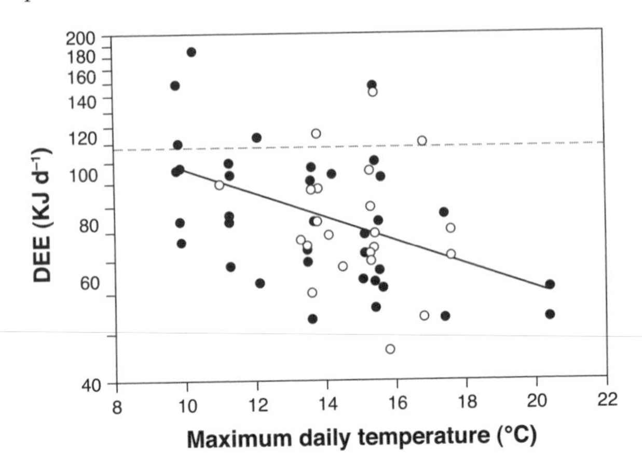

Fig. 3.15. Daily energy expenditure (DEE) in egg-laying great tits in relation to ambient temperature in two years: 1997 (filled circles) and 1998 (open circles). (Reprinted by permission from Macmillan Publishers Ltd: [*Nature*] Stevenson and Bryant 2000.)

Meijer et al. (1999) heated and cooled nest boxes by 2°C–3°C for 2–3 weeks before laying and found that temperature had a direct effect on the timing of laying and that the effect was independent of food supply and photoperiod. Cool temperatures retarded reproductive development, and most birds started laying 1 week after temperatures rose by 5°C. Neither of these studies directly measured energy expenditure. Salvante et al. (2007b) showed that female zebra finches producing eggs at 7°C had basal metabolic rates that were 218% higher than females producing eggs at 21°C (105.9 vs. 48.4 ml O2 per hour; this rate includes the costs of egg production, cold acclimation, and heat production). Although females took longer to initiate egg-laying at 7°C, the difference was relatively small (6.5 days vs. 6.1 days at 21°C), perhaps because females greatly increased their energy intake via greater seed consumption (a situation that is probably unrealistic in nature).

In conclusion, there is only limited evidence that temperature has a direct effect on hypothalamus-pituitary function, gonadal growth, or energy expenditure in birds of either sex (that is, during the maturation phase; but see A. Dawson and Sharp 2010 for temperature effects on gonadal regression). However, most studies have used temperature ranges that greatly exceed those which birds normally experience, and most of the existing studies are correlational. In females, photoinduced gonadotropin release appears to be independent of temperature, and there is little evidence for any temperature dependence of ovarian function and follicle development. Thus, despite the wealth of correlational data from field studies linking temperature and timing of egg-laying, exactly how temperature might directly influence *female* reproduction remains unknown (Visser et al. 2011) and virtually unexplored. Given the small number of experimental studies with female birds, it might be premature to conclude unequivocally that temperature is unimportant in the timing or rate of female gonadal development (cf. Bourgault et al. 2010), and regardless, that conclusion would be very hard to reconcile with data from field studies.

3.7.4. Where Would Temperature Response Mechanisms Reside?

Most studies discussed in section 3.7.1 strongly suggest that Ta in the several weeks to several months before egg-laying is initiated can affect the timing of laying, and this effect could involve two potential mechanisms with different sensory pathways. First, birds might be responding to visual cues (e.g., bud burst), in which case, environmental information would probably be encoded by the retinal photoreceptors and then would pass to the midbrain region (optic tectum) via the optic nerve, and from it via projections to other brain regions. Second, birds might directly integrate sensory information about Ta itself. Birds are known to possess several types of thermoreceptors, which are transmembrane proteins located in the plasma membrane of sensory neurons that project to the skin. These receptors belong to the transient receptor potential ion-channel family (Dhaka et al. 2006), and both cold receptors and heat receptors with activation temperatures below and above 25°C have been identified in pigeons and chicks, respectively (Myers et al. 2009; Schäfer et al. 1989). Sensory neurons with these receptors form part of the peripheral nervous system, which has neural inputs passing along the dorsal root ganglion to the spinal cord and then to the brain. Very little work has been conducted on these avian thermoreceptors, and none in the context of temperature-dependent timing of breeding. Therefore it remains unknown if, or how, birds might use these receptors to detect and integrate temperature information over long periods of time (e.g., to perform some form of warmth sum calculation?) or with the sensitivity required to detect annual differences in average temperatures of only a few degrees. Presumably, either visual or thermal information eventually must be integrated with, and affect the functioning of, GnRH and/or GnIH neurons in the hypothalamus (or potentially downstream targets) in a similar way as photoperiodic information. How this occurs remains a major challenge for understanding the timing of breeding in birds.

3.8. Physiological Mechanisms Associated with Food Availability as a "Proximate" Factor

Perrins (1970) was the first to highlight the role of food availability in timing of egg-laying of individual females, and it has been suggested that "Perrins was clearly thinking primarily of a *proximate* limitation by food" (Drent 2006; my italics), although Perrins himself never used this term (Perrins 1970, 1996). Perrins recognized that birds initiated reproductive development *before* the onset of any seasonal increase in food availability, a fact that would preclude food itself from being used as a long-term predictive cue. Food availability cannot therefore serve as a proximate factor as first defined by J. Baker (1938) or, as subsequently defined by others, as a cue "which changes in a predictable way during the year" (A. Dawson 1986) and which "stimulates the neural substrates that lead to reproductive cascades" (G. Ball and Ketterson 2008). Perrins (1970, 1996) did suggest that an individual female initiates egg-laying on the date on which she is able to find enough food to form eggs "without risk to herself" (Perrins 1970), that is, when she is in positive energy balance, including both maintenance and the costs of egg production. Thus, low food availability might provide the "constraint," or counteracting disadvantage, that prevents many individuals from breeding at the optimal time for having their young in the nest coincident with the seasonal peak in food supply. As with temperature, greater availability of food might therefore "lift a constraint" on earlier timing of reproduction (Visser et al. 2009).

The ideas proposed by Perrins (1970, 1996) have remained very influential and have routinely been cited in other influential reviews (Drent 2006; Wingfield 1983). However, none of these papers deals explicitly with the physiological mechanism underlying the effects of food availability. This mechanism requires that females can directly assess (a) food availability in the environment and/or (b) their own "nutritional plane," body condition, or endogenous resource level, and that they can (c) integrate this information to *directly* affect timing of onset and rate of gonadal development and egg formation. I will return to this issue in section 3.8.4 below. However, much of the evidence for the hypothesis that food availability has a direct effect on the timing of egg-laying, and for the required corollary that egg production is nutritionally and/or energetically demanding, comes largely from field studies and is correlational, and often circumstantial, even if the hypotheses themselves are intuitively attractive (see below). This body of studies has been extensively reviewed elsewhere (Drent 2006; Nager 2006), and several reviews have highlighted the limitations of these data in the context of egg production

(T. Williams 2005) and timing of laying (Verhulst and Nilsson 2008), so only the main arguments will be summarized here.

3.8.1. Argument 1: Food Availability Is a Constraint Because Supplemental Food Advances the Timing of Laying

Drent (2006) argued that a major strength of Perrin's hypothesis (1970) was that it was directly amenable to testing by feeding experiments in the field, which would be predicted to lift any food-dependent "constraint" and advance the timing of laying. Nager (2006) reviewed 57 food-supplementation studies in 38 species and showed that timing of egg-laying was significantly advanced in only 57% of the cases. Numerous explanations for the different outcomes of supplemental feeding experiments have been presented, including the idea that most studies have considered the wrong "currency" and that *specific* resources or nutrients, rather than gross energy or nutrient availability, might be limiting (Nager 2006; this is a common argument made in light of inconsistent results [see chapter 7]). However, providing birds with specific foods, for example, protein, lipids, or calcium, also has inconsistent effects on laying date: some studies have shown an advancement of laying (Nager et al. 1997; Ramsay and Houston 1997), whereas others have shown no effect (Bolton et al. 1992; Graveland and Drent 1997). Food supplementation can also indirectly manipulate other aspects of territory or individual quality; for example, high-quality birds might selectively occupy food-supplemented sites, and this situation could confound interpretation of the results (Nager 2006). The majority of supplemental feeding experiments have provided birds with additional food for weeks, and sometimes months, starting before egg-laying, and in most studies an "enormous" amount of food has been supplied (Meijer and Drent 1999). Despite this, food supplementation has a relatively modest effect in advancing the laying date in single-brooded species (≤5 days), but it has a greater effect on multi-brooded species (c. 15 days; Drent 2006; Svensson 1995), as would be predicted if multi-brooded species are "mistiming" their first brood relative to peak food availability to match their overall timing of breeding. Drent (2006) concluded that food *is* one of the proximate drivers of laying date "but that given the relatively modest response to food there must be other constraining factors as well." However, based on the currently available data, a devil's advocate faced with this intuitively logical argument might suggest that these inconsistent results reflect the fact that we have been looking at entirely the wrong currency (i.e., "resources") rather than considering alternative, potentially non-resource-based, mechanisms (T. Williams 2005).

The experimental design used in most supplemental feeding experiments, which use chronic provisioning of food over a long time period, is not well suited to testing Perrin's hypothesis (1970, 1996) that increased food availability releases females from constraint in the short-term, during the period immediately prior to onset of gonadal development or "at the time of laying." Rather, long-term food supplementation might provide birds with a predictive cue that could actually produce inappropriate or maladaptive timing decisions. In support of this idea, the few studies that have looked at the long-term effects of supplemental feeding have shown that fed birds that advance their laying date often do not breed more successfully and can actually have lower fitness (Nager et al. 1997; Nilsson 1994). An intriguing example is provided by a recent comparison of life-history traits of urban and rural bird populations (Chamberlain et al. 2009). Urban birds have access to greater food availability due to human-provided food and generally lay earlier than rural conspecifics, but on average they actually have smaller clutch sizes, lower nestling masses, and lower productivity per nesting attempt in urban landscapes. Similarly, chronic supplemental feeding of blue tits advanced the laying date but *decreased* the clutch size and breeding success (T. Harrison et al. 2010). These observations suggest that although birds can be "tricked" (Verhulst and Nilsson 2008) into laying earlier, the response to food supplementation might be maladaptive and does not reflect ecological, evolutionary, or physiologically relevant modulation of endogenous physiological response mechanisms.

3.8.2. Argument 2: Food Availability Must Be a Constraint Because Egg Production Is Energetically Expensive

It is considered a truism that "egg formation is a demanding process both in terms of energy and nutrient requirements" (Nager 2006), and this idea in itself has gone a long way to supporting the hypothesis that energy and/or nutrients can directly constrain egg production. However, there is actually very little direct evidence to support this statement, as highlighted by Nager (2006) himself, who went on to state, "we know little about how egg formation in wild birds is constrained by energy [or] other nutrients." It seems incredible that 40 years on from Perrin's seminal study (1970) directing our focus on the role of food availability as a constraint, such statements as "resource demand to prepare for reproduction, including growth of the gonads and accumulation of sufficient endogenous nutrient sources, may be substantial *but have largely been ignored* [my italics] and would deserve more attention in the future" are still manifestly true (Nager 2006; see also T. Williams 2005). This issue will be discussed further in chapters 4 and 5 in relation to quantitative aspects of egg production (egg size and number), so here I will simply highlight a few key problems. Recent empirical studies in both the field and laboratory have shown that the metabolic cost of egg production results in a 16%–27% increase in resting metabolic rate (Chappell et al. 1999; Nilsson and Raberg 2001; Vézina and Williams 2002, 2005), values that are much lower than earlier theoretical predictions. Few studies have measured DEE in egg-producing females, but those that have generally show that DEE is similar to, not higher than, that for other life-history stages. In other words, sustained DEE appears to be more or less constant during all phases of reproduction, and perhaps even throughout the year (I. Stevenson and Bryant 2000; Ward 1996; T. Williams and Vézina 2001). However, it is important to note here that average values of DEE might mask biologically significant inter-individual variations in energy "management" associated with egg production; that is, any energetic cost of egg production is not simply additive (Vézina et al. 2006; T. Williams et al. 2009; see chapter 4). If this is generally true, then the problem of identifying any additive energetic cost of egg production itself in individual females becomes very challenging. Overall, however, the data suggest that if the metabolic demands of egg production "constrain" the onset of egg-laying, then this constraint is not related to the total amount of energy (or resources) available, since the energy expenditure of laying females does not appear to markedly increase. Perhaps instead the costs and benefits of how this energy is allocated, and mechanisms that facilitate this allocation, are more important; but those ideas have been less well studied (see chapter 7).

3.8.3. Argument 3: Food Availability Is a Constraint Because Pre-breeding "Body Condition" Determines Laying Date

Although "body condition" itself is rarely defined, or even understood, it is commonly argued that high-quality individuals in good "condition" lay earlier because their better condition reflects the fact they have more resources for egg-laying. This argument is intuitively appealing, which explains why it has persisted almost as dogma in the literature. However, for income-breeding species that rely on daily food intake to support egg production, we would not actually predict *any* relationship between body mass or body "condition" and timing of egg-laying. The rate of food intake might increase to meet the greater demands of egg production, but this higher intake could be matched by the greater "demand" from reproductive processes such that endogenous reserves, and body condition, would remain constant. This idea has important implications for potential mechanisms mediating the effects of food (see below). It probably also explains why few studies have found strong relationships between *pre-breeding* body mass or size-corrected mass and the timing of laying in small-bodied passerines. In female European starlings there were no consistent differences in any component of non-reproductive body composition, organ mass, or nutrient reserves when pre-breeding females were compared with egg-laying ones over 3 years (Vézina and Williams 2003). Similarly, in migratory American redstarts (*Setophaga ruticilla*), although fat mass at arrival was positively correlated with arrival date and with some aspects of reproduction (e.g., egg mass), females arriving with more fat did not breed sooner than females in poor condition at arrival (R. Smith and Moore 2003). For larger-bodied "capital" breeders, which rely more on stored, endogenous reserves, pre-breeding body mass or condition can be positively correlated with reproductive success (e.g., Ebbinge and Spaans 1995; Madsen 2001), but relatively few studies have shown that females in better pre-breeding condition actually lay earlier than poor-condition females. Devries et al. (2008) showed that in female mallards (*Anas platyrhynchos*), the date of nest initiation was negatively correlated with size-corrected body mass, so that females in the best condition laid approximately 15 days earlier than the females in worst condition. Similarly, Bety et al. (2003) showed that in snow geese, the laying date was correlated with pre-migration body condition, controlling for arrival date. These birds will delay the onset of laying to accumulate more nutrients even at the expense of the survival of juveniles (Gauthier et al. 2003), an observation that provides more compelling evidence for food or resource availability acting as a constraint in capital breeders (but see chapters 4 and 5).

3.8.4. Where Would Food (Resource) Response Mechanisms Reside?

The foregoing sections should not, at this point, lead to the conclusion that food availability has *no* effect on timing of egg-laying; rather, the intent is to highlight the relative lack of *direct*, consistent empirical data in support of this hypothesis, which has instead largely relied on less-than-critical acceptance of intuitively attractive suppositions. The strongest support for the idea that food availability constrains timing of egg-laying in individual females would come from identification of the *physiological mechanisms* through which these effects are mediated. Most of the seminal papers on food and timing of laying (e.g., Drent 2006; Drent and Daan 1980; Perrins 1970) highlight the importance of this issue, and they deal with resource acquisition and resource allocation, which are inherently physiological concepts. For example, Drent and Daan (1980) suggested that the timing of egg-laying is determined by the female's "energy and nutrient balance," but they also articulated much more explicit ideas about *how* food availability might affect timing of laying, including the importance of rates of daily (food) intake, endogenous energy or nutrient stores, and "threshold values" of body condition. Nevertheless, they acknowledged that the "physiological mechanism by which food intake affects laying date [and clutch size] has so far hardly been touched upon." That statement remains true today. What do concepts such as "threshold body condition" or "rate of daily income" translate into at the mechanistic level? And how are these processes themselves integrated into the complex machinery regulating ovarian function (described in chapter 2) to modulate reproduction?

The first of these questions would appear to reside firmly within the domain of nutritional physiology and involve phenotypic flexibility of the digestive system, that is, mechanisms underlying food selection, food intake, digestion, processing, and nutrient uptake (Karasov 1996; Karasov and McWilliams 2005). In other situations where increased food intake is required, for example, during hyperphagia associated with low ambient temperatures or migration, there is considerable evidence that the size of the digestive tract (mass and/or length) and its function (e.g., upregulation of digestive enzymes and nutrient transporters, retention time) can be modulated in relation to variation in required food intake (Battley and Piersma 2005; Karasov and McWilliams 2005). These changes can occur rapidly (20%–40% increases within 2–3 days of a diet switch, Starck 1998), and there is some evidence for considerable inter-individual variation in at least some of these traits (e.g., digestive efficiency, Shuman et al. 1989). Although a large number of studies have demonstrated this marked phenotypic flexibility in body composition, organ sizes, nutrient uptake and transport, and fuel use in non-breeding birds, relatively few studies have investigated similar changes associated with reproduction (Vézina and Salvante 2010). Karasov (1996) suggested that egg production might involve increases in feeding rate (hyperphagia) of 17%–70%, but this idea was based on estimated increases in the metabolic rate of laying birds rather than on empirical measurement of food intake. Salvante et al. (2007b) directly measured food intake in zebra finches at different ambient temperatures and showed that seed consumption was 45% higher in egg-laying birds at 7°C than at 21°C. If we assume that egg production requires an increase in the total amount of food (energy), and that variation in the ability to obtain this higher level of food constrains reproduction (sensu Perrins 1970), then we should *expect* to see an adaptive phenotypic flexibility in the digestive system that regulates the acquisition of the energy and nutrients used during egg-laying. There is currently only very limited evidence for such adaptive phenotypic responses. In wood ducks (*Aix sponsa*, Drobney 1984) and cowbirds (Ankney and Scott 1988), laying females had heavier (+ 47%) and longer (+ 10%) intestines compared with pre-laying birds. Similarly, in geese

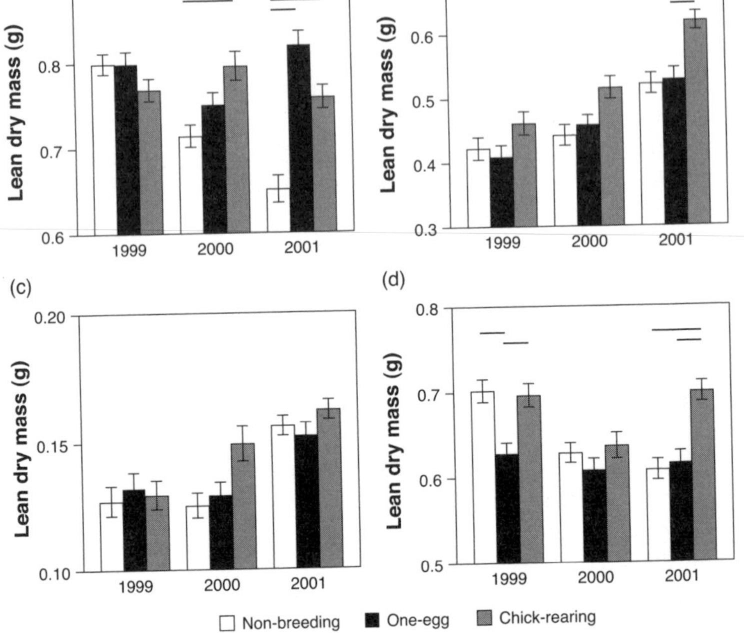

Fig. 3.16. Variation in residual lean dry mass of (a) liver, (b) small intestine, (c) pancreas, and (d) gizzard in relation to year and breeding stage in European starlings in 1999–2001. Bars indicate significant differences between columns.(Based on data in Vézina and Williams 2003.)

and eider ducks pre-laying and laying females have larger digestive tracts than birds at later stages of breeding though, at least in geese, these differences occur in males and females (Ankney 1977; Parker and Holm 1990). Therefore, these changes might in part reflect changes in diet quality associated with breeding rather than with reproductive state per se. In a 3-year study of European starlings comparing pre-breeding females with egg-laying females at the 1-egg stage, there were no consistent changes in the relative size of food-processing organs (liver, small intestine, pancreas, gizzard; fig. 3.16). Some digestive organs increased gradually in mass through the cycle of follicle development and egg-laying (fig. 3.17), but this pattern was not consistent among years and it did not match the predicted temporal pattern of energetic demand of follicle development (Vézina and Williams 2003). Given the central importance and influence of Perrins' food limitation hypothesis (1970, 1996; Drent 2006), it would clearly be valuable to apply the concepts and techniques used

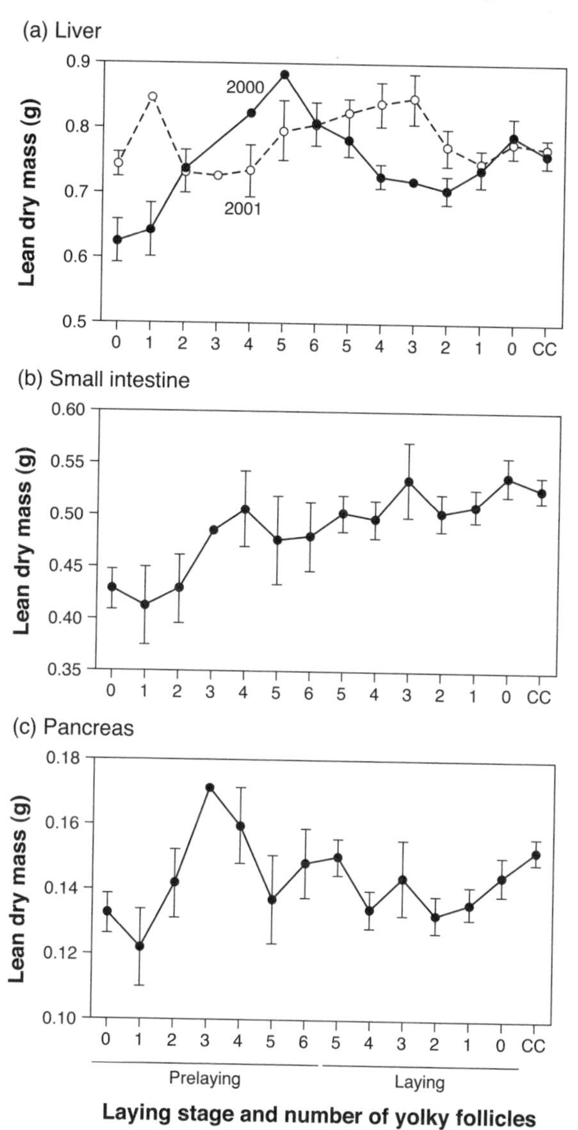

Fig. 3.17. Changes in residual lean dry mass of (a) liver, (b) small intestine, and (c) pancreas relative to the number of developing yolky follicles in female European starlings that ultimately developed a full six-follicle ovary. (Based on data in Vézina and Williams 2003.)

widely in studies of phenotypic flexibility of the digestive tract in non-breeding birds (e.g., enzyme analysis, nutrient uptake measurement) to breeding birds, specifically in the context of nutritional constraint of egg production.

Let us turn to the second question: assuming that greater food intake around the time of onset of egg-laying translates into an increase in endogenous energy and nutrient balance, how is this information integrated with the HPG axis to regulate ovarian function? In mammals a wide range of hormonal signals have been identified that regulate "hunger," satiety, intake rate, meal size, feeding frequency, and plasma metabolite homeostasis (Badman and Flier 2005; K. Murphy and Bloom 2006; Stanley et al. 2005). These include many "brain-gut" peptides, for example, ghrelin, glucagon-like peptide 1, and peptide YY released by enteroendocrine cells throughout the gastrointestinal tract and other peripheral hormones secreted by adipocytes (leptin, adiponectin) and the pancreas (insulin, glucagon). Hormone secretion occurs in response to metabolic or physical stimuli associated with feeding, for example, changes in blood metabolite levels and gut distension, and these hormones play a key role in regulating digestive tract function. However, many of these peripheral signals are also relayed to the hypothalamus or brain stem, either humorally or via paracrine effects on vagal nerve stimulation. Anorexigenic (appetite-inhibiting) and orexigenic (appetite-stimulating) populations of neurons in the arcuate nucleus of the hypothalamus—the melanocortin system—then mediate the central regulation of food intake and energy expenditure. These central mechanisms involve neuropeptides such as agouti-related peptide, neuropeptide Y, and α-melanocyte stimulating hormone. The arcuate nucleus receives sensory input from other brain regions (e.g., the mesolimbic system in humans) and sends connections to other hypothalamic nuclei including the paraventricular nucleus and the ventromedial nucleus of the hypothalamus Stanley et al. 2005), potentially regulating GnRH- and GnIH-neuron function and reproduction (J. Hill et al. 2008). Recent evidence also suggests that the hypothalamic-pituitary-thyroid axis interacts with neurons in the arcuate nucleus (Decherf et al. 2010).

Very little work has been conducted on any of these potential regulatory mechanisms of feeding and energy balance, especially in non-domesticated, free-living birds. Birds differ from mammals in a number of ways with regard to gastrointestinal tract structure and function: for a given body size, birds tend to eat more food but have smaller digestive tracts, and they retain food in their gut for shorter periods of time, despite generally higher mass-specific energy demands (McWhorter et al. 2009). Nevertheless, most brain-gut peptides identified in mammals have also been identified in birds, and they have been shown to have similar inhibitory or stimulatory effects on food intake in experimental studies, although largely in poultry and in experiments using injection into either central or peripheral sites (Richards 2003). Furthermore, the neuroanatomical architecture underlying central regulation of food intake and energy expenditure, and in particular the hypothalamic melanocortin system, is well conserved between birds and mammals, although relatively little is known about how this system functions in birds (Boswell and Takeuchi 2005; Kuenzel et al. 1999). Most recent and current studies on regulation of food intake have focused on tissue-specific changes in gene expression, but few studies have investigated it in relation to hyperphagia associated with *reproductive* development. In terms of peripheral hormone signals, leptin is an obvious candidate that might be predicted to play a key role in regulating energy status, food intake, and reproduction, perhaps functioning as a "lipostat" indicating nutrient or energy status (but there are many other candidates). The existence and function of leptin in birds, or more precisely the avian leptin gene, has proved controversial (Sharp et al. 2008), but several studies have reported systematic variation in leptin-like immunoreactivity in free-living birds in relation to egg production (Kordonowy et al. 2010) and chick-rearing (Quillfeldt et al. 2009). Other studies, in ring doves, have shown that agouti-related peptide and neuropeptide Y immunoreactivity within the infundibular nucleus of the hypothalamus (the avian homolog of the arcuate nucleus) is highest during pre-laying and chick-rearing, and lowest during early incubation (Ramakrishnan et al. 2007; Strader and Buntin 2003). Further investigation of both peripheral and central hormone signaling—specifically in the context of hyperphagia, food intake, and nutrient balance during seasonal gonadal recrudescence and egg production, and especially in free-living birds—would clearly be valuable to improve our understanding of the role of food in timing of breeding.

3.9. Conclusion

The central tenet that day length plays a key role in regulating timing of breeding has been widely accepted for more than 80 years, since the pioneering work of Rowan (1926), and has been very influential in driving neuroendocrinological research on the photoperiodic control of seasonality. However, it remains unclear if day length and the associated photoperiodic response mechanism(s) determine local, population-level, or individual variation in the timing of breeding, which has such important consequences for fitness. One general view is that day length simply provides a relatively broad "window" within which breeding can occur if other environmental cues are present. However, the alternative view—that adaptive, intraspecific variation in egg-laying date *is* attributable to among-population differences in photoresponsiveness—persists (Coppack and Pulido 2004) despite few examples (Lambrechts et al. 1997). In reality it is not clear, even for males, *how* timing of breeding is affected by changing day length, that is, whether timing is dependent on the absolute duration of photoperiod or on the direction and/or rate of change of photoperiod, or even if photoperiod serves to entrain an endogenous rhythm (A. Dawson 2007). In the context of fitness-related variation in timing, further studies on population-level and/or individual variation in the photoperiodic response mechanism would be warranted.

The other central tenet of seasonal reproduction—that precise fine-tuning of laying date in a given year and locality necessarily involves a set of supplementary environmental cues, such as food availability or ambient temperature, and that these are especially important in female birds—has also been widely accepted for over 50 years (Farner et al. 1966; Wingfield 1983; Wingfield and Farner 1993). Here, the idea that food availability and timing of breeding are causally linked has predominated in driving largely ecological, long-term population studies (Perrins 1991, 1996), and there has been a persistent disconnect between this evolutionary or ecological work and the neuroendocrinological work primarily focusing on photoperiod. Despite the widely accepted importance of supplemental cues there is very little experimental evidence that variation in temperature or food availability, within an ecologically relevant range, have either indirect or direct effects on timing of onset or rate of gonadal development in female birds. As others have highlighted recently, we still know surprisingly little about the exact nature of the environmental cues that individuals use to regulate the precise timing of breeding or the physiological mechanisms underlying plasticity in the processing of environmental information (Bourgault et al. 2010; Visser 2008; Visser et al. 2010). By far the best, most robust, evidence to unequivocally support a role for environmental cues in timing of breeding would come from identification of the specific mechanisms in females that mediate detection and integration of these cues with the HPG axis to ultimately regulate onset of vitellogenesis, ovarian and oviduct development, and egg-laying. For both temperature and food availability we have considerable knowledge of the *potential* mechanisms that might be involved, in some cases mainly from mammalian studies, and application of this knowledge and associated techniques to breeding decisions in free-living birds could help advance what has become a rather static field of research. Finally, given that males and females are under very different selection pressures in terms of the cost of sex (A. Hayward and Gillooly 2011), we should expect to see the evolution of more "conservative" regulatory mechanisms in females downstream of the HPG axis.

3.10. Future Research Questions

- What environmental factors and physiological mechanisms explain individual variation in the timing of egg-laying? How does this variation differ in single- versus multiple-brooded species?
- Are there sex-specific differences in the timing and duration of the period of physiological "preparation" that must precede onset of egg production and egg-laying? How important are differences in cue values, or in the prevailing environmental cues themselves, that birds use to time these sex-specific reproductive "decisions"?
- Is the photoperiod response mechanism variable among individuals within populations and plastic within individuals? Is there heritable, individual variation in timing mechanisms (the clock), specific day lengths required for rapid gonadal growth, or rate of gonadal growth in relation to day length that can respond to selection?
- Do supplemental cues (temperature, food) within an ecologically relevant range have an effect on plasma estradiol levels, vitellogenesis, or follicle development in females on *intermediate* long day lengths?
- Are transient-receptor-potential thermoreceptors used in birds to detect and integrate temperature information over long time periods? Do these receptors have the sensitivity required to detect annual differences in average temperatures of only a few degrees? How is temperature information encoded by these receptors and integrated with the HPG axis?
- Is hyperphagia associated with the increased demands of egg production, and does increased food intake "lift the constraint" of food availability? Does this involve phenotypic flexibility of the digestive tract or central regulation of food intake versus expenditure at the hypothalamic level? How are mechanisms of energy or nutrient homeostasis integrated with hypothalamic control of reproduction in the context of food availability and timing of breeding?
- Many individual females that initiate laying more than a few days away from some "optimum" date appear to have zero fitness (no recruitment of offspring), and many individuals that fledge young fail to recruit any offspring. What are the implications of these observations for experimental field studies that use fledging success or offspring "quality" at fledging as end points?

CHAPTER 4

Egg Size and Egg Quality

VARIATION IN EGG SIZE appears to contribute relatively little to variation in lifetime fitness (see below), so why devote a chapter to egg size or egg quality? First, a major goal of this book is to explain *individual variation* in reproductive traits at both the evolutionary and physiological level and there is marked individual variation in egg size that remains largely unexplained (Christians 2002; T. Williams 2005). Second, variation in egg quality has received more attention in the last 10–15 years than most other reproductive traits (certainly compared with mechanistic work on clutch size or incubation; see chapters 5 and 6), so this should be one area where we are not data poor in attempting to explain the functional significance of, and mechanisms underlying, variation in a life-history trait. Ironically, this abundance of research has occurred, in part, as a result of the problems of demonstrating the predicted strong, positive relationship between egg *size* and offspring fitness (Clutton-Brock 1991; T. Williams 1994). Much of this work has been at the interface between ecology and physiology, and there has been an explosion of studies looking at other "minor" egg components, such as yolk hormones, antimicrobial proteins, and antioxidants (Blount et al. 2000; Grindstaff et al. 2003; Groothuis, Müller, et al. 2005), that might potentially provide better measures of egg quality. Since these egg components are maternally derived, they have also been considered an important example of maternal effects (nongenetic contributions of the mother to her offspring) that have been a major focus of life-history studies (Badyaev and Uller 2009; Bernardo 1996; Mousseau and Fox 1998). Maternally derived egg components might therefore provide individual females with additional mechanisms to fine-tune offspring phenotype to the prevailing environmental conditions independently of egg size per se.

4.1. Individual Variation in Egg Size

Early theoretical studies postulated and modelled the evolution of a single optimum egg size (McGinley et al. 1987; C. Smith and Fretwell 1974), an idea that has remained influential. However there is, in fact, marked

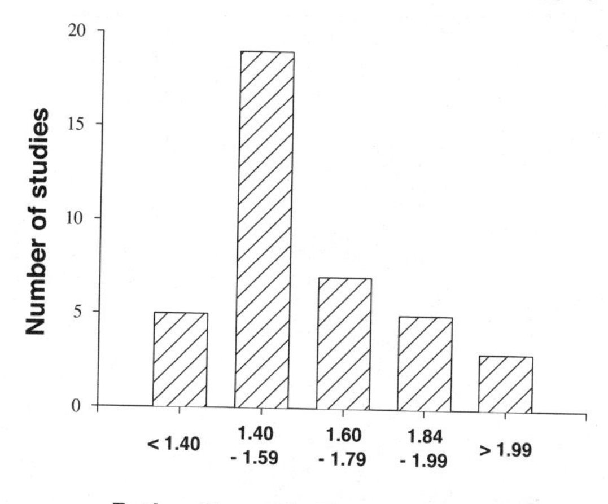

Fig. 4.1. Individual variation in egg size given as the ratio of the size of largest to smallest egg laid by different females within a population (n = 39 studies). (Based on data in Christians 2002.)

individual variation in egg size *within populations*: the largest eggs laid by individual females are generally at least 50% larger, and sometimes twice as large, as the smallest eggs laid by other individuals in the same population (fig. 4.1). Approximately 70% of the total variation in egg size is due to differences between clutches, that is, between individual females, not within clutches (Christians 2002). As an example, in the lesser snow goose, the largest eggs within a clutch are on average 8.1 g heavier (6.7%) than the smallest, last-laid eggs (fig. 4.2a), whereas egg mass varies from 86 to 166g (93%) among females (Cooke et al. 1995). Similarly, in the European starling, the largest eggs are on average 0.13–0.15 g (<2%) larger than the smallest eggs within a clutch (fig. 4.2b), but egg mass varies from 5.64 to 8.86 g (57%) among females. In general, variation in intra-clutch egg-size is smaller in precocial species compared with altricial species (0.68% vs. 3.91%) and is smaller in hole-nesting passerines compared with open-nesting species (3.56% vs. 0.05%; Slagsvold et al. 1984). However, intraclutch egg-size variation is *very* small relative to among-individual variation in all avian taxa, even though much more research has been directed at the former source of variation than the latter (Christians 2002).

These data suggest that egg size is an inherent characteristic of an individual female's phenotype and has relatively little intra-individual plasticity (see below). So what components of the female's phenotype might explain individual variation in egg size from one female to the next? In general, individual factors (age, experience, mass) and environmental

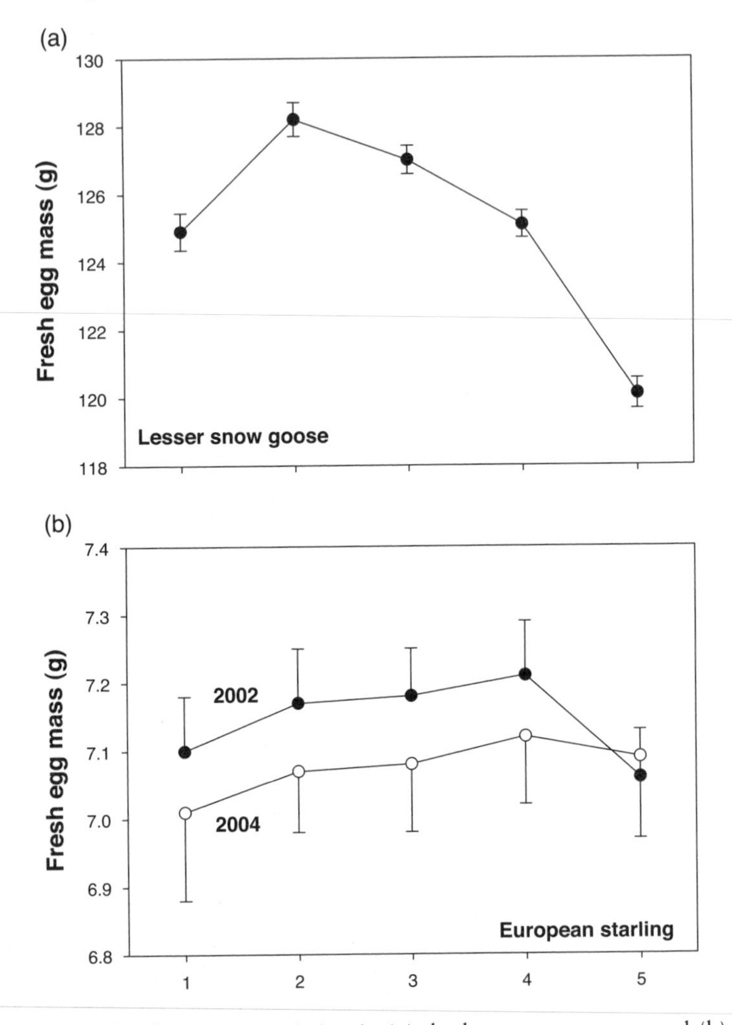

Fig. 4.2. Intra-clutch egg-size variation in (a) the lesser snow goose and (b) the European starling.

factors (temperature, food, date) continue to be the focus of most research, even though it is clear that these explain only a small proportion (10%–20%) of individual variation in egg size (Christians 2002), leaving 80%–90% unexplained. Increases in egg size with age or experience were found in almost half (17 of 37) of the studies reviewed by Christians (2002), but the mean difference between the most extreme age classes was only 5.9% (range 2%–17%). Positive correlations between egg size and either body mass, body size, or size-corrected body mass (as an index of "condition") were reported in 33 of 50 studies reviewed by Christians

(2002), but these traits explained less than 20% of individual variation in egg size (e.g., body mass: passerines, 10%, $n = 17$ studies; shorebirds, 15.7%, $n = 10$; ducks and geese 17.3%, $n = 3$). One problem here is that in almost all of these studies, body mass was measured after completion of egg-laying and thus does not provide a reliable measure of pre-breeding "condition" in terms of resources potentially available for egg production. In European starlings, individual variation in mean egg mass (4.8–8.1 g) was independent of non-reproductive body mass in females at the 1-egg stage of laying (fig. 4.3a). Similarly, in captive-breeding zebra finches, pre-breeding body mass explains only 6% of the variation in mean egg mass over a wide range of body masses (12–20 g, fig. 4.3b). Thus, body mass is a poor predictor of individual variation in egg size, even when mass is recorded early in the laying cycle. Several experimental studies have directly manipulated body mass or condition, for example, "handicapping" females by removing or clipping feathers, and looked for effects on reproductive investment in replacement clutches following egg removal (Slagsvold and Lifjeld 1988, 1990; Winkler and Allen 1995) or in natural second clutches (Love and Williams 2008a). None of these studies found any treatment effect on egg size even though the experimental manipulation did significantly change body mass or body condition.

It is widely assumed that female birds might adjust egg size in response to changes in environmental or ecological factors, such as temperature, food availability, or laying date, as has been demonstrated in other taxa (Mousseau and Fox 1998). However, individual plasticity in egg size is very limited in birds, and egg size is not very variable within individuals *even under different environmental conditions* (Christians 2002). Some correlational studies have reported positive relationships between egg size and Ta, with females laying larger eggs at warmer temperatures (Jarvinen and Pryl 1989; Ojanen 1983a; Saino et al. 2004), but other studies have reported negative correlations (T. Williams and Cooch 1996) or no effect (Robertson 1995). However, in all studies temperature explains less than 10% of the variation in egg size, and often much less (Christians 2002), and significant correlations are often found only for later-laid eggs, not for all eggs (Magrath 1992; Saino et al. 2004). In the lesser snow goose, the average range of ambient temperatures encountered would predict only a 3%–5% difference in mean egg size (T. Williams and Cooch 1996), and in great tits, egg volume changes by only 0.5% per 1°C change in temperature (Nager and Zandt 1994). Few studies have experimentally manipulated temperature during egg-laying. Nager and Noordwijk (1992) manipulated nest-box temperatures during the pre-laying and laying period in blue tits. Females in cooled nest boxes laid smaller eggs (–0.11 cm^3, or –7%), and females in heated nest boxes laid

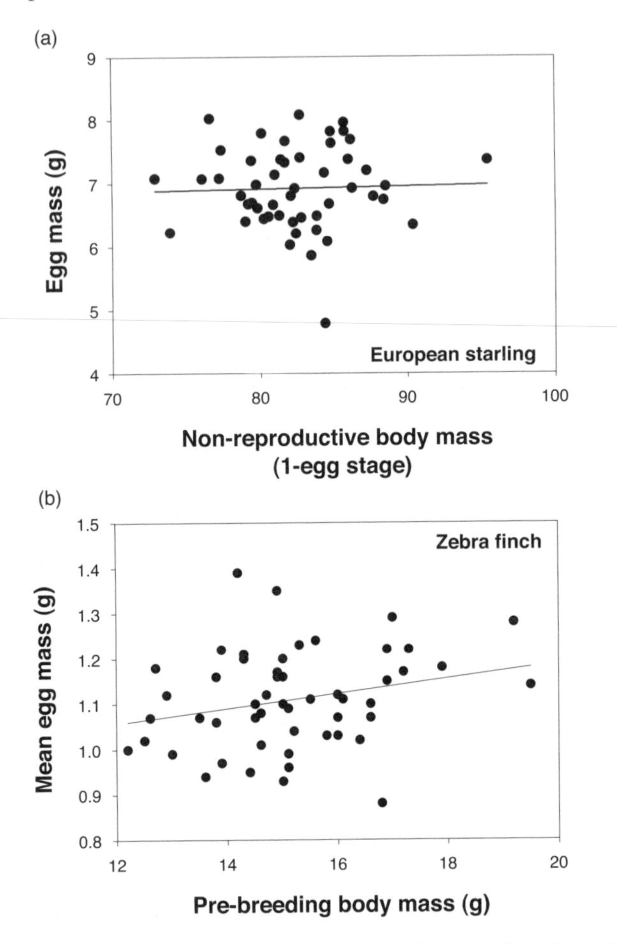

Fig. 4.3. Relationship between mean egg mass of individual females and (a) body mass at the one-egg stage in European starlings and (b) pre-breeding body mass in the zebra finch.

larger eggs (+0.12 cm^3 or +8%) compared with control females (mean egg volume 1.53 cm^3), a result that Nager and van Noordwijk (1992) interpreted in terms of energy limitations on egg size. Salvante et al. (2007b) found no difference in egg mass in captive-breeding zebra finches at 7°C and 21°C, although the birds increased their food consumption by 45% at the lower temperature. These latter studies are more consistent with an indirect effect of temperature on egg size acting through energy or resource availability; so how does food availability or diet quality affect egg size? Christians (2002) reviewed food-supplementation experiments and found that only 10 of 28 studies (36%) reported an increase in egg

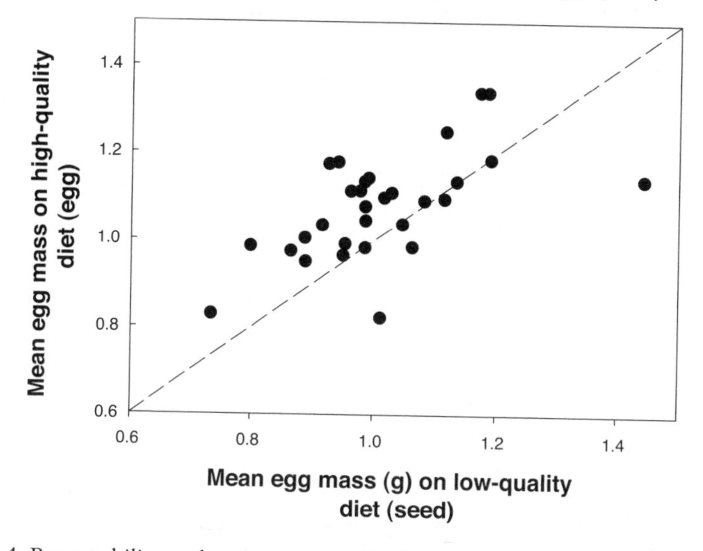

Fig. 4.4. Repeatability and maintenance of individual variation of mean egg mass in female zebra finches breeding on a high-quality (egg-supplemented) and low-quality (seed) diet.

size, far fewer than those reporting positive effects of food on clutch size (63%) and laying date (57%), and mean egg size increased only by 8.1% on average (range 4%–13%). More importantly, food supplementation does not reduce individual variation in egg size; that is, "small-egg" females do not lay large eggs simply with access to more or better food (Arcese and Smith 1988). As an example, individual female zebra finches breeding on a high-quality diet laid eggs that were on average 8%–10% larger than when they bred on a low-quality diet (fig. 4.4), but egg size varied between 0.8 and 1.4 g (75%) among individuals *on both diets* (T. Williams 1996). A smaller number of food-restriction experiments have also found no effect on egg size (Giuliano et al. 1996; Meijer and Langer 1995).

Finally, what about seasonal variation in egg size? As we will see in chapter 5, clutch size varies systematically with laying date; however, most studies reviewed by Christians (2002) found no relationship between egg size and laying date (40 of 69 studies). In 20 studies a seasonal decline in egg size was observed, and as with clutch size, there is evidence that it can be explained both by quality and date effects (see section 3.2). In some species, younger or lower-quality birds lay later in the season and lay smaller eggs, an observation that supports a role for individual quality in seasonal variation in egg size (Hipfner et al. 1997; Vinuela 1997). However, in other species, individual females laying eggs in replacement clutches, after egg removal, lay smaller eggs, which suggests an effect of laying date or time per se (e.g., Hipfner et al. 2003).

In conclusion, there is marked individual variation in egg size among females within populations (up to twofold, or 100%), but egg size is relatively inflexible within individuals (generally varying at most ±10%): "individual" factors (age, experience, mass) and environmental factors (temperature, food) explain little of this variation in egg size. Now I will consider if this individual variation in egg size, and perhaps also the more limited intra-individual variation in egg size, is related to fitness.

4.2. Fitness Consequences of Variation in Egg Size

Given that there can be up to a twofold difference between the smallest and largest mean egg size of individual females within a population, does this variation influence offspring phenotype and, ultimately, offspring or maternal fitness? In the early 1990s it was widely assumed that there *was* abundant, consistent evidence for a positive relationship between egg size and offspring growth and/or survival in birds (Amundsen and Stokland 1990; Clutton-Brock 1991). However, in reviewing over 40 studies on intraspecific variation in egg size and offspring fitness T. Williams (1994) concluded that "there is little unequivocal evidence to date for a positive relationship between egg size and offspring fitness in birds." The primary reason for this contradictory conclusion was that few studies at the time had controlled for the potential confounding effects of parental quality. Putative high-quality females might lay large eggs, and their offspring might show high rates of growth and survival, but the latter could be due to these females also maintaining high provisioning rates rather than the offspring effects being the result of larger egg size per se. Cross-fostering experiments, where "large-egg" females are given small eggs to rear and vice versa, or where eggs are randomized among nests, have to be used to resolve the independent effects of parental quality and egg size. The fact that there is still surprisingly little empirical, and especially experimental, evidence for strong, positive effects of egg size on offspring phenotype (described further below) has recently led some authors to suggest that egg size is a neutral trait that is under no selection pressure (Jager et al. 2000; van de Pol, Bakker, et al. 2006). For a number of reasons this conclusion, and this line of reasoning, is probably premature or, at least, unhelpful from a mechanistic viewpoint. An equally plausible explanation is that we have simply not asked the right questions or done the right experiments (see Love and Williams 2011; E. Wagner and Williams 2007). Indeed, a recent meta-analysis of egg-size effects from 283 studies (Krist 2011) concluded that egg size does indeed have positive effects on "nearly all studied offspring traits" (but see below). The equivocal nature of support for strong effects of egg *size* on offspring phenotype likely explains the rapid growth in studies focusing on other components of egg quality (yolk hormones, antibodies) and I shall return to the question of whether these might be more important for chick quality than egg size.

At the time of T. Williams' review (1994), only 6 of 41 studies had used an experimental approach to investigate egg-size effects. Only 4 studies reported some evidence for a significant residual effect of egg size on offspring growth (Magrath 1992; Nisbet 1978) or survival to fledging (Bolton 1991; Nisbet 1973) when controlling for female quality and 3 of these studies were in seabirds. There was more consistent evidence for a positive relationship between egg size and chick growth and survival *early* in the chick-rearing period, suggesting that the primary benefit of larger eggs might be to increase offspring survival in the first few days after hatching (T. Williams 1994). Most studies have also failed to find an effect of among-clutch variability in egg mass on hatchability (Sanchez-Lafuente 2004; H. Smith et al. 1995; T. Williams 1994). More recent experimental work has generally confirmed these findings. Egg size is often correlated with offspring mass and size in the first 7–10 days after hatching (Amundsen 1995; Amundsen et al. 1996; Bize et al. 2002) but only a few studies have reported longer-lasting effects (H. Smith and Bruun 1998; Styrsky et al. 1999), and then only for certain morphological traits (e.g., wing length, fig. 4.5), not for body mass in general (Hipfner and Gaston 1999). Most studies have failed to find an effect of egg size on offspring growth or survival at fledging (Blomqvist et al. 1997; W. Reed et al. 1999; Risch and Rohwer 2000; H. Smith et al. 1995). As would be predicted, larger egg size confers a greater advantage to offspring in situations where resources are generally low and where selection on offspring development is likely to be stronger. In European starlings, a positive effect of egg size on early nestling survival was only apparent in habitats where the availability of pasture, the preferred foraging habitat for starlings was low, suggesting that egg size has a more pronounced effect on survival when food availability is low (H. Smith and Bruun 1998). Similarly in house wrens (*Troglodytes aedon*), nestling mass in early-season broods was positively correlated with egg mass only through to brood day 6, but nestling mass was positively correlated with egg mass until the nestlings achieved asymptotic mass in late-season broods when preferred food resources were less available (fig. 4.6; Styrsky et al. 1999).

It is notable that there are still few data available on the long-term consequences of egg size on offspring fitness, for example, effects on post-fledging survival, recruitment, or female fecundity traits (laying date, clutch size). In their long-term study of the lesser snow goose, Cooke et al. (1995) found no relationship between the fresh mass of eggs and

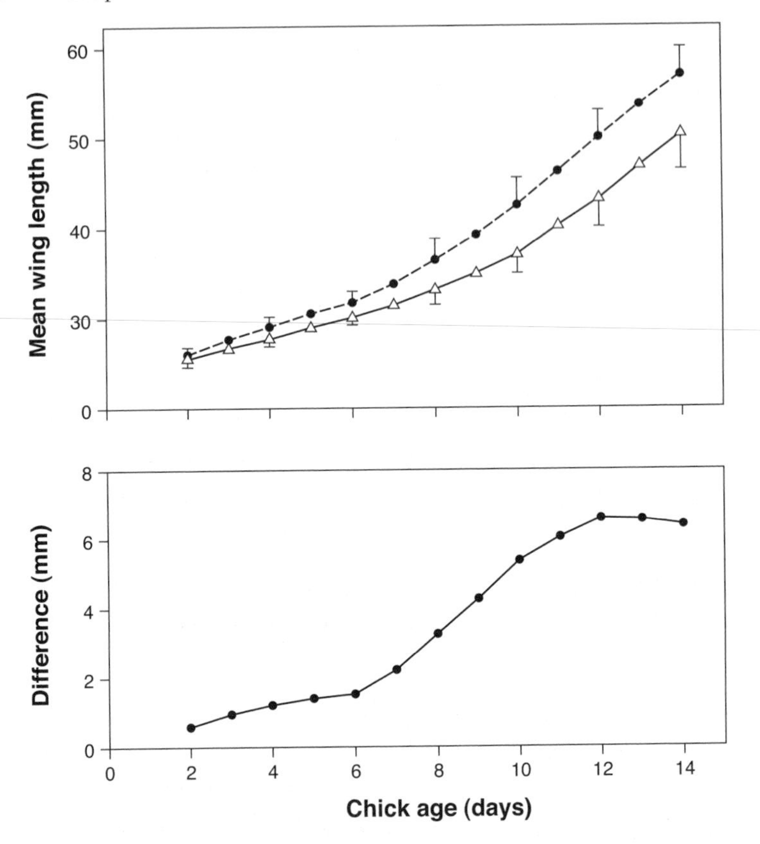

Fig. 4.5. (Top) Mean wing length (± 1 SD) against age for thick-billed murre chicks, from the largest one-third of eggs (circles) and the smallest one-third of eggs (triangles). (Bottom) Difference in mean wing length between the two groups. (Reproduced from Hipfner and Gaston 1999, with permission of the Ecological Society of America.)

probability of female offspring from those eggs recruiting to the breeding population. Krist (2009) used a cross-fostering design in the collared flycatcher and followed young until recruitment to estimate the long-term effects of egg size. Egg-size had the greatest effect (largest effect size, r = 0.2–0.4) on offspring morphology, and effect sizes were positive at 6 days post-hatching (nestling mass) and at adulthood (mass and wing length of recruits), though, curiously, effect sizes for fledging mass and size were not different from zero. However, egg-size effects were absent for three traits describing offspring survival at various life stages (hatchability, fledging success, and recruitment probability), and the effects were weak

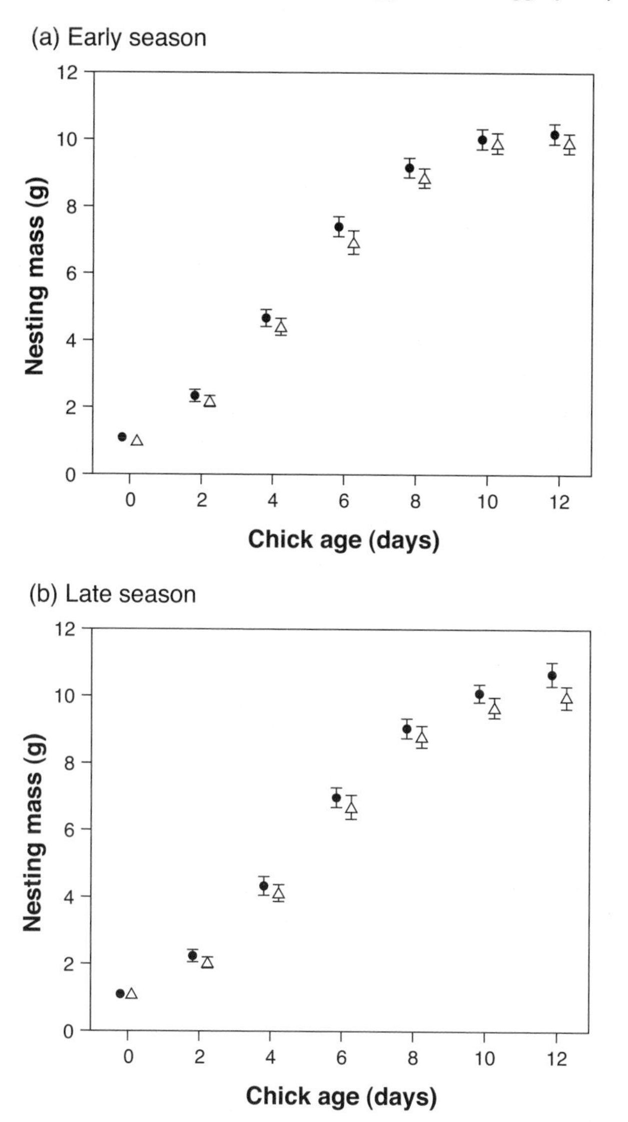

Fig. 4.6. Effect of egg size on chick growth in house wrens from cross-fostered clutches of small and large eggs in (a) early season and (b) late season broods. Nestling mass was significantly correlated with original egg mass for days 0–6 in early-season brood and for days 0–10 in late-season broods.(Redrawn from Styrsky et al. 1999, with permission of the Royal Society.)

(≤0.2) for laying date and clutch size of recruits (95% CI overlapped zero for these fecundity traits; Krist 2009).

Krist's meta-analysis (2011) of egg-size effects confirms the main conclusions from these earlier studies described above, although he argued "that egg size had a positive effect on nearly all studied offspring traits up to fledging," based on positive effect sizes in his analysis. Effects of egg size were largest at hatching (effect size, r = 0.4–0.6) but were lower during the nestling and post-fledging stage ($r \leq 0.2$; 95% CI included zero for many post-fledging offspring traits although small sample sizes generated large CIs). Morphological traits were most strongly influenced by egg size, but significant effects of egg size were also found on hatching success, nestling growth rate, and survival ($r \leq 0.2$). However, effect sizes were not significantly different from zero for egg fertility, offspring immune function, behavior, and life-history or sexual traits (although the sample size was $n = 4$ for the latter analysis, representing a very mixed set of male/female traits). Effect size did not depend on species-specific traits such as developmental mode, clutch size, or relative size of the egg. However, this meta-analysis also generated two counterintuitive results: (a) the egg-size effect did not depend on whether cross-fostering was used, suggesting that parental quality was not a confound, despite the fact that analysis of cross-fostering studies did reveal a positive effect size for parental quality, and (b) the effect size was larger when tested in captive compared with wild populations.

The most direct, causal test of fitness consequences of egg-size variation while controlling for confounding factors should be the experimental manipulation of egg size itself. However, indirect methods such as manipulation of temperature, diet quality, or photostimulation are limited by a general lack of specificity in targeting the underlying mechanism of interest (Zera and Harshman 2001) and can be confounded by potential effects of the manipulation on parental condition (Bernardo 1996). An alternative approach is direct hormonal manipulation of the female's reproductive system to take advantage of the specific mechanisms that generate variation in egg size and to experimentally uncouple any correlation between maternal condition and egg size (Sinervo 1999; Sinervo and Licht 1991; T. Williams 2000). Few studies have taken this approach in birds, but Williams (T. 2000, 2001) and E. Wagner and Williams (2007) used the anti-estrogen tamoxifen to manipulate egg size in captive-breeding zebra finches by disrupting the hormonal signal (estrogen action) that drives egg production. Tamoxifen treatment decreased circulating yolk-precursor levels by up to 50% (T. Williams 2000), decreased relative yolk mass and yolk protein content (T. Williams 2001), and decreased egg size in a systematic, repeatable manner while maintaining laying-sequence specific variation in egg mass (fig. 4.7). At a low

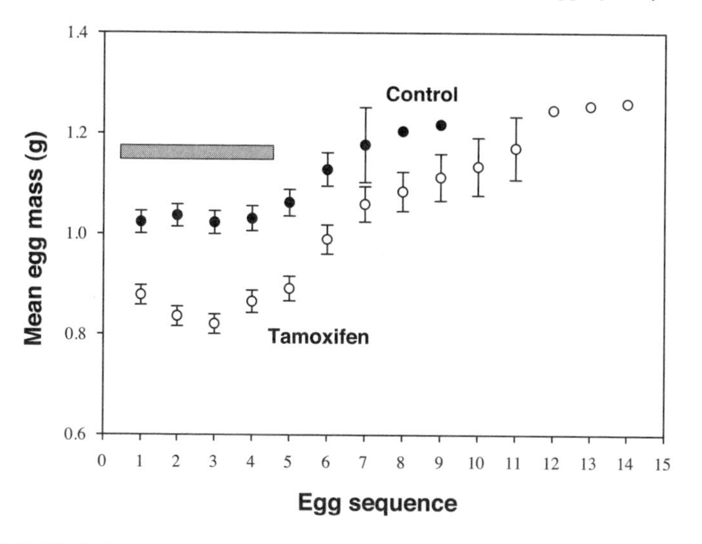

Fig. 4.7. Variation in mean egg mass with laying sequence in anti-estrogen (tamoxifen)–treated treated female zebra finches (open circles) and sham-treated females (filled circles). The shaded bar indicates the period of tamoxifen treatment. (Based on data in E. Wagner and Williams 2007.)

dose, tamoxifen decreased egg size by approximately 8%, and there were few effects on offspring survival, body size and mass at fledging, and reproductive performance at sexual maturity (T. Williams 2001). In contrast, at a slightly higher dose, tamoxifen treatment decreased the egg size of individual females by approximately18%, and this reduction had marked effects on offspring development, including a significant decline in embryo viability, slower post-hatching growth, lower fledging mass (fig. 4.8a), and higher nestling mortality (fig. 4.8b; E. Wagner and Williams 2007).

E. Wagner and Williams (2007) suggested that a *change* in egg size of *individual* females might have a greater effect on the development of their offspring than absolute egg mass if individual females produce eggs of an "optimum" size, or composition, for the growth of their chick(s) and if any manipulation moves the egg size away from this optimum. Given that embryo growth and metabolism are genetically determined by the maternal/parental genotype (Starck and Ricklefs 1998), each egg might contain the exact amount of nutrients required to maintain expression of this phenotype in individual chicks. Thus, negative effects on offspring development might be observed only when the egg size is experimentally manipulated such that it deviates significantly from each individual female's phenotypic optimum. This individual optimization of egg size

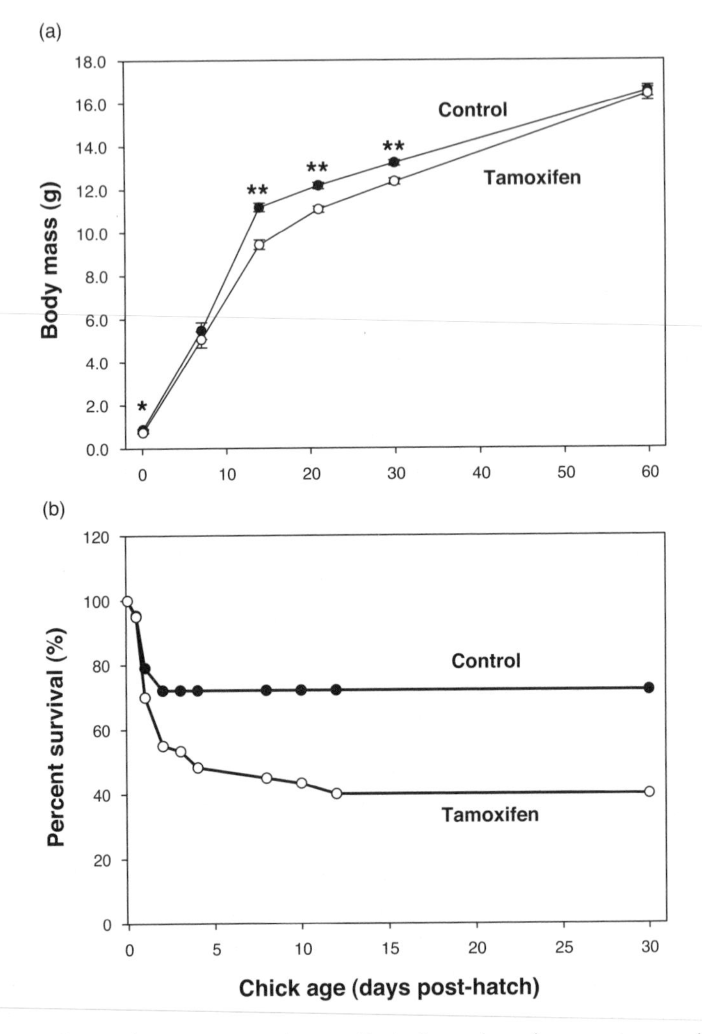

Fig. 4.8. Effects of anti-estrogen (tamoxifen)–dependent decrease in egg size on (a) chick growth and (b) chick survival in zebra finches. (Based on data in E. Wagner and Williams 2007.)

matched to offspring phenotype might explain why, even with robust cross-fostering experiments, egg size appears to have little effect on offspring growth or survival: offspring development is not uncoupled from female- or chick-specific egg size or other maternal effects and is still "optimum" for the growth of that particular chick regardless of the foster environment. This could also explain why studies more commonly find effects of within-clutch variation in egg size (i.e., variability around some individual-specific optimum) on offspring development even though the magnitude of variation at this level is much smaller: deviations from a female's mean egg size are more significant than absolute variation in egg size among females.

4.2.1. Egg-Size–Clutch-Size Trade-Offs and Fecundity Costs of Large Egg Size

One potential fitness consequence of large egg size that has not been explicitly addressed so far is that, *given finite resources*, a female that lays larger eggs *must* lay fewer eggs; that is, large-egg females should have reduced fecundity. This trade-off between egg (or offspring) size and number is a central and long-standing component of life-history theory (Clutton-Brock 1991; Roff 2002; C. Smith and Fretwell 1974; Stearns 1992). However, comparative studies have found mixed support for this trade-off among species (P. Bennett and Owens 2002; Blackburn 1991; Christians 2000b; T. Martin et al. 2006; Rohwer 1988). Although this trade-off should exist within species as well as among species, direct evidence to support an intraspecific trade-off between egg size and clutch size in oviparous vertebrates is very limited (Bernardo 1996). This lack of data is particularly true for birds, where intraspecific analyses have commonly failed to find evidence for a trade-off. Christians (2002) reviewed 63 studies, and only 5 (8%) reported a negative relationship between egg size and clutch size, 15 (23%) reported a positive relationship, and 46 (69%) found no relationship. One commonly proposed explanation for these findings is that if individual females differ in the absolute amount of resources they have to allocate to reproduction, then *phenotypic correlations* between traits would not be expected, or might even be positive, because females with access to lots of resources can lay large clutches *and* large eggs (Reznick 2000; van Noordwijk and de Jong 1986; fig. 4.9). Physiological manipulations are especially useful for analysis of the relationship between egg size and egg number in part because they can uncouple any correlation between the mother's condition or resources and reproductive investment. In a series of classic studies Sinervo and colleagues (Sinervo 1999; Sinervo and DeNardo 1996; Sinervo and Licht 1991) utilized two physiological manipulations to demonstrate an egg-size–clutch-size trade-off in the oviparous lizard *Uta stanburiana*. Yolk ablation, or "yolkectomy," of gravid females reduced clutch size, and hormone treatment (FSH) raised clutch size. In each case these treatments resulted in a compensatory increase and decrease in egg size, respectively, exactly as predicted if there is an obligate trade-off between these two traits. This approach only appears to have been used in one instance in birds: T. Williams (2001) used the anti-estrogen tamoxifen to decrease

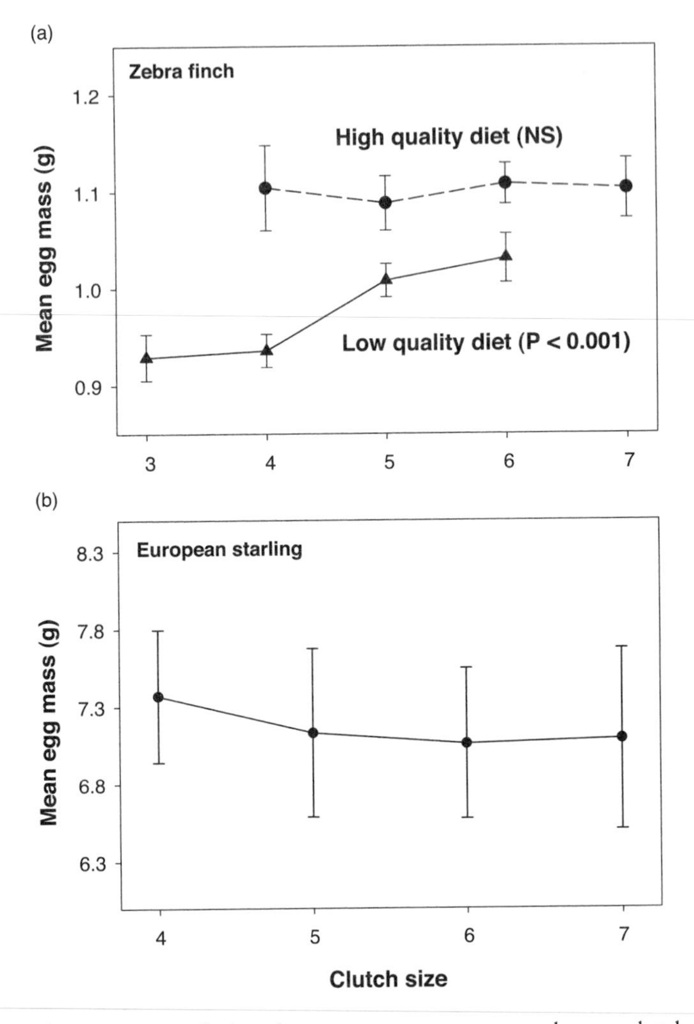

Fig. 4.9. Phenotypic correlations between mean egg mass and mean clutch size in (a) zebra finches on a high- and low-quality diet and (b) in European starlings.

egg size and measured the correlated response for clutch size in captive-breeding zebra finches. Among individual tamoxifen-treated females, birds that laid the smallest eggs early in the laying sequence also laid the largest number of additional eggs; that is, there *was* a negative correlation between the treatment-induced change in egg size and clutch size. In contrast, sham-manipulated females showed a positive phenotypic correlation between these traits (fig. 4.10). In contrast, E. Wagner and Williams (2007) actually found a positive phenotypic correlation between egg size

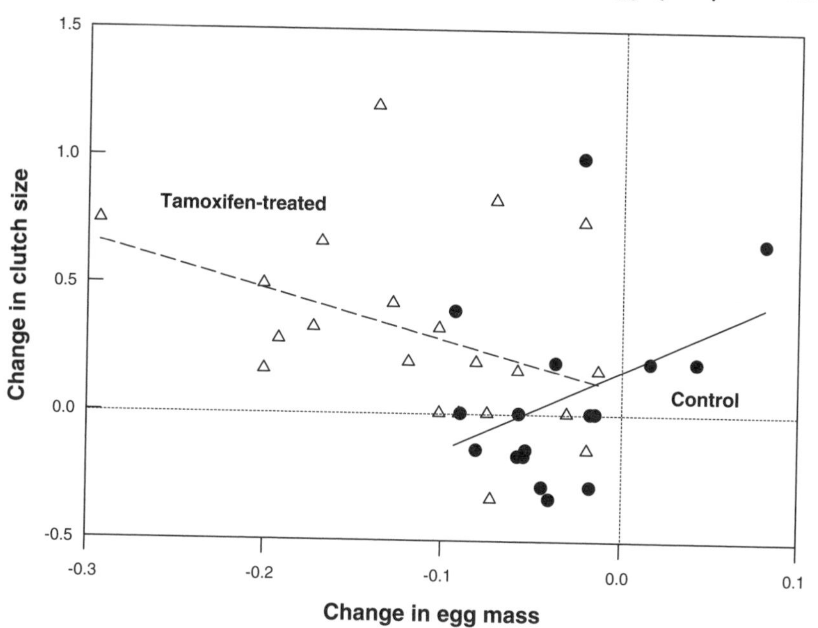

Fig. 4.10. Experimental evidence for an inverse relationship (trade-off) between change in clutch size and change in egg mass for early laid eggs (eggs 1–6) of anti-estrogen (tamoxifen)–treated females (triangles) and control females (circles). (Based on data in T. Williams 2001.)

and clutch size and a strong positive relationship between the change in egg and clutch size in response to tamoxifen treatment, although this experiment used a different protocol and a higher tamoxifen dose, which caused a greater decrease in egg mass (see above, this section). They suggested that the more marked inhibition of egg formation associated with more frequent tamoxifen treatment might have uncoupled the compensatory relationship between egg size and number. Further experimental work of this sort in a wider range of species would be valuable.

Although there is still surprisingly little data to support the idea of an egg-size–clutch-size trade-off within species in birds (and we know almost nothing about the mechanisms underlying such a trade-off), it is clear that *individual females* do pay a "fecundity cost" (smaller clutch sizes) when they make the reproductive "decision" to lay larger eggs. There is substantial overlap in the variance of total clutch mass among different clutch sizes both in free-living birds and in captive-breeding birds (fig. 4.11). Females with the same level of reproductive investment (i.e., the same total clutch mass) either produce a clutch of n large eggs or a clutch of n + 1 smaller eggs (Cooke et al. 1995; Flint and Sedinger

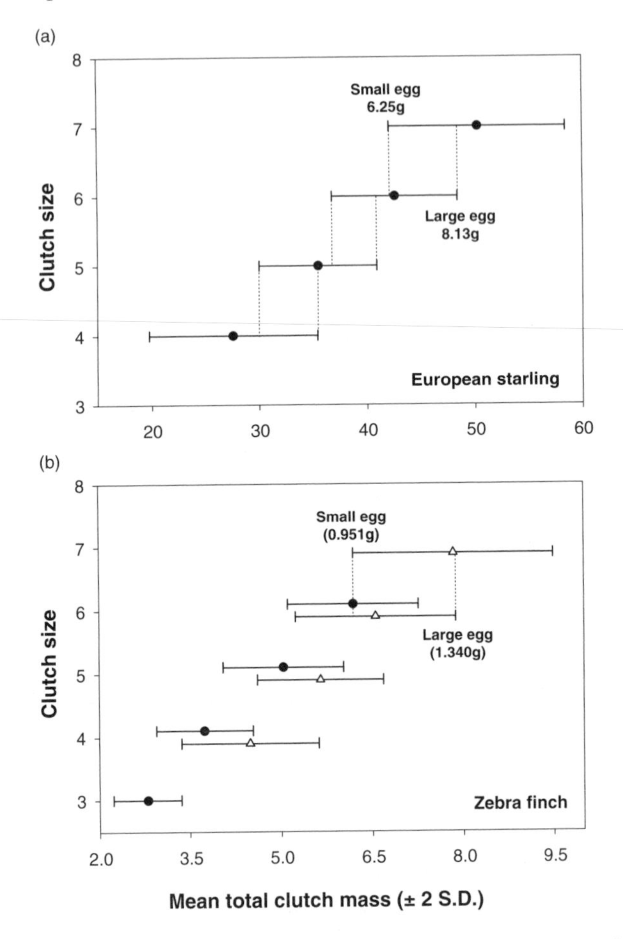

Fig. 4.11. Egg-size clutch-size trade-offs as reproductive "decisions" of individual females in relation to total clutch mass (± 2 S.D), clutch size, and individual egg mass in (a) European starlings and (b) zebra finches on both high- (triangles) and low-quality (circles) diet. Vertical dotted lines indicate overlap in total investment between large-egg birds with clutch size n and small-egg birds with clutch size $n + 1$.

1992). In the European starling, females producing a six-egg clutch can lay eggs with a mean mass of up to 8.13 g, for a total clutch mass of 48.8 g (fig. 4.11a). In terms of total investment this is equivalent to a mean egg mass of 6.96 g for a seven-egg clutch, which is only slightly smaller than the actual mean egg mass of seven-egg birds (7.18 g). Similarly, in zebra finches, although variance in total clutch mass is slightly less on a low-quality (seed) diet compared with a high-quality (egg food) diet, there is still considerable overlap among different clutch sizes on both diets (fig. 4.11b). Given that females do pay a fecundity cost for large egg size, there ought to be compensating benefits for large egg size. Thus, the significance, functional consequences, and mechanisms underlying this observed reproductive decision about egg and clutch size in females still need to be resolved.

4.3. Selection on Egg Size

Most of the studies described in section 4.2 considered the fitness consequences of egg size from the offspring's perspective. However, given that selection optimizes maternal, not offspring, fitness (Marshall and Uller 2007), another way to ask this question is, Do females that lay larger eggs have higher fitness? As we have seen, egg size is a highly repeatable trait, even in different environments, and egg size also shows high heritability. The mean heritability was 0.66 in nine studies reviewed by Christians (2002), and both repeatability and heritability of egg size are much higher than for laying date and clutch size (see chapters 3 and 5). As highlighted in chapter 3, most heritability estimates are calculated from single-generation mother-daughter regressions, which may inflate heritability due to common environmental effects. However, Garant, Hadfield, et al. (2007) estimated a constant high heritability of egg size ($h^2 = 0.45$–0.51) in great tits at Wytham Woods, Oxford (UK) over a 40-year period using an animal model. Therefore, given that egg size is heritable, is there any evidence for directional selection on egg size through higher fitness of females laying larger eggs?

In fact, fewer selection studies have analyzed variation in egg size compared with variation in laying date (chapter 3) and clutch size (chapter 5). McCleery et al. (2004) found no evidence for an effect of egg size on a female's lifetime recruitment of offspring in great tits at Wytham Woods. In a subsequent analysis of the same data, Garant, Kruuk, et al. (2007) confirmed that egg size had no effect on the number of offspring a female recruits to the breeding population from each breeding attempt. They concluded that this "constitutes something of a puzzle" and highlighted the idea that there might be differences in optimum egg size between individuals while acknowledging that "the aspects of a female phenotype that would determine such optima are unknown." Alternatively, Garant, Hadfield et al. (2007) reported a significant negative genetic correlation between egg weight and clutch size (−0.016) and suggested that egg weight might influence the "multivariate evolutionary response" indirectly via its negative link with clutch size, which itself is usually under selection (see chapter 5). Garant, Hadfield, et al. (2007) did find evidence for positive viability selection on egg size, which they suggested was related to variation in female quality, with high-quality individuals producing larger eggs and also having higher survival. Hõrak et al. (1997) found a clear and consistently positive selection differential for egg size in a population of great tits in Estonia. However, because there was no effect of egg size on embryo or nestling survival, they also suggested that some aspect of parental quality (as yet unknown), rather than egg size per se, explained why females with large eggs recruited more fledglings.

4.4. Variation in Egg Composition or Egg Quality

Eggs contain three different "macro" components—shell, yolk and albumen—each of which contributes to offspring development. The yolk supplies protein and energy-rich lipids to the developing embryo, and in the form of the retained yolk sac, to the neonate (Carey 1996). Albumen provides protein and water as well as having anti-bacterial properties (B. Palmer and Guilletter 1991), and the shell provides physical protection and is important in regulating water and gas exchange (Carey 1996). If the yolk, albumen, and shell mass vary independently of egg mass, then the composition of the egg might affect offspring phenotype, and egg composition might be a better indicator of egg quality than egg size per se. In addition, during egg formation many substances, such as steroid, thyroid, and protein hormones, antibodies (immunoglobulins), antioxidants, and RNA, are transferred from the mother to the egg in either the yolk or albumen. Both macro and micro components of eggs have been viewed as potential mediators of "maternal effects": non-genetic, or epigenetic, contributions to offspring phenotype, by which mothers might "adaptively" adjust the development and the phenotype of their offspring to match the prevailing environmental conditions in order to increase fitness (Groothuis and Schwabl 2008; Mousseau and Fox 1998; Uller 2008). This idea assumes that variation in egg composition or amounts of the minor egg components will have positive effects on offspring growth and development, and ultimately on offspring or maternal fitness, a key issue that will be addressed below in sections 4.4.1–4.4.4.

4.4.1. Egg Macronutrient Composition

Among-female variation in egg composition (yolk, albumen, and shell mass) is largely a function of variation in egg size (W. Hill 1995; T. Williams 1994). Larger eggs contain absolutely more shell, dry albumen, and dry yolk, not just higher water content (fig. 4.12); egg composition varies much less within clutches than among females (Carey 1996), and in poultry, there is heritable, genetic variation in egg composition (Hocking et al. 2003). If egg component mass is *directly* proportional to egg mass, then the slope of this relationship should be $b = 1$ based on log-log plots for albumen and yolk, and $b = 0.75$ for shell (i.e., isometry). In many species, the slopes of the log10 egg component × log10 egg mass relationship do not differ significantly from these predicted exponent values (T. Williams 1994). However, in some passerines and seabirds, larger eggs contain disproportionately more albumen ($b > 1$), which largely reflects variation in water content, and less yolk ($b < 1$, e.g., see fig. 1 in Hipfner et al. 2003). In contrast, in *some* precocial species, such as some waterfowl, larger eggs contain relatively more yolk, largely reflecting variation in lipid content. Therefore, different egg components can show positive or negative allometry with egg size in different taxa (W. Hill 1995; T. Williams 1994). Among individuals, those females laying large eggs will, on average, lay eggs with different relative composition, for example, large eggs with relatively more albumen (seabirds) might provide more protein, perhaps enhancing offspring structural size (Dzialowski et al. 2009), whereas large eggs with relatively more yolk (waterfowl) would provide more energy, which might enhance growth rate or post-hatching survival due to a larger retained yolk sac.

In addition to systematic variation in egg composition with increasing egg size, there is some residual, individual variation in egg composition when controlling for egg mass (fig. 4.12). On average, egg size explains 57% (a range of 31%–86%) and 46% (3%–77%) of variation in dry albumen and dry yolk content, respectively ($n = 11$ studies cited in T. Williams 1994). The functional significance of this egg-mass-independent variation in egg composition is poorly understood, in large part because of the problems of obtaining egg composition data on intact eggs that can then hatch. A more direct approach is, therefore, to experimentally manipulate egg composition to investigate the effects on offspring development. Early studies in poultry suggested that removal of 8%–20% of yolk had little effect on hatchling mass, but removal of similar amounts of albumen reduced late-stage embryo and hatchling mass (although these studies were characterized by high total mortality (>75%) of the manipulated eggs (Finkler et al. 1998; W. Hill 1993). Ferrari et al. (2006) manipulated the albumen content of barn swallow (*Hirundo rustica*) eggs, reducing the proportion of albumen relative to egg content by approximately one standard deviation (2.8%) or 2 standard deviations (5.6%) of the mean original amount of albumen. Albumen removal had no effect on hatching success (85%–91%), although it delayed hatching, and chick survival to fledging was lower (0.85) in the eggs with 2×SD albumen removed compared with the controls (0.95). Nevertheless, the original egg mass had positive effects on all nestling morphological traits at all ages after controlling for albumen manipulation. Bonisoli-Alquati

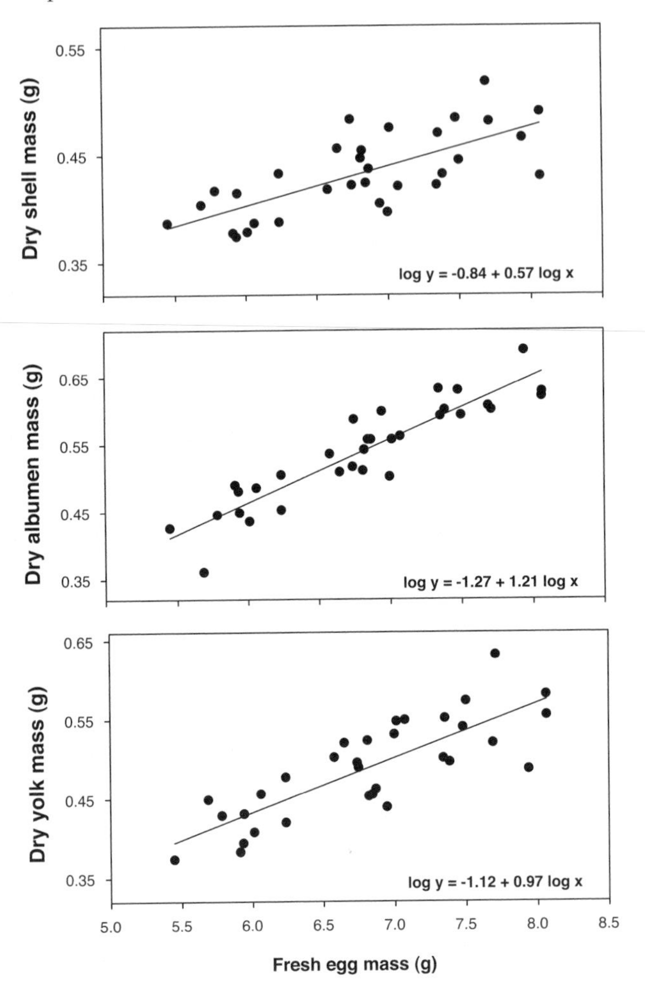

Fig. 4.12. Relationships between fresh egg mass and dry shell, dry albumen, and dry yolk mass in European starlings. Equations are derived from log10:log10 plots.

et al. (2007) also removed albumen equal to 2.3% of the estimated albumen content in the eggs of yellow-legged gulls (*Larus michahellis*), which decreased hatching success for a- and b-eggs from approximately 80% to 50%. However, this loss of albumen did not negatively affect body size or mass in the first days after hatching, and the overall egg mass strongly predicted chick size and mass at all ages irrespective of albumen removal. These studies therefore provide some support for the idea that individual variation in egg composition within the normally observed range can affect offspring phenotype independently of egg size. However, they also confirm the predominant effect of egg size per se on offspring development, with relatively minor effects of egg-size-independent egg composition. Nevertheless as Carey (1996) pointed out, many questions with regard to the importance of intraspecific variation in egg composition for offspring phenotype remain unresolved. It is also plausible that the specific content or composition of freshly laid eggs is less important than individual variability in how the components are metabolized and utilized by the developing embryo (Carey 1996; Hamdoum and Epel 2007).

4.4.2. Yolk Hormones

Since the first report of the transfer of maternal hormones to yolk in birds (Schwabl 1993), hormonally mediated maternal effects have become one of the more active fields of avian reproduction, but one largely driven by behavioral ecologists, not endocrinologists or physiologists (Groothuis, Müller, et al. 2005). Much of this work has been reviewed (D. Gill 2008; Groothuis, Müller, et al. 2005; Groothuis and Schwabl 2008; Groothuis and von Engelhardt 2005), so here I will focus on *individual variation* in yolk hormone levels and the *long-term* fitness consequences of this variation (mechanisms will be considered in section 4.5). As Groothuis and Schwabl (2008) pointed put, until recently there has been an almost "myopic" focus on yolk androgens in the context of hormonally mediated maternal effects. An increasing number of studies have investigated maternally derived yolk glucocorticoids (Hayward et al. 2006; Love and Williams 2008a; Saino et al. 2005), but far fewer studies have considered effects of yolk estrogens (von Engelhardt et al. 2004; T. Williams 1999) or thyroid hormone (McNabb and Wilson 1997). In addition, the field has been dominated by a focus on "extrinsic" factors that correlate with variation in yolk hormones and on the relatively short-term effects of yolk androgens on nestling growth and behavior. Little is known about the mechanisms by which hormones accumulate in the egg, how they are regulated (or not), and how maternal hormones affect embryo target tissues (but see Groothuis and Schwabl 2008; Moore and Johnston 2008; and Navara and Mendonça 2008). There have still been very few studies that have unequivocally demonstrated long-term fitness consequences of variation in maternally derived yolk hormones for either the offspring or the mother. Yolk hormone concentrations have been shown to correlate with social factors (Love, Wynne-Edwards, et al. 2008; Mazuc et al. 2003), maternal condition or quality (Love et al. 2005; Tschirren et al. 2004), mate quality or attractiveness (Gil et al. 2004; Tanvez et al. 2004), and food availability (Verboven et al. 2003). However, as Groothuis, Müller, et al. (2005) concluded, results from the many studies are less consistent than widely assumed, often contradictory, and for each extrinsic factor different studies have reported either positive or negative relationships, or no effect.

There is large, systematic variation in yolk androgen (Groothuis and Schwabl 2002) and glucocorticoid (Love, Wynne-Edwards, et al. 2008) levels within clutches in relation to laying sequence and among clutches in relation to clutch size. However, there is even more marked variation in yolk hormone levels among clutches laid by *different* females, independent of laying sequence or clutch size (T. Williams 2008). In black-headed gulls, *average* yolk testosterone (T) levels increase with laying sequence from 14.1 to 16.9 to 18.1 pg/mg in a-, b-, and c-eggs, respectively (Groothuis et al. 2006), but for any given laying sequence, yolk T varies more than twofold among individual females (fig. 4.13). In Cassin's auklets (*Ptychoramphus aleuticus*), which lay a single-egg clutch, yolk androstendione (A4) and yolk T varied eightfold, from 141.5 to 1091.5 ng/g yolk (95% CI) and from 3.6 to 28.6 ng/g respectively, among females, and in rhinoceros auklets (*Cerorhinca monocerata*), yolk A4 and yolk T varied from 486.3 to 1386.9 ng/g and from 6.7 to 39.0 ng/g, respectively, that is, three- to sixfold (Addison et al. 2008). There is accumulating evidence that this individual variation in maternally derived yolk hormones is repeatable (Eising et al. 2008; Tobler et al. 2007), and Tschirren et al. (2009) recently showed that yolk T, but not yolk A4, was highly heritable ($h^2 = 0.746$) based on mother-daughter regression in collared flycatchers. Although these estimates might have been inflated by common-environment effects, heritable genetic variation in maternally derived yolk hormones is consistent with the fact that yolk hormones show correlated responses to artificial selection for behavior, personality, and stress response (Gil and Faure 2007; Groothuis et al. 2008; Hayward et al. 2005). This suggests that yolk hormone levels might be a component of individual female phenotype, reflecting differences in individual "quality," with only relatively minor variation due to environment (Eising et al. 2008).

Yolk hormone concentrations are usually reported on a per mass basis (pg/mg or ng/g yolk mass), and the variation in yolk hormones might simply be determined by that in the egg or yolk mass (as is the case for the major egg components; see section 4.4.1, and for maternal transfer of antibodies and carotenoids, see sections 4.4.3 and 4.4.4). This dependence of the amount of yolk hormones on yolk mass might be predicted, for example, if the large lipid-rich yolk simply acts as a "sink" for lipophilic hormones, and so the amount of hormone transferred is proportional to the size of the yolk. Pilz et al. (2003) and Safran et al. (2008) specifically tested the hypothesis that higher-quality females would lay larger eggs with more yolk androgens, and they showed that larger eggs or eggs with larger yolks do contain more total T and A4 but do not have higher yolk

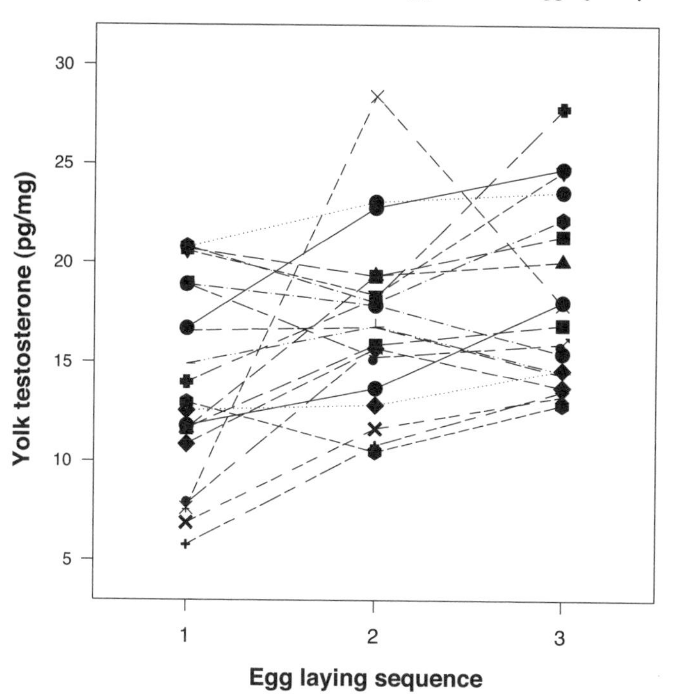

Fig. 4.13. Individual and laying-sequence-specific variation in yolk testosterone in eggs of the black-headed gull. Different individual females are indicated by different symbols. (Based on data provided by T.G.G. Groothuis, pers. comm.)

hormone *concentrations*. This observation suggests that larger embryos developing in eggs with larger yolks might be exposed to the same relative amounts of yolk hormone as smaller embryos developing in small-yolked eggs. In contrast, in European starlings, total yolk corticosterone is also positively correlated with yolk or egg size, but this relationship explains only 6%–7% of the total variation in yolk corticosterone. For any given yolk size, total corticosterone varies fourfold (10–40 ng per yolk) considering all eggs and twofold (10–20 ng per yolk) considering only second-laid eggs (fig. 4.14). So, similar-sized embryos developing with similar-sized yolks could be exposed to markedly different amounts of hormone. Other studies have shown that the effects of manipulated yolk hormones are independent of egg mass (Gil et al. 2004; Pilz et al. 2003) and that natural variation in yolk hormones is also independent of egg mass (Addison et al. 2008).

Given that maternally derived yolk hormones can vary markedly among individual females, and to a large extent independently of clutch size and egg size, this phenotypic variability would potentially provide

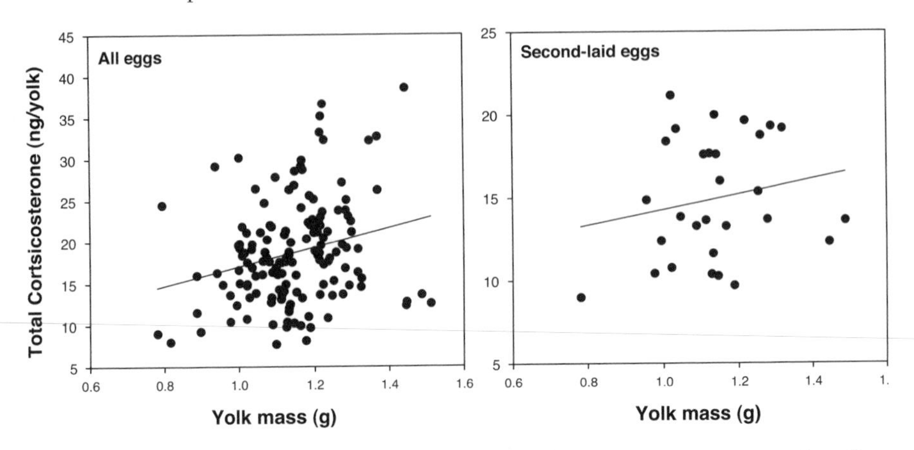

Fig. 4.14. Variation in total yolk corticosterone in relation to yolk mass for all eggs (left) and second-laid eggs only (right) in European starlings. (Based on data provided by O. Love, pers. comm.)

females with additional mechanisms to modify offspring phenotype, perhaps matching it to environmental conditions, and thus increasing offspring and maternal fitness. Early *in ovo* exposure to endogenous hormones produced by the embryo's own gonads is known to have long-lasting organizational effects on the brain and behavior, especially on sex-specific differentiation of the morphological, physiological, and behavioral phenotype (Adkins-Regan 2005; Balthazart and Adkins-Regan 2002). This leads to the a priori prediction that maternally derived hormones *will* have significant effects on offspring phenotype. However, as Carere and Balthazart (2007) point out, much of the work on hormonally mediated maternal effects, which has been driven by behavioral ecology, developed as "an independent line of research," and an unresolved problem is how exposure to maternal hormones affects an individual offspring's phenotype without interfering with the normal process of sexual differentiation. Evidence for significant, short-term effects of maternally derived yolk androgens on offspring phenotype is, in fact, rather contradictory. Some studies have reported positive effects of experimentally elevated yolk T during the early nestling period, including earlier hatching, enhanced begging and growth, enhanced immune function (Navara et al. 2006), and lower nestling mortality (Groothuis, Müller, et al. 2005; Gil (2008); and references therein). However, other studies have reported negative effects, including *delayed* hatching, *higher* nestling mortality (Sockman and Schwabl 2000; von Engelhardt et al. 2006), *suppressed* immune function (Groothuis, Eising, et al. 2005), and *decreased* growth (Rubolini et al. 2006; Rutkowska et al. 2007).

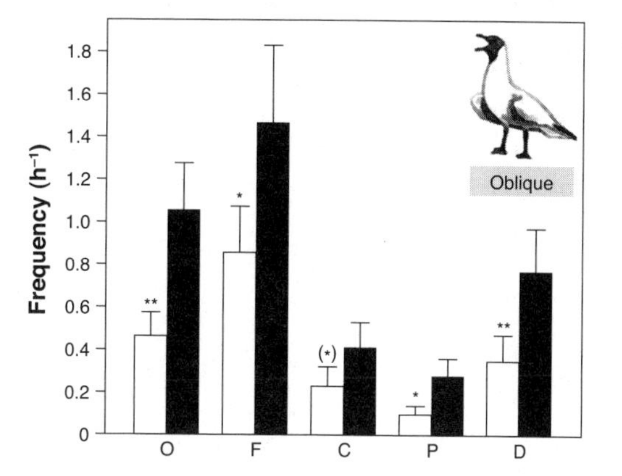

Fig. 4.15. Enhanced social behaviors and dominance in 10-month-old gulls from eggs injected with androgens (filled bars) compared to birds from eggs injected with oil (open bars). O, oblique display; F, forward display; C, charge; P, aggressive peck; D, displacement. (Reproduced from Eising et al. 2006, with permission of the Royal Society.)

A general assumption in all of these studies is that *short-term* effects on nestling growth, physiology, or behavior translate into *long-term* effects on offspring fitness (Eising et al. 2006), but few studies have directly demonstrated such long-term effects. In studies where the magnitude of the reported effects of yolk androgens were small, these effects were no longer apparent in older-aged nestlings (Navara et al. 2006; Pitala et al. 2009). In other studies, yolk androgens have been reported to increase, decrease, or have no effect on the size of sexually selected traits in males at reproductive maturity (Müller and Eens 2009; Rubolini et al. 2006; Strasser and Schwabl 2004). Partecke and Schwabl (2008) showed that house sparrows from testosterone-treated eggs had a greater frequency of aggressive, dominance, and sexual behavior as 1-year-old, reproductively competent adults. Similarly, in black-headed gulls (fig. 4.15), experimentally elevated maternal yolk androgens enhanced the development of nuptial plumage and the frequency of aggressive and sexual displays almost 1 year after hatching (Eising et al. 2006). However, none of these studies actually measured the reproductive success or fitness of these birds as adults. In females, experimentally elevated yolk T either had no effect on reproductive traits of female offspring (Rutkowska et al. 2007; von Engelhardt et al. 2004), negative effects on egg size (Uller et al. 2005), or positive effects on clutch size but not egg mass (Müller et al. 2009). As with egg size, most studies have addressed the question of whether yolk hormones enhance offspring fitness, not whether females that lay eggs with higher (or lower) yolk hormone levels have better fitness (but see Hinde et al. 2009 for an example of yolk androgens matching offspring demand and maternal effort). In this regard, the studies by Love and colleagues provide the most comprehensive example to date of fitness consequences of variation in maternally derived yolk corticosterone (see chapter 6; Love et al. 2005; Love and Williams 2008a; Love, Wynne-Edwards, et al. 2008).

In summary, there is marked individual variation in maternally derived yolk hormones that appears to be repeatable and might reflect differences in individual "quality" but that is independent of other measures of female quality such as egg or clutch size. However, at present there is inconsistent, equivocal evidence even for short-term effects of yolk hormones on offspring phenotype. It is possible that the effects of maternal hormones might differ with the sex of the offspring (Love et al. 2005; Love and Williams 2008a; Rutkowska and Cichon 2006; Rutkowska et al. 2007; Sockman et al. 2008), which was not considered in many earlier studies. However, a more generic explanation is that maternally derived yolk hormones would be *expected* to have complex positive and negative effects, because of the inherent pleiotropic nature of hormones. Females should therefore balance the costs and benefits of maternal "deposition" into their eggs for themselves and their offspring in terms of their lifetime reproductive success (Groothuis et al. 2006; Groothuis, Eising, et al. 2005), assuming females can regulate this deposition (Groothuis and Schwabl 2008). Moreover, the relative costs and benefits of yolk hormones would be expected to change with individual female quality or condition at the time of laying, with ecological context, and, among species, with evolutionary or life-history context (Love et al. 2009). For example, in species laying single-egg clutches, hormonally mediated maternal effects might be more important for modulation of the maternal reproductive effort through offspring phenotype, because clutch size is univariate. In contrast, in species with large clutch sizes, females have many potential mechanisms to adjust reproductive effort (timing, clutch size, total clutch mass), and yolk hormones might play a very minor role. Since ecological context (e.g., natural food availability, Pilz et al. 2004) or life histories have rarely been used to generate explicit a priori predictions about the likely magnitude and direction of effects of yolk hormones, this omission could explain the many seemingly contradictory results of studies to date. Nevertheless, at present it would appear that a female's reproductive "decision" to lay earlier or later (chapter 3) or to lay larger or smaller clutches (chapter 5) is likely to have far more important and far-reaching fitness consequences than any "decision" about variation in yolk hormones.

4.4.3. Egg Immunoglobulins and Antimicrobial Proteins

During egg formation, maternal antibodies can be transferred from an immunocompetent female to the immunologically naive embryo via deposition in the yolk or albumen, and maternally derived antibodies are thought to provide the primary form of humoral immune defense for offspring early in life, prior to the offspring producing their own antibodies (Boulinier and Staszewski 2008; Grindstaff et al. 2003; Hasselquist and Nilsson 2009). In birds, IgY (the avian equivalent of mammalian IgG) is the principal immunoglobulin transferred into eggs in the egg yolk. Other antibodies (IgA, IgE, and IgM) are deposited in egg albumen at low concentrations and are transferred via ingested allantoic or amniotic fluid to the embryonic gut, where they can be utilized for local intestinal protection (Bencina et al. 2005; Boulinier and Staszewski 2008). The type and amount of antibodies transmitted to the offspring reflect the circulating antibodies in the laying female (Gasparini et al. 2001; Hasselquist and Nilsson 2009), both of which can include antibodies generated by the mother over a relatively long time period before egg-laying (up to 1 year, Staszewski et al. 2007).

There are positive relationships between the concentration of specific antibodies circulating in mothers (but not fathers) and the antibody levels in eggs or offspring (fig. 4.16), but IgY concentrations in the yolk are similar to or lower than those in the maternal serum (Gasparini et al. 2002; Grindstaff 2008; Kowalczyk et al. 1985). This observation suggests that antibodies are transferred from females to the eggs in proportion to plasma antibody concentrations with no evidence of mechanisms for accumulation of higher IgY concentrations in the yolk (Grindstaff 2008; Grindstaff et al. 2003; Staszewski et al. 2007). However, there is marked individual variation in plasma immunoglobulin levels among different females (e.g., 3.3–5.3 mg/ml for IgG in the chicken), which might be related to female "condition" or food availability during laying (Gasparini et al. 2007; Karell et al. 2008; Pihlaja et al. 2006). These differences could then account for the marked individual variation in the amount of antibodies transferred to the eggs in clutches of different females. There is currently only very limited evidence to suggest that females can facultatively adjust the concentration of maternal antibodies that they transfer to eggs in order to make "adaptive" adjustments to egg quality (Hasselquist and Nilsson 2009; Saino et al. 2001).

Because of the rapid endocytosis of IgY by the developing oocytes, it has been estimated that a laying chicken "loses" 10%–20% of her steady-state IgY to the oocytes every day (Kowalczyk et al. 1985). This calculation has led to the suggestion that the transfer of antibodies to

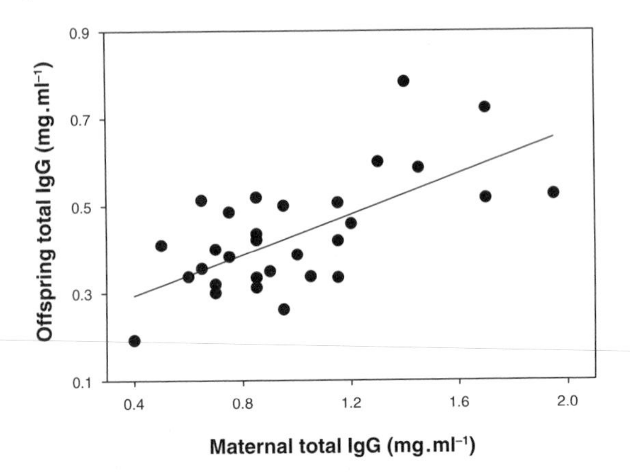

Fig. 4.16. Relationship between total plasma IgG concentrations in maternal circulation and total IgG concentrations measured in offspring on day 6 post-hatch in Japanese quail. (Reproduced from Grindstaff 2008, with permission of the Company of Biologists.)

eggs might represent a cost to laying females through suppression of their own immune function (Boulinier and Staszewski 2008). However, Saino et al. (2001) showed in female barn swallows that there was no depletion of total immunoglobulin proteins in plasma during egg-laying, because egg-laying females raised their production of antibodies compared with non-laying females (Saino, Dall'ara, et al. 2002). While this enhancement could still intensify the demand for resources, even in the domestic hen, which sustains high laying rates, the daily total synthesis of immunoglobulin is less than 0.05% of body weight per day (Klasing 1998). At present, therefore, there is little evidence to support the idea that the transfer of antibodies to eggs is costly to females.

The direct, positive effects of maternally derived antibodies should be of short duration because antibodies are rapidly catabolized: most maternal antibodies disappear from the offspring within 5–14 days of hatching (Grindstaff et al. 2003; Staszewski et al. 2007). In some non-domesticated birds, the chicks develop their own immune system and start to produce endogenous antibodies around 10–14 days post-hatch (Grindstaff et al. 2006; Pihlaja et al. 2006; Staszewski et al. 2007), so maternally derived antibodies should be important for the protection of the developing embryo and the newly hatched chick. The positive effects of maternal antibodies on offspring growth and survival after a disease or immune challenge have been demonstrated in poultry and other domesticated species (Grindstaff 2008; Grindstaff et al. 2003), and immunization of breeding flocks is routinely used in the poultry industry for this purpose (Etches 1996). However, evidence for the positive effects of maternal antibodies on offspring phenotype in free-living birds is limited and mixed. Early studies by Heeb et al. (1998) showed a positive effect on nestling growth rate and recruitment in broods where mothers were exposed to an ectoparasite (hen fleas) during egg-laying. Follow-up studies suggested that this maternal effect was due to immunoglobulins transferred via the egg (Buechler et al. 2002; Gallizzi et al. 2008). In contrast, experimental manipulation of maternal immunoglobulin levels had no effect on offspring growth or survival to fledging in other studies (Grindstaff et al. 2006; Pihlaja et al. 2006).

Longer lasting, indirect effects of maternal antibodies have been proposed, largely from studies in mammals, via a positive or negative effect on development of the offspring's own (endogenous) humoral immune response (Hasselquist and Nilsson 2009). In birds there is some evidence that maternal antibodies might enhance the endogenous humoral immunity of offspring up to the end of the nestling period (14–20 days post-hatching; Gasparini et al. 2006; Grindstaff et al. 2006). J. Reid et al. (2006) showed in free-living song sparrows that vaccination of parents with a novel antigen (tetanus toxoid) influenced the humoral antibody response mounted by fully grown offspring up to 1 year later. The offspring of vaccinated mothers (but not vaccinated fathers) mounted substantially stronger antibody responses than offspring of unvaccinated mothers, although the success or fitness of these offspring was not reported. In contrast, Staszewski et al. (2007) found evidence that maternal antibodies suppressed the development of humoral immunity of offspring during the nestling period (25 days post-hatching) in the black-legged kittiwake (*Rissa tridactyla*). These contradictory results might be due to the fact that studies are currently constrained by methodological limitations and lack of knowledge about natural host-parasite systems (i.e., ecological context) in which protective antibodies are known to play a role (Boulinier and Staszewski 2008). Several experimental studies have shown that maternal antibodies or early immune responses in neonates do not predict adult immune responses (Addison et al. 2010; Love, Salvante, et al. 2008).

A number of other immunoactive proteins, including lysozyme, ovotransferrin, and avidin, are transferred to the albumen during egg production and are thought to provide innate, non-specific antibacterial immunity to the embryo and newly hatched chick (B. Palmer and Guilletter 1991). Once again there is individual variation in the amounts of antimicrobial proteins transferred to albumen among clutches laid by different females, but there is little evidence for a systematic relationship with laying sequence within clutches (D'Alba, Shawkey et al. 2010; Saino, Dall'ara, et al. 2002). Although Saino, Dall'ara, et al. (2002) reported a positive, but weak, correlation between the mean within-brood lysozyme activity of nestling barn swallows and the residual lysozyme activity in mothers, this relationship would actually *not* be expected because egg lysozyme is produced by tubular gland cells in the oviduct rather than being sequestered from the plasma pool (Burley and Vadehra 1989; Mandeles and Ducay 1962). Mean lysozyme activity in the albumen can be up to 20-fold higher than in the maternal plasma, a rate that demonstrates the existence of active mechanisms for the transfer of lysozyme to, and accumulation in, albumen during egg formation in the oviduct (Saino, Dall'ara, et al. 2002). Nevertheless, there are currently no data to suggest that transfer of antimicrobial proteins to eggs is costly for mothers or that the differential allocation of these proteins has fitness consequences for offspring or the mother (D'Alba, Shawkey et al. 2010). In fact, Shawkey et al. (2008) suggested that the production cost of antimicrobials in albumen appears to be low, and behavioral regulation of their function may make their differential allocation unnecessary, for example, incubation itself inhibits the growth of bacteria on egg shells (Shawkey et al. 2009).

4.4.4. Egg Antioxidants

Egg yolk is yellow due to the presence of carotenoids, biologically active pigments that can function as antioxidants and in immunity (Blount et al. 2000). Carotenoids are transferred from the yolk to the tissues of the developing embryos (Surai and Speake 1998), and these antioxidants might be important in reducing lipid peroxidation and tissue damage from oxidative stress, thus protecting cell membranes and/or cell function in developing embryos and chicks, which have high rates of oxidative metabolism and free radical production (Blount et al. 2000; Catoni et al. 2008). The availability of yolk-derived carotenoids appears to be essential for embryos to develop an ability to acquire and use carotenoids post-hatching. In chickens, embryonic exposure to maternally derived carotenoids determines the subsequent efficiency of carotenoid metabolism and the capacity of offspring to assimilate dietary carotenoids and to use and accumulate these at the cellular level in different tissues (Koutsos et al. 2003). Most work on antioxidants in birds has focused on carotenoids, but the antioxidant system is complex, comprising numerous exogenous antioxidants of dietary origin (including the carotenoids lutein, α- and β-carotenes, and lycopene, and vitamin E) and endogenous antioxidants (including vitamin C, uric acid, superoxide dismutase; Catoni et al.2008; Cohen and McGraw 2009). Recent work, however, has concluded that the importance of carotenoids as antioxidants might have been overestimated and that carotenoids appear to contribute little to total antioxidative capacity (Costantini and Møller 2008; Hartley and Kennedy 2004;

Isaksson et al. 2007). In addition, measurement of antioxidant capacity (concentration) alone is not sufficient to assess oxidative stress without measurement of oxidative damage (Costantini and Verhulst 2009). It is possible that carotenoids mediate their effects through alternative functions, such as their role in enhancing immune function (Costantini and Møller 2008), which highlights our need for an improved understanding of the physiology of the antioxidant system (Costantini 2008).

As with the other "minor" egg components, there is marked individual variation in the total amount of carotenoids deposited in egg yolks by different females (Royle et al. 2001); for example, yolk-carotenoid deposition among female great tits varied from 2.5 to 75 μg/g yolk (Isaksson et al. 2008). Some studies have also reported systematic variation in total yolk carotenoids within clutches in relation to laying sequence, with higher concentrations in earlier-laid eggs (Royle et al. 2001; Royle et al. 1999; Safran et al. 2008), although more detailed analysis suggests that the different yolk carotenoids can show very different patterns of variation across the laying sequence (Newbrey et al. 2008). These patterns have led to the widespread conclusion that carotenoids might mediate maternal effects, with mothers influencing offspring phenotype by varying the amounts of carotenoids deposited in eggs (Romano et al. 2008; Royle et al. 1999; Saino, Bertacche, et al. 2002). Some studies have suggested that high-quality females invest more carotenoids in their clutch than low-quality females (Isaksson et al. 2008). However, unlike most other components of egg quality, yolk carotenoids do not co-vary with measures of individual female quality. For example, in barn swallows the most colorful females that have the earliest laying dates and lay the largest eggs deposited lower concentrations of yolk carotenoids (Safran et al. 2008). Similarly, in collared flycatchers there was no effect of female condition, body size, or age on yolk-carotenoid concentration (Török et al. 2007). Various studies have also reported positive, negative, or no relationship between carotenoid concentrations and egg size (Cassey et al. 2005; Fenoglio et al. 2003; Newbrey et al. 2008; Royle et al. 2001), although in general, larger eggs contain more total carotenoids (Isaksson et al. 2008; Safran et al. 2008). Carotenoids cannot be synthesized de novo by birds or stored for long periods in the liver and so must be obtained in the diet (Surai 2002). Therefore, variation in yolk-carotenoid deposition might mainly be determined by variation in dietary intake, perhaps reflecting differences in foraging ability of or food availability for egg-laying females (Hargatai et al. 2006; Török et al. 2007). Yolk carotenoids are highly correlated with the carotenoid content of the diet (Bortolotti et al. 2003); carotenoid supplementation of laying females increases yolk-carotenoid levels (Blount et al. 2002; McGraw et al. 2005; Remes et al. 2007); and plasma and yolk carotenoids are positively correlated (Bortolotti et al. 2003).

Are carotenoids limiting during egg production in females, and might females pay a cost of carotenoid deposition in eggs? The first question is long-standing but remains unresolved (Olson and Owens 1998). A general approach has been to supplement laying females with carotenoids (mainly lutein), and an early study concluded that "egg-laying capacity is limited by carotenoid pigment availability" (Blount et al. 2004). However, even Blount et al.'s study (2004) found very limited effects of carotenoid supplementation on female reproductive traits, with no effects on clutch mass or re-laying interval, a result that has been confirmed by subsequent studies. Carotenoid supplementation before laying had no effect on laying date, egg mass, variation in intra-clutch egg size, or clutch size in a wide range of species (Blount et al. 2002; Bortolotti et al. 2003; Remes et al. 2007; Royle et al. 2003). More complex "costs" have been proposed that involve trade-offs between the allocation of carotenoids and other physiological systems, mainly in males in relation to sexual signaling (Alonso-Alvarez et al. 2009; Alonso-Alvarez et al. 2008; Bertrand, Faivre, and Sorci 2006). For example, Groothuis et al. (2006) proposed that maternal effects might be mediated by an interaction of yolk carotenoids and yolk T in relation to laying sequence. However, there is currently only very limited evidence that egg production causes a reduction in the female's own antioxidant capacity (Morales et al. 2008; cf. Surai and Speake 1998), and there is no evidence that this effect has long-term costs.

Does carotenoid supplementation have effects on offspring phenotype? Again, in general, experimental studies that have reported unequivocal, positive effects of carotenoids on offspring performance have either been conducted in captivity (McGraw et al. 2005) or involved injection of carotenoids directly into eggs in the wild (Saino et al. 2003). Most studies have shown no effect or only a very weak effect of carotenoid supplementation of offspring phenotype, and then only in the early nestling phase. Carotenoid supplementation of laying females had no effect on hatching success, nestling growth, fledging mass, or survival to fledging stage in free-living great tits (Remes et al. 2007), blue tits (Biard et al. 2007), or Eurasian kestrels (*Falco tinnunculus*) (De Neve et al. 2008), or in captive red-legged partridges (*Alectoris rufa*; Bortolotti et al. 2003). Maternally derived carotenoids increased nestling begging behavior in great tits, but there was no effect on the body mass of 6-day-old nestlings (Helfenstein et al. 2008). De Ayala et al. (2006) found that vitamin E supplementation of laying females enhanced the mass and size of nestling barn swallows, but only up to 10–12 days post-hatching, not in older nestlings. Similarly, Romano et al. (2008) reported that lutein supplementation via egg injection enhanced the growth of sons from first eggs but depressed that of sons from last eggs; enhanced the survival of daughters, but only late in the season; and promoted the immunity of male chicks and chicks from small eggs but not female chicks or chicks from large eggs (see also Berthouly et al. 2008). Therefore, at present there is little consistent evidence to support strong effects of yolk carotenoids on offspring growth and development, and no evidence for long-term effects.

4.5. Physiological Mechanisms Underlying Individual Variation in Egg Size and Egg Quality

Larger eggs contain absolutely more major egg components (shell, albumen, yolk) and absolutely more of most of the minor egg components (total amounts of maternally derived antibodies and antioxidants). Assuming that high-quality females lay larger eggs they would provide their offspring with greater *total* amounts of most major and minor egg components, and thus egg size *would* be a good measure of egg quality. If we can identify the mechanisms by which individual females regulate egg size itself, they will probably also provide insight into the mechanisms that regulate other aspects of egg quality. For example, if yolk size determines egg size (T. Williams et al. 2001), then we need to identify mechanisms underlying the differences in yolk uptake, and if antibodies and antioxidants are transported "piggy-back" via receptor-mediated endocytosis of yolk precursors, these same mechanisms could also explain variations in the minor components of egg quality (MacLachlan et al. 1994). It would also seem parsimonious to assume that the same mechanisms that generate the large individual variation in egg size also generate smaller age-, mass- or environment-dependent variation, but these mechanisms should be much easier to identify if we focus on the major source of variation, that is, differences among individual females. There appear to be two main exceptions to this line of reasoning: the amounts of both yolk hormones and albumen antimicrobial proteins can vary independently of yolk or egg size. Therefore, we also need to identify regulatory mechanisms for these two egg components that might be *independent* of mechanisms underlying yolk formation per se.

4.5.1. Mechanisms Regulating Egg Size and Egg-Size-Dependent Egg Quality

Theoretically, the maximum egg size might be set by *peripheral* effectors, such as the rate of synthesis of yolk precursors in the liver or their uptake at the ovary, or by the *central* physiological "machinery," such as the digestive, pulmonary, cardiovascular, or excretory organs that supports these peripheral effectors. These represent the *peripheral limitation hypothesis* and the *central limitation hypothesis*, respectively (Peterson et al. 1990). As an example of central limitation, given the proposed importance of food supply in regulating egg production, one might predict that in egg-producing females, *increases* in the size of the digestive system or of nutrient stores would maximize the supply of nutrients for egg formation. If the up-regulation of "essential" supporting physiological machinery is costly, this expense might necessitate reallocation of resources such that the size of other "non-essential" components of body composition; for example, flight muscles, might actually *decrease* in egg-producing females compared with other breeding stages. This line of reasoning suggests that (a) there should be predictable, systematic variation in body composition in egg-producing females: organs specifically adjusted to the demands of egg production should be largest (and "non-essential" organs smallest) in egg-laying females compared with non-breeders and chick-rearing individuals; (b) among egg-producing females, non-reproductive organ mass should show a parallel (or inverse) pattern of mass change to gonadal development and regression, matching the demands of follicle development (fig. 4.17); and (c) individual variation in the size of key "limiting" organs should correlate positively with individual variation in egg size or egg quality.

Only a few studies have tested these hypotheses through detailed analysis of changes in body composition in relation to known breeding state (egg-producing versus non-egg-producing) and individual variation in egg, yolk, or follicle size. As described in chapter 3, there is little evidence for consistent, systematic up-regulation of the food-processing organs (gizzard, intestine) in relation to the greater demands of egg production, even though body composition and organ masses can vary markedly between breeding stages and among years (see fig. 3.17). With the exception of the kidney, the "metabolic machinery" (sensu Daan, Masman, and Groenwald1990: heart, lungs, liver) also did not change systematically between egg-laying and non-laying female European starlings (Vézina and Williams 2002, 2003). Furthermore, individual variability in egg size was independent of the relative mass-corrected size of each of these organs, and the changes in yolk protein and yolk lipid were independent of individual variation in flight-muscle mass (the main protein store) and total body fat, respectively (Christians and Williams 2001). In captive-breeding zebra finches, although there was marked individual variation in the residual organ masses for these same organs, it did not explain individual variation in reproductive effort as measured by egg size, F1 and F2 follicle mass, or oviduct mass (T. Williams, unpublished data; T. Williams and Martyniuk 2000). Organ *mass* might not be a reliable index of physiological or metabolic capacity. However, Vézina and Williams (2005) measured tissue-specific metabolic variation via citrate synthase (CS) activity (a putative "pace-making" enzyme in the Krebs cycle) in

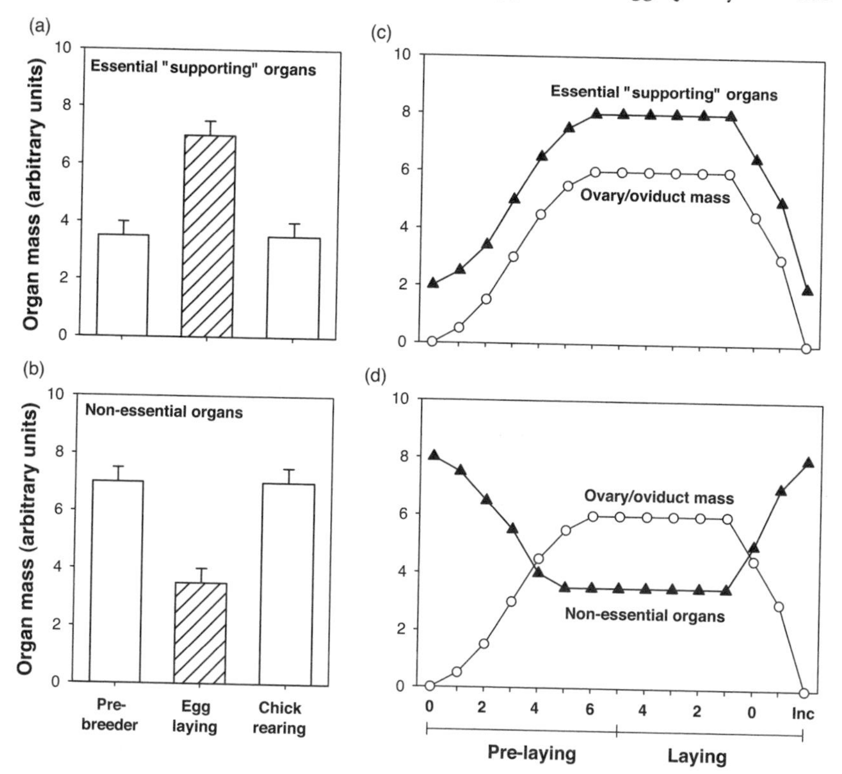

Fig. 4.17. Examples of organ mass changes that support the hypothesis of "central" limitation of egg size: (a) mass of essential "supporting" organs is highest in egg-laying females; (b) mass of non-essential organs is lowest in egg-laying females, consistent with reallocation; (c) mass of essential "supporting" organs parallels that of reproductive organs to support demands of egg formation; and (d) mass of non-essential shows an inverse relationship to that of reproductive organs, reflecting reallocation.

pectoral muscle, heart, kidney, and liver in free-living female European starlings during egg production. Citrate synthase activity varied in relation to the breeding stage and/or year, but this variation did not support the hypothesis that egg production was associated with up-regulation of organ-mass-specific enzyme activity. For example, even though the liver is actively involved in yolk-precursor production, no evidence was found for up-regulation of mass-specific CS activity in this organ during egg formation, and in fact total CS enzyme activity was lower at the one-egg stage than in non-breeding females (fig.4.18). In summary, there is currently no strong evidence for the central limitation of egg production, in which individual variation in egg size is related to the relative size or

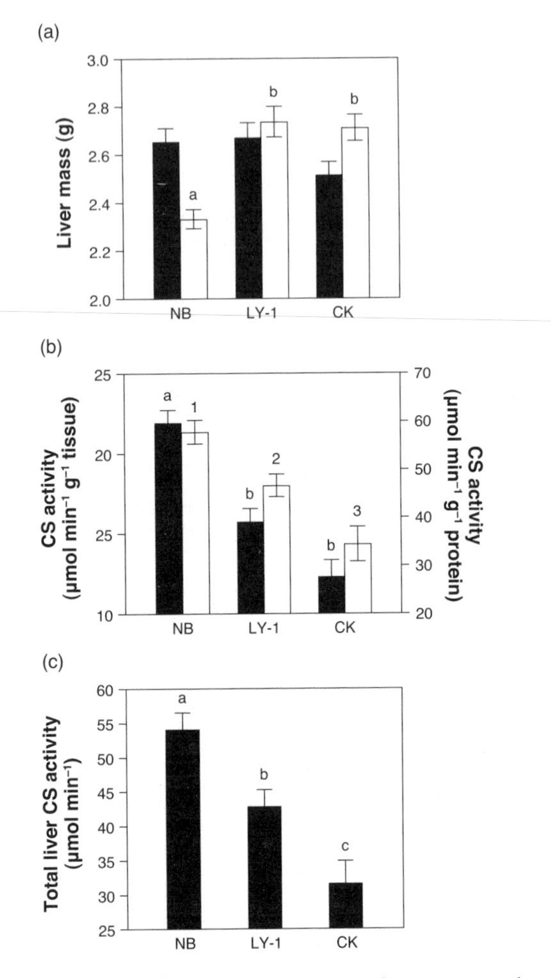

Fig. 4.18. Variations in (a) fresh mass (b) mass-specific citrate synthase (CS) activity, and (c) total CS activity in the liver of female European starlings at different breeding stages. NB, non-breeding; LY–1, one-egg; CK, chick-rearing. Letters and numbers indicate significant differences. Black bars in (b) are for activity per gram of wet tissue, and white bars indicate activity per gram of protein. (Based on data in Vézina and Williams 2005.)

metabolic activity of digestive, pulmonary, cardiovascular, or excretory organs that must support the demands of egg production.

Is there evidence that individual variation in egg size might be determined by the capacity of peripheral effectors, for example, the rate of synthesis, transport, and/or uptake of yolk precursors (described in detail in chapter 2)? Individual variation in the rate of synthesis of yolk precursors in the liver has not yet been directly measured, but several studies have measured individual variation in steady-state plasma yolk-precursor concentrations and, more recently, variation in rates of yolk-precursor uptake at the ovary. In European starlings, follicle mass was *negatively* correlated with plasma VTG and VLDLy levels (Challenger et al. 2001; fig. 4.19), and variation in yolk-lipid and yolk-protein mass was negatively correlated with plasma VTG (Christians and Williams 2001). In captive-breeding female zebra finches, egg mass was negatively correlated with the levels of plasma VTG in females on a high-quality diet, but these traits were positively correlated in females on a low-quality diet (Salvante and Williams 2002). Salvante and Williams (2002) suggested that females that laid either very small or very large eggs were driving this relationship: individual females that lay small eggs might do so because egg size is limited by the size of the yolk-precursor pool in their plasma; that is, their circulating levels of VTG may fall below the level required to maintain high levels of receptor-mediated yolk uptake. Conversely, in females that produce the largest eggs, yolk formation may actually deplete the plasma VTG pool, with the uptake of VTG exceeding the rate of VTG synthesis and secretion by the liver. Overall, however the relationships between the circulating levels of yolk precursors and egg, yolk, or follicle size are relatively weak (r^2 = 10%–38%), and in all studies there is a large degree of individual variation (8- to 10-fold) in yolk-precursor levels for any given egg or follicle size (see figs. 4.19 and 4.20). Thus, individual variation in *circulating* yolk-precursor levels might contribute to, but does not fully explain, variation in egg size. This conclusion is supported by data from experimental and selection studies. In female zebra finches, a 30%–50% reduction in the circulating levels of VTG, caused by administration of the anti-estrogen tamoxifen, reduced egg mass only by 10% (T. Williams 2000, 2001). In Japanese quail, 32 generations of divergent selection for high and low yolk-precursor production resulted in a 10-fold increase and 4-fold decrease in circulating yolk-precursor levels, respectively, in the high- and low-selection lines. However, despite these differences in yolk-precursor levels, mean egg and yolk size varied little among the selected quail (Chen et al. 1999).

Other components of yolk-precursor dynamics also fail to fully explain individual variation in egg, yolk, or follicle mass. In chickens, females with higher laying rates had a higher proportion of smaller-diameter VLDLy particles that were available to be incorporated into developing yolks (Walzem 1996), but in the zebra finch, individual variation in egg and clutch size and the timing of laying was unrelated to individual variation in VLDLy particle-size distribution (Salvante et al. 2007a). Estrogens provide the hormonal signal that drives most aspects of vitellogenesis and follicle and oviduct development in laying females, and on average,

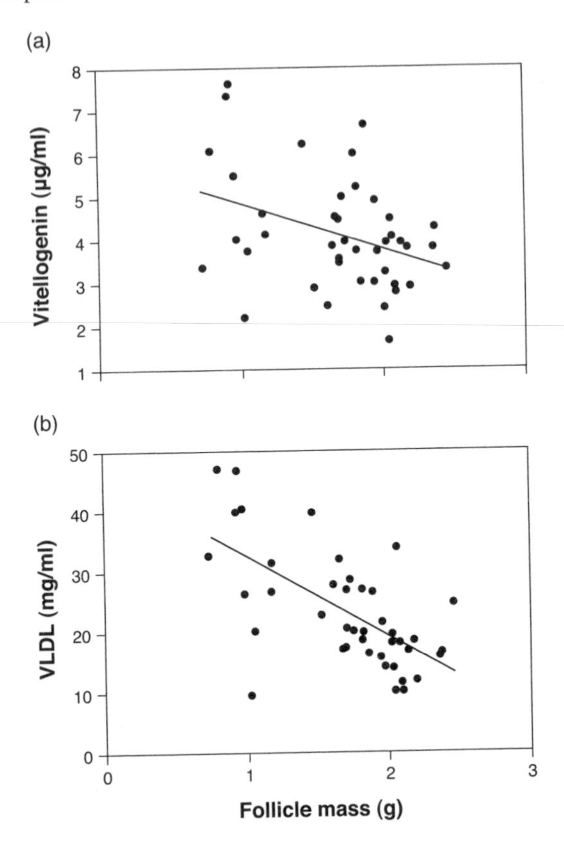

Fig. 4.19. Relationship between total vitellogenic follicle mass and plasma yolk-precursor levels during egg production in European starlings for (a) plasma vitellogenin (µg/ml zinc) and (b) plasma VLDLy (mg/ml, triglyceride).

the levels of plasma estrogen parallel changes in reproductive function as follicles develop and are ovulated (see fig. 2.13). However, although there is up to a fourfold variation in plasma E2 levels among individual egg-laying females, it is unrelated to the plasma levels of the main yolk precursor VTG or to the total mass of yolky follicles developing at the time of blood sampling (T. Williams, Kitaysky, and Vézina 2004; fig. 4.20).

Thus, the picture that emerges is one of marked individual variation in each component of the "peripheral" mechanisms underlying egg production (plasma estradiol, yolk-precursor levels, VLDLy particle size) that is only weakly related, or is unrelated, to individual variation in the measures of female reproductive output. This is counterintuitive since there *should* be functional relationships between these traits, based on our knowledge of the mechanisms underlying egg production (chapter 2).

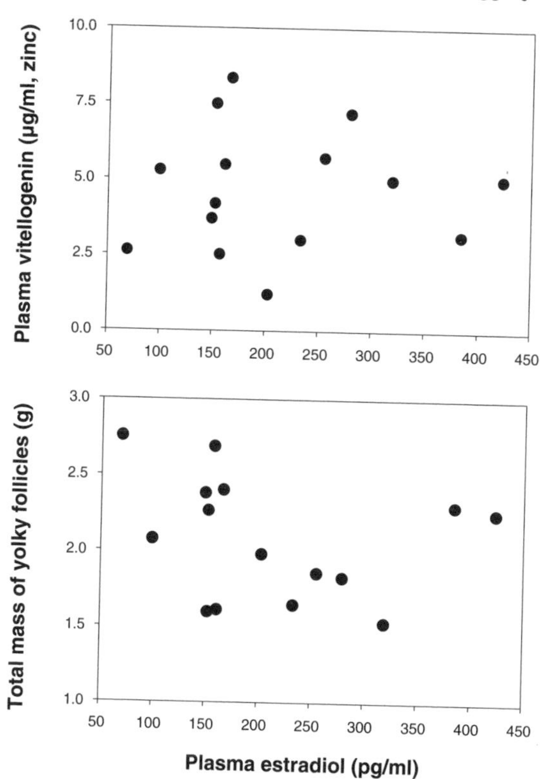

Fig. 4.20. Individual variation in and lack of relationship between plasma vitellogenin (top) and total vitellogenic follicle mass (bottom) and plasma estradiol levels measured at the one-egg stage of laying in female European starlings. (Based on data in T. Williams et al. 2005.)

What is intriguing, however, is that individual variation in these physiological traits is not *random* but, rather, is systematic. Plasma yolk-precursor levels are highly repeatable among breeding attempts ($r > 0.80$, Salvante and Williams 2002) and even across years ($r = 0.50$, Challenger et al. 2001; fig. 4.21). However, it remains unclear why, from a functional and adaptive perspective, some birds maintain 10-fold higher plasma estradiol or VTG levels compared with other birds when this has no apparent benefit in terms of increased reproductive output. Within a cost-benefit framework, this relationship presents a special paradox, especially if there are costs, such as osmoregulatory perturbation, or transient inhibition of erythropoiesis, to maintaining high hormone or yolk-precursor levels (Challenger et al. 2001; E. Wagner, Prevolsek, et al. 2008). This variability would be consistent with individual females adopting different physiological "strategies" in which different combinations of traits

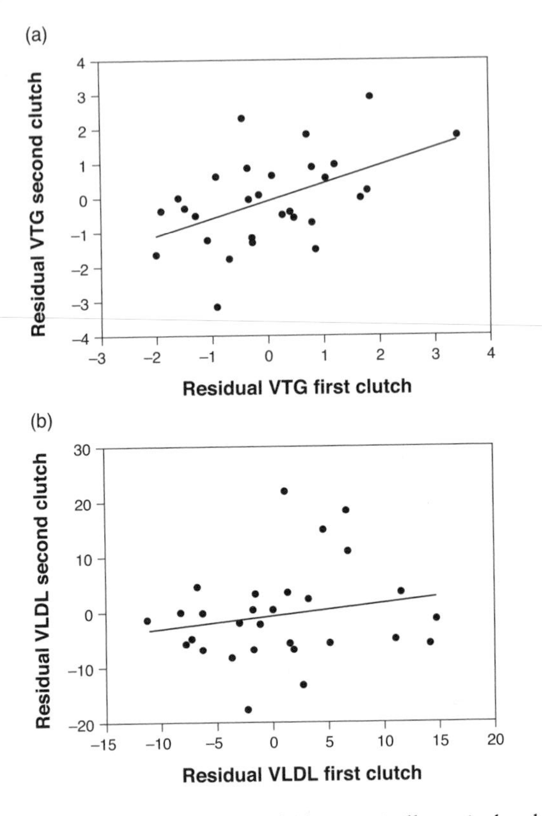

Fig. 4.21. (a) Within-year repeatability of plasma vitellogenin levels (μg/ml zinc) and (b) lack of repeatability of plasma VLDLy (mg/ml, triglyceride) between first and second clutches in female European starlings. (Based on data in Challenger et al. 2001.)

or trait values determine egg size. For example, might some individuals have very different sensitivities to circulating E2 levels, such that in different individuals very different plasma E2 levels are required to support the same level of physiological function? Or might some females require high concentrations of yolk precursors to compensate for a low efficiency of receptor-mediated uptake, for example, low VTG/VLDL-R number or activity? Resolving the functional consequences of this large-scale, individual physiological variation in the "peripheral" mechanisms of egg production represents an exciting challenge for the future in terms of our understanding of costs of egg production, both from a physiological and an evolutionary perspective.

In contrast to mechanisms underlying the *supply* of yolk precursors to developing ovarian follicles, there is evidence that receptor-mediated

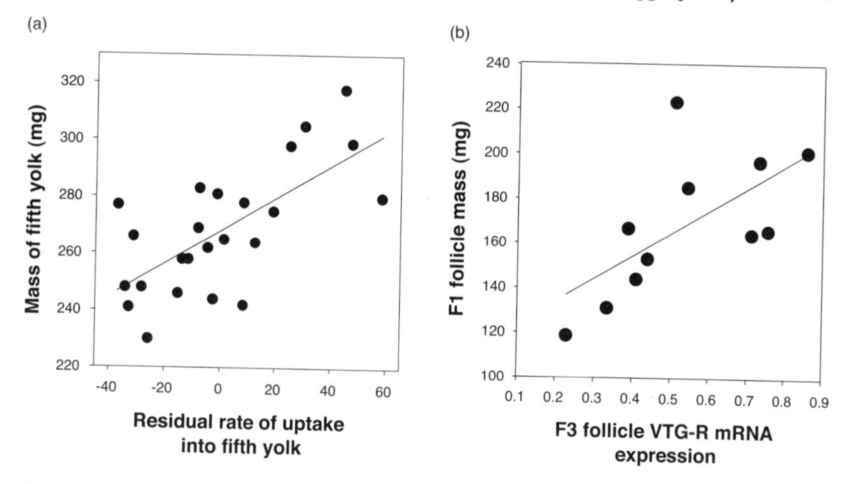

Fig. 4.22. Positive relationships between individual variation in (a) yolk mass and rate of yolk precursor uptake (based on data in Christians and Williams 2001) and (b) F1 follicle mass and F3 follicle VTG/VLDL-R mRNA expression (based on data in Han et al. 2009).

uptake of yolk precursors is a determinant of individual variation in egg size. Christians and Williams (2001) measured the variation in yolk uptake rates indirectly by injecting laying female zebra finches with a radiolabelled amino acid (serine) and then measuring the radioactivity in subsequently laid eggs after in vivo incorporation of amino acids into yolk precursors and uptake into the yolk. The rate of uptake of radiolabel by the yolks was positively related to yolk mass ($r^2 = 0.24$, 0.35, and 0.50 for the yolks of the third-, fourth- and fifth-laid eggs, respectively; fig. 4.22a), suggesting that individual variation in yolk mass is due, at least in part, to variation in the *rate* of follicle growth and yolk-precursor uptake. Furthermore, the uptake of radiolabel was repeatable between breeding attempts in ($r = 0.23$–0.44), as was mean yolk mass ($r = 0.35$), suggesting that these traits are characteristic of the individual female's phenotype. This strongly suggests that receptor-mediated yolk uptake, and potentially the expression level or the functional activity of the VTG/VLDL receptor, might be a key determinant of phenotypic variation in follicle, yolk, or egg size in birds. In support of this idea, Han et al. (2009) found that individual variation in VTG/VLDL-R mRNA expression in F3 follicles was positively correlated with individual variation in egg and F1-follicle mass in female zebra finches (fig. 4.22b). In addition, VTG/VLDLR mRNA expression in the ovary was positively correlated with individual variation in clutch size and laying interval. Individuals with larger clutches had higher levels of mRNA expression in the ovary, which

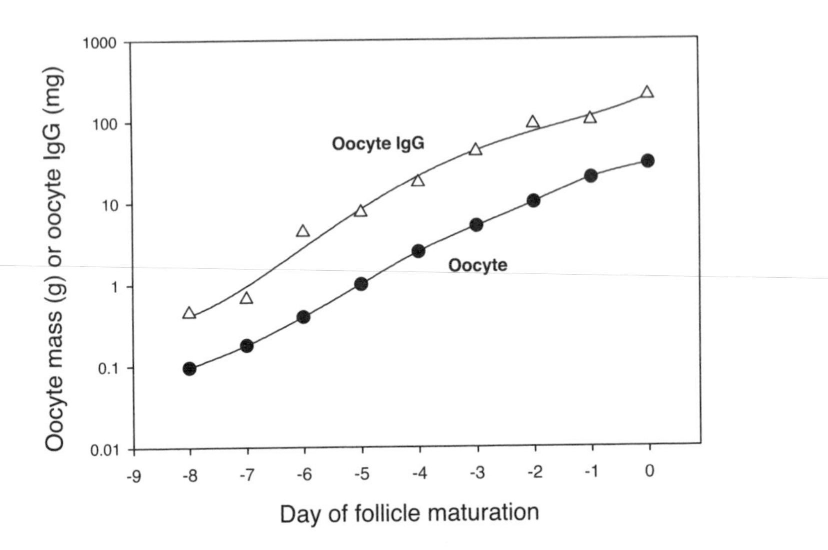

Fig. 4.23. Timing and rate of accumulation of chicken IgG in maturing vitellogenic follicles in relation to timing and rate of oocyte growth. (Redrawn using data in Kowalczyk et al. 1985.)

would be consistent with the idea that higher mRNA levels are required to support greater overall yolk uptake by a greater number of follicles.

This idea that VTG/VLDL-receptor-mediated yolk uptake is a key determinant of individual variation in egg size and other aspects of reproductive phenotype seems a long way removed from ideas about food availability, condition-dependent individual quality, and resource-allocation trade-offs. Why would receptor number or activity be limiting? What prevents "small-egg" females from expressing high levels of VTG/VLDL-R to support high rates of yolk uptake? At present we know very little about how the VTG-VLDL receptor itself is regulated, but T. Williams (2005) speculated that there might be developmental pathways or constraints that link lipoprotein receptor expression, such that up-regulation of VTG receptors causes up-regulation of receptors in other tissues, with potentially negative consequences. These ideas await further study, but the centrality of receptor-mediated yolk uptake is consistent with the observation that some components of egg quality (e.g., antibodies, antioxidants) correlate with, and therefore might be functionally linked to, the amount of yolk transferred to developing follicles. A striking example is provided in hens by the parallel between the uptake of IgG (now IgY) and the increase in oocyte mass during the exponential vitellogenic growth phase (Kowalczyk et al. 1985), a relationship that supports the idea that IgG uptake occurs via receptor-mediated endocytosis (fig. 4.23). Bortolotti et al. (2003) also suggested that the liver incorporates carotenoids into VLDL and is the site where they are delivered to the egg yolk and that VLDLy "is the main delivery system for carotenoids into the developing oocyte" (although no citation is provided for this point). If so, then the VTG/VLDL-R must also play a key role in regulating variation in yolk carotenoids.

4.5.2. Mechanisms Underlying Egg-Size-Independent Variation in Egg Quality

The concentrations of yolk hormones and antimicrobial albumen proteins vary markedly in clutches laid by different females, and to a lesser extent within clutches, and the sizes of the yolk or eggs appear to explain relatively little of this variation (see section 4.4). Variation in, and the potential regulation of, antimicrobial proteins must involve secretory mechanisms in the tubular gland cells in the oviduct. As described in chapter 2, the secretion of albumen proteins is thought to involve mechanical stress from the yolk descending the oviduct, which regulates gene expression in the uterus (shell gland), perhaps with hormonal "priming" of mechanical sensitivity involving estrogens or progesterone (Lavelin et al. 2002; Lavelin et al. 2001). However, it is currently not known if, or how, this basic mechanism can mediate the differential regulation of albumen components.

Until recently there has been only limited work on the physiological mechanisms underlying variation in yolk hormones, and as Groothuis and Schwabl (2008) concluded, the "lack of knowledge about the underlying mechanisms hampers further progress" in this field. Several reviews have provided excellent summaries of the unresolved mechanistic issues (Groothuis and Schwabl 2008; Moore and Johnston 2008), so here I will highlight some points relevant to maternal physiology, egg formation, and ovary function. In short, important gaps in our knowledge include (a) *how* hormones are transferred from mother to eggs; (b) to what extent females can *regulate* or "control" this process, and to what extent this might be constrained by hormonal requirements for the female's own physiological processes; (c) what mechanisms mediate exposure to, and effects of, yolk hormones in the embryo; and (d) how the effects of maternally derived hormones are integrated with, or differentiated from, the critical developmental effects of the embryo's endogenous hormones (Carere and Balthazart 2007).

The most parsimonious mechanism to account for the occurrence of maternal hormones in yolk is that highly lipophilic hormones partition passively down a concentration gradient into the lipid-rich yolk (Moore and Johnston 2008). In other words, physical chemistry alone might be sufficient to explain the transfer of hormones between mother and yolk. Groothuis and Schwabl (2008) termed this concept the *physiological epiphenomenon hypothesis*. For glucocorticoids and thyroid hormones, which are secreted by extra-gonadal endocrine glands, transfer must occur from the general circulation to the yolk across the basal lamina, perivitelline layer, and oocyte plasma membrane (see fig. 2.2). However, for gonadal steroids it has been suggested that, rather than the general circulation, the main source of yolk hormones might be steroids produced locally in the follicle cells of the ovary and that those hormones move directly from the thecal or granulosa cells into the yolk (Hackl et al. 2003). This is an important point since plasma hormone levels from gonadal sources will reflect the integrated secretion of all developing follicles, and this integration might limit the potential for follicle-specific yolk hormone deposition (sensu Badyaev et al. 2008; see section 4.7). If yolk hormones are derived from the general circulation, then the physiological epiphenomenon hypothesis predicts that the amount of hormone transferred to the yolk will be positively correlated with hormone levels in the maternal circulation This prediction is consistent with studies that have shown that treatment of laying females with exogenous hormone elevates levels in the plasma and results in elevated yolk hormone levels (Adkins-Regan et al. 1995; Love et al. 2005; T. Williams et al. 2005), and it also explains why environmental factors that are assumed to elevate *plasma* steroid levels in females also elevate yolk hormone levels (see section 4.4.2). However, studies that have investigated the relationship between endogenous gonadal steroids in the female's circulation and those in the yolk have reported inconsistent results, with positive, negative, and no correlations (10 studies in seven species; Groothuis and Schwabl 2008). If yolk hormones are derived mainly from plasma, then the "availability" of these hormones should reflect, and perhaps be dictated by, hormonal regulation of the female's own physiological processes, such as egg formation, reproduction-related behavior (ovarian steroids), and energy homeostasis (corticosterone, thyroid hormones, Groothuis and Schwabl 2008). If, for example, the development and functioning of the female reproductive tract during egg formation dictates a fixed or constrained hormonal milieu, which is reflected in the female's own plasma hormone levels, then this constraint might limit a female's ability to adjust the hormone levels in yolks independently of the variations in plasma levels (T. Williams et al. 2005). In other words, the mother's functional hormonal requirements might not match those of her offspring, and physiological trade-offs will determine if hormone transfer can be "costly" to mothers (Groothuis and Schwabl 2008). This maternal perspective has rarely been incorporated in work to date, which has instead focused mainly on the benefits of yolk hormones to offspring. Might this explain why estrogens, which are critical for female reproductive function, occur in yolks at lower concentrations than androgens? Can females "utilize" androgens with far more flexibility without compromising their own physiological function?

Can females regulate or control the transfer of hormones from the circulation or follicle cells to the yolk? The literature on hormonal maternal effects commonly states that females "allocate." "manipulate," or "invest" yolk hormones, all of which imply a level of female control for "adaptive" manipulation of offspring phenotype, but current data are inconclusive on this point (Groothuis and Schwabl 2008). Numerous *possible* mechanisms exist for the differential regulation of yolk and maternal hormone levels (Groothuis and Schwabl 2008; Moore and Johnston 2008). These include (a) follicle-specific regulation of follicle development, assuming that steroid levels in the yolk are influenced mostly by hormone production in the follicle cells adjacent to each yolk; (b) selective uptake of hormones into the oocyte or barriers to passive diffusion (Lovern and Wade 2003; Painter et al. 2002), which would uncouple yolk and plasma hormone levels; (c) retention and accumulation of passively derived steroids to concentrations greater than in plasma, which would occur passively based on partition coefficients since at equilibrium much more steroid would be dissolved in the yolk than in the aqueous plasma or cytosol (Moore and Johnston 2008), although Groothuis and Schwabl (2008) also suggested that steroid-binding globulins could be involved in this process; and (d) differential metabolism of the steroids in yolk versus those in plasma. The selective transport of hormones, in either direction across the plasma-yolk boundary, could involve P-glycoproteins, membrane-associated transporters belonging to the ATP-binding cassette superfamily of proteins, which are known to regulate tissue distribution of endogenous compounds, including steroid hormones (Dean and Annilo 2005). Although P-glycoproteins are expressed in birds (Barnes 2001; A. Green et al. 2005), nothing is currently known about their expression or function in developing follicles. Nevertheless, it is important to realize that a simple, passive mechanism of hormone transfer to yolks can still generate adaptive variation in offspring phenotype and the matching of offspring phenotype to prevailing environmental conditions. For example, Love et al. (2005) and Love and Williams (2008a) showed that environmental factors such as parental body condition, food, or social factors generate variations in maternal plasma levels of corticosterone that are translated into variations in yolk corticosterone and that in turn affect offspring phenotype. At no point is there any requirement for females to "control," "regulate," or otherwise determine the amount of hormone they transfer to eggs. Given that there is marked individual variation in plasma hormone levels (T. Williams

2008), a passive mechanism would still generate marked intra- and inter-clutch variation in yolk hormone levels that could be functionally important for the development of offspring phenotype (but see section 4.4.2).

Finally, what mechanisms mediate exposure to, and effects of, yolk hormones in the embryo? Although many studies suggest that maternally derived yolk hormones *do* affect offspring growth and development, yolk steroid levels actually decrease rapidly during early incubation (Elf and Fivizzani 2002; Pilz et al. 2004). A number of possible explanations have been proposed for the fate of these maternally derived steroids (L. Gilbert et al. 2007; Moore and Johnston 2008): (a) embryo formation might cause early mixing of yolk and albumen, diluting the yolk steroids, although these would then still be available to the embryo; (b) yolk hormones could diffuse into other compartments of the egg (e.g., albumen); (c) yolk hormones could be metabolized to inactive forms, which might protect the embryo from maternal influence; or (d) they could be metabolized into other biologically active compounds that could still affect the developing embryo. Ultimately, maternally derived hormones must get from the yolk to the developing embryo if they are the cause of the many reported effects on offspring phenotype, but this movement in itself is not a trivial problem since steroids must pass from the lipid-rich yolk to the aqueous embryo circulatory system against the lipid solubility gradient. Once again, mechanisms regulating the utilization of maternal hormones are unknown, especially in birds. Paitz and Bowden (2008) proposed that in turtles steroids are converted to water-soluble metabolites via the sulfotransferase pathway and can then be transported to the embryo, where they could serve as precursors for subsequent steroid production via sulfatase enzyme activity. More recently, similar mechanisms for embryonic modulation of maternally derived hormones and of the embryo's local steroid environment have been demonstrated in birds (Paitz et al. 2010). It is probably naive, therefore, to assume that the embryo is simply a passive recipient of maternal hormones. Rather, embryonic modulation of maternal signals (via embryonic enzyme activity) might significantly affect offspring responses (Paitz and Bowden 2009). Hormonally mediated maternal effects could therefore more correctly be viewed as an example of parent-offspring conflict *if* both mothers and offspring possess mechanisms to manipulate hormone signals and *if* their evolutionary interests differ (Müller et al. 2007).

4.6. Variation in the Primary Sex Ratio and Sex-Specific Follicle Development

If the relative "adaptive value" of sons and daughters differs, then mothers might maximize their own fitness by differentially investing in one sex over the other. Differential reproductive investment in relation to the sex of offspring, which could ultimately lead to biased sex ratios, might be achieved in a number of ways including (a) adjustment of the primary sex ratio at fertilization, (b) adjustment of the secondary sex ratio at hatching or fledging, through differential investment in offspring growth and/or sex-specific embryo or nestling mortality, and (c) sex-specific post-fledging survival (Clutton-Brock 1991). Some of these issues are dealt with elsewhere in this book, so here I will focus on maternal adjustment of primary sex ratio and sex-specific follicle development during egg production. In birds, females are the heterogametic sex (ZW) and therefore they have the potential to adjust the primary sex ratio via non-random segregation of the sex chromosomes: depending on whether the ovum retains the W or Z chromosome, the resulting offspring will be female or male, respectively (Pike and Petrie 2003). An extension of this idea is that females might differentially provision "male" or "female" follicles or eggs with yolk, hormones, carotenoids, etc., to manipulate offspring phenotype in a sex-specific manner. This second idea is more problematic since sex is determined 1–2 hours before ovulation at completion of the first meiotic division, when either the W or the Z chromosome is retained in the ovum and the other sex chromosome is drawn into the inactive polar body (Rutkowska and Badyaev 2008). By this time, all of the yolk, and yolk-associated minor egg components, have been deposited into the fully developed follicle. This idea of differential provisioning of specific follicles, in turn, has led to the suggestion that one of these yolk components, the maternally derived yolk steroids, may influence the segregation of the sex chromosomes at meiosis (Petrie et al. 2001) to determine offspring sex (see below).

Sex-ratio variation, or sex-specific reproductive investment, can be favored whenever the relative fitness of sons and daughters differs due to factors other than sex ratio per se (Daan, Dijkstra, and Weissing 1996; Trivers and Willard 1973). For example, if the reproductive success of sons increases more with higher parental investment than does the reproductive success of daughters (e.g., due to condition-dependent male-male competition), then the theory predicts that mothers in good condition should bias the sex ratio of their offspring toward males and mothers in poor condition should bias sex ratio toward females. The development of molecular sexing techniques in the late 1990s (Griffiths et al. 1996; Griffiths et al. 1998) allowed for collection of empirical data to test these ideas, and the field of adaptive sex-ratio manipulation has expanded rapidly in the last 10–15 years. Many current papers start with statements to the effect that "adaptive" maternal sex-ratio adjustment has been convincingly demonstrated in birds, but in fact there is still considerable debate and controversy over how widespread the facultative maternal adjustment of the primary sex ratio is (Ewen et al. 2004; Komdeur and Pen 2002; Maddox and Weatherhead 2009; Postma et al. 2011). Indeed, there are broad parallels between this field of research and that of hormonally mediated maternal effects: (a) many studies are correlational, with few experimental studies; (b) results tend to be inconsistent among different studies, with a general lack of replicability (Maddox and Weatherhead 2009 and references therein), perhaps due to technical (K. Arnold et al. 2003) and analytical problems (e.g., sample sizes, potential publication bias); (c) most studies simply report differences in sex ratio without directly demonstrating short- or long-term fitness consequences either to the offspring or the mother of these sex-differences in reproductive investment; and (d) the physiological mechanisms underlying "facultative" manipulation of the sex ratio remain completely unknown. Although many empirical studies test clear, intuitive, theoretically based a priori predictions, it has been suggested that the life histories of birds violate a number of assumptions of the standard theoretical models of sex allocation (Komdeur and Pen 2002). As just one example, theory predicts that females will bias the sex of offspring sired by extra-pair mates toward males if extra-pair mating increases the variance in fitness of sons relative to daughters, but only 2 of 23 studies (in 15 different bird species) have found support for this hypothesis (L. Johnson et al. 2009 and references therein).

A handful of studies are widely cited for reporting marked non-random sex allocation at conception or fertilization (Heinsohn et al. 1997; Komdeur et al. 1997; West and Sheldon 2002). Numerous other studies have reported variation in sex ratio in relation to season (date), parental quality (e.g., mate attractiveness), social environment (e.g., cooperative breeding), and sexual size dimorphism, although the direction of these relationships often varies among different studies (reviewed in Komdeur and Pen 2002). Among the most extreme sex-ratio biases are those seen in the Seychelles warbler (*Acrocephalus sechellensis*), and Komdeur et al. (2002) concluded that these *were* the result of pre-ovulation control of sex ratio (although this conclusion was not based on any *physiological* data). Females on high-quality territories produce eggs that are 88% female, whereas females on low-quality territories produce eggs that are 77% male (Komdeur 2003). In contrast, in the majority of other studies, the deviation from a 1:1 sex ratio is relatively small. One meta-analysis of 40 studies of primary sex-ratio adjustment concluded that overall these studies did not exhibit *any* significant variability beyond that expected due to sampling error and that the studies to date "did not provide any evidence that facultative adjustment [of sex ratio] is a characteristic biological phenomenon" (Ewen et al. 2004; also see Postma et al. 2011). Reconciling these differing interpretations of the existing data is clearly an important challenge for the future. However, beyond the question of whether the primary sex ratio does vary non-randomly, very few studies have addressed the question of how, and to what extent, facultative sex allocation contributes to variation in offspring or maternal fitness (sensu fig. 1.1; but see Appleby et al. 1997; Badyaev et al. 2002). Lessells et al. (1998) experimentally manipulated brood sex ratio in great tits within days of hatching, creating all male, all female, or mixed-sex broods, and found no differences in provisioning behavior, nest defense, changes in parental mass, occurrence of second broods, or overwinter survival of parents. In the Seychelles warbler the short-term benefit of biased sex allocation is thought to be the recruitment of female helpers that increase reproductive success in high-quality territories, whereas sons disperse from low-quality territories. However, even in this well-studied system, the long-term fitness consequences of sex allocation have not been quantified, and the adaptive nature of this variation has yet to be confirmed (Komdeur 2003; Komdeur and Richardson 2007). This is especially important in light of the fact that some experimental studies have shown that the "preferred,", or more common sex, might actually have lower fitness (Rutkowska and Cichon 2006). In any case, how important is the adjustment of sex allocation likely to be for maternal lifetime fitness compared, for example, to decisions on when to breed or how many eggs to produce? The answer to this question is likely to be very different for the 1-egg-clutch Seychelles warbler compared with a species that might lay between 5 and 15 eggs in a single breeding attempt. Once again the ecological-, life-history-, or condition-dependent context of sex allocation needs to be taken into account (and might go a long way to explaining inconsistent results to date, Rutkowska and Badyaev 2008).

What physiological or molecular mechanisms might females use to manipulate offspring sex? Some early papers suggested that birds might have a limited ability to manipulate offspring sex, given the chromosomal nature of sex determination (Krackow 1995). However, several potential mechanisms have subsequently been proposed, including sex-specific follicle recruitment or the development of follicles "pre-destined" to give rise to either males or female offspring, pre-ovulatory sex-specific follicle atresia, selective abortion of ovulated ova, and non-random chromosome segregation (meiotic drive; Pike and Petrie 2003; Rutkowska and Badyaev 2008). Given our lack of knowledge of general mechanisms underlying follicle recruitment (see chapter 2), it is not surprising that we have not resolved the mechanisms underlying the *sex-specific* recruitment of follicles "pre-destined" to become male or female. In general, the mechanisms involving sex-specific follicle atresia and selective abortion are likely to be costly in terms of delays to laying, potentially smaller clutch sizes, and larger laying gaps. Furthermore, the empirical data on atresia of pre-ovulatory vitellogenic follicles suggest that it occurs at too low a frequency, and with too limited a variation in timing, to explain the myriad of patterns of sex-ratio bias that have been reported (Challenger et al. 2001; Goerlich, Dijkstra, and Groothuis 2010). This leaves non-random segregation of chromosomes as perhaps the most likely mechanism for sex-ratio manipulation, assuming one exists. Rutkowska and Badyaev (2008) provide an extensive review of molecular and cytological mechanisms of avian meiosis and suggest a multitude of potential targets for selection on non-random segregation of sex chromosomes (meiotic drive) that they suggest "may reflect the diversity of mechanisms and levels on which such selection operates in birds to facilitate sex-ratio bias." Their paper also highlights several features of avian meiosis that might be influenced by maternal hormones in response to environmental stimuli and "may account for the precise and adaptive patterns of offspring sex ratio adjustment observed in some species" (see also Uller and Badyaev 2009). Direct molecular evidence for any of these mechanisms is lacking, and experimental evidence demonstrating the effects of maternal hormones on offspring sex ratio is mixed. In several studies, elevation of maternal testosterone induced a male-biased sex ratio that persisted in subsequent breeding attempts even though no additional hormone treatment was given (Goerlich et al. 2009; Rutkowska and Cichon 2006; Veiga et al. 2004). However, samples sizes in these studies have generally been small, and in a follow-up study Goerlich, Dijkstra, et al. (2010) were unable to replicate this effect (and see Pike and Petrie 2006). Correa et al. (2005) reported that progesterone treatment of laying female chickens caused a female-biased sex ratio (75%), although only a single egg, the first egg laid following injection, was collected from each hen. To my knowledge this study has not been replicated. Finally, no bias in primary sex ratio was observed in female zebra finches injected with E2 (von Engelhardt et al. 2004). So, in summary, there is no empirical physiological evidence to date to support the idea that sex-biased meiotic drive underlies sex-ratio manipulation in birds, and there is no consistent evidence that maternal hormones regulate this process—although Rutkowska and Badyaev's paper (2008) provides numerous hypotheses that could be tested experimentally via direct hormonal manipulations of growing follicles or follicles cultured in vitro.

4.7. Extreme Flexibility in Reproductive Investment: The House Finch

Few studies have attempted to integrate the multiple components of egg quality, offspring phenotype, and follicle development, even though these most likely form a complex, interacting whole (but see, e.g., Alonso-Alverez et al. 2008; De Neve et al. 2008; Groothuis et al. 2006; Romano et al. 2008). However, a series of studies by Alex Badyaev and colleagues (reviewed in Badyaev 2009, 2011; Uller and Badyaev 2009) have proposed an unprecedented level of complexity and flexibility in female reproductive investment in the house finch. Their hypothesis involves integration of the hormonal mechanisms underlying oogenesis, interactions among oocytes, facultative sex-determination, local adaptation of differences in sex ratio, and sex-specific offspring phenotype across recently established populations in response to novel environments. This model proposes that oocytes destined to become male or female ova differ in their timing of recruitment to the ovulatory sequence and/or their pattern and duration of oocyte development (Badyaev, Acevado Seaman, et al. 2006; see fig. 2.6). Differences in the timing of oocyte development in relation to temporal variation in ovarian steroidogenesis (Bahr et al. [1983]) result in the exposure of individual follicles to different hormonal milieus during development. This difference in "local" hormonal environment, along with growth-inhibiting interactions among growing oocytes, leads to development of temporal clusters of oocytes that will develop into the same sex, minimizing developmental overlap between the sexes (Badyaev et al. 2005) and ensuring the sex-specific accumulation of different yolk components in different follicles (Badyaev, Oh, and Mui 2006; Badyaev et al. 2008; Young and Badyaev 2004). This follicle-specific provisioning, in turn, drives adaptive patterns of offspring phenotype in relation to laying order (Badyaev, Hamstra et al. 2006; Badyaev et al. 2002). The formation of these same-sex oocyte clusters can apparently be induced by distinct environmental cues or different environments (Badyaev and Oh 2008). Finally, this model proposes that the same hormone, prolactin, is involved in all stages of this process (Badyaev et al. 2005) and therefore provides a *proximate link* between sex-determination and sex-specific accumulation of maternally derived products in the follicles or eggs (Uller and Badyaev 2009). It is to be hoped that this exciting body of work can be confirmed through experimental studies and can be extended and generalized to other species, in the near future.

4.8. Conclusions

Egg size is a highly variable life-history trait, with up to twofold differences in egg mass among individual females within a population. Larger eggs contain absolutely more major egg components (shell, albumen, yolk) and absolutely more of several minor egg components (maternally derived antibodies and antioxidants), and, in this regard, egg size *is* a good proxy for egg quality. The only exceptions to this generalization appear to be for yolk hormones and antimicrobial albumen proteins, which might allow for some egg-size-independent variation in offspring phenotype. Many studies assume that high-quality females produce large, high-quality eggs, but it is equally plausible that individual females produce eggs of the optimum size and quality for their phenotype or genotype. Either way, the aspects of a female's phenotype that would determine maximum or optimum egg size are unknown, although these are *not* primarily factors such as age, experience, body condition, or mate quality that continue to be the focus of much current work. Christians (2002) discussed the possibility that several of these factors together might explain a substantial amount of the variation in egg size between females, and certainly such a multivariate approach would be valuable in future studies. However, these factors are not completely independent, and it seems unlikely that the cumulative effect of these multiple factors would equal the sum of their individual effects.

As a consequence, identifying the specific *physiological mechanisms* that generate repeatable, individual variation in yolk or egg size will probably also answer the question of what mechanism(s) underlie variation in egg quality and what components of an individual's phenotype determines egg size. At present the most likely candidate mechanism appears to be receptor-mediated endocytosis and yolk-precursor uptake. However, the identification of a putative cell-level, receptor-mediated process as *the* component of phenotype that determines individual quality, at least in terms of egg size, raises a number of interesting, conceptual questions that contrast with how we have previously thought about factors that might underpin individual variation. How does a central effect of variation in receptor expression or activity fit with traditional, long-standing ideas such as resource or condition dependence, nutritional constraint, or resource-allocation trade-offs mediating reproductive investment? Receptor-mediated mechanisms could still form the basis of physiological trade-offs, but a more productive approach for future research would be to include a broader focus on putative non-resource-based costs of egg production, including those mediated by the pleiotropic effects of reproductive hormones (see chapter 7).

Considering the fitness consequences of individual variation in egg size, the most parsimonious conclusion, based on a very large number of studies and analytical approaches (experimental manipulation, meta-analysis), is that egg size has at most a weak positive effect on offspring phenotype, and then, mainly early in development. Egg size determines hatching mass and can have significant effects on chick growth or survival early in chick-rearing. Although slower growth immediately post-hatch, due to hatching from a smaller egg, may be compensated for later in the rearing period, early chick mortality cannot be compensated for.

There is currently only limited evidence for the longer-term effects of egg size, either at fledging (in most species; but see Love and Williams 2011) or post-fledging, in terms of recruitment and future fecundity. There is also little evidence for a strong positive selection on egg size, independent of individual female quality. The extensive research effort that has gone into investigating other components of egg "quality" (yolk hormones, carotenoids, antibodies, etc.) in the last 10–20 years has not helped answer this question of the effects of egg quality on offspring phenotype: there is virtually no evidence that variation in these factors has significant, long-term effects on offspring or maternal fitness independent of any effects of egg size per se (but see Love, Salvante, et al. 2008). Certainly the potential fitness consequences of variation in egg composition would seem to be far less obvious than the fitness consequences of the female's decision on when to lay or how many eggs to lay.

If egg size is only weakly related to fitness, this might resolve the paradox of high heritability of egg size, since traits closely related to fitness should generally have low heritability (McCleery et al. 2004; Mousseau and Fox 1998). However, a weak relationship between egg size and offspring or maternal fitness presents another, unresolved, paradox: if there are few benefits of large egg size, why do some females continue to produce very large eggs, especially when this "decision" results in a clear "fecundity cost" (see fig. 4.11)? If the benefits of large egg size do not outweigh the costs (reduced clutch size), then selection should favor all females laying a minimum viable egg size and maximizing fecundity. However, would a large-egg female laying a five-egg clutch have higher fitness if she were actually to lay a six-egg clutch of medium-sized eggs? From this perspective, identifying the physiological and/or evolutionary processes that generate and maintain individual, phenotypic, variation in egg size remains an important, but still enigmatic, issue for the future.

4.9. Future Research Questions

- Why have female birds not retained the ability to facultatively adjust egg size, adaptively tuning allocation to eggs in response to varying ecological conditions (cf. other oviparous taxa); that is, why is egg size "inflexible" in individual females?
- In what ecological, environmental, or life-history contexts does egg size have the greatest effect on offspring phenotype? Can experimental studies provide support for the idea that variation in egg size has *long-term* effects on offspring survival, recruitment, and future fecundity?

- Why do large-egg females pay a substantial fecundity cost in terms of decreased clutch size if there are no clear benefits of large egg size?
- Are natural or experimentally induced *deviations* in egg size from an individual's overall (lifetime) mean egg size more important in determining offspring phenotype or fitness than absolute variation in egg size (following E. Wagner and Williams 2007)?
- Is there an egg-size–clutch-size trade-off among individuals within species in birds? How are egg and clutch size coupled mechanistically?
- If relatively small intra-clutch variation in yolk hormones has significant effects on offspring phenotype, why does the much larger inter-individual variation not generate even larger effects among chicks of different females; for example, why do females producing eggs with such high yolk T levels not give rise to "super-chicks" and females producing eggs with very low T not give rise to "super-duds"?
- Why do some individual females have 10-fold higher plasma E2 and yolk-precursor levels than other females when this has no apparent benefit in terms of egg size and number? What other functions might E2 and VTG have, or are there costs of maintaining these elevated plasma levels? To what extent do the functional demands of the female's own hormonal *milieu* constrain flexibility in the partitioning of hormones between the mother and yolks?
- Is receptor-mediated yolk uptake (which can be measured simply in intact birds using radiolabelled amino acids) a main determinant of variation in the amounts of maternally derived antibodies or antioxidants in yolks?

CHAPTER 5

Clutch Size

CLUTCH SIZE is generally considered to be one of the most important determinants of reproductive success and lifetime fitness in birds (Lack 1947; McCleery et al. 2004; Rockwell et al. 1987; Sockman et al. 2006). Clutch size is also perhaps the most well-studied life-history trait in birds, though many reviews have focused on interspecific variation (Jetz et al. 2009; E. Murphy and Haukioja 1986; Winkler and Walters 1983). Clutch size sets an upper limit on brood size, and in single-brooded species, clutch size will therefore determine the maximum annual fledgling productivity. In short-lived, single-brooded species, clutch size will also be the main determinant of lifetime fledgling productivity, the main predictor of variance in lifetime fitness (fig. 1.1). In multiple-brooded species, annual fecundity is also related to clutch size, although the *number* of broods can be a stronger predictor of fecundity among (T. Martin 1995) and within species (Weggler 2006). Sockman et al. (2006) suggested that estimating reproductive effort based only on the *number* of eggs a bird lays is "oversimplified and imprecise" and highlighted the potential importance of fine-tuning of reproductive investment through continuous variation in egg size or egg quality. Sockman et al. (2006) argued that a female laying a modal 4-egg clutch "could only" adjust reproductive effort in increments of 25% (one egg) by adjusting egg number alone. Although the implication here was that clutch size is a limited way for females to adjust reproductive investment, laying one less or one more egg immediately adjusts female fecundity, and potential fitness, by 25%. As I argued in chapter 4, there is currently little evidence that continuous variation in egg size or egg quality has anywhere near this magnitude of effect on fecundity or fitness. Therefore, from an evolutionary perspective, identifying the physiological mechanisms underlying individual, phenotypic variation in clutch size would seem to be one of the worthier and more significant challenges in avian reproduction.

5.1. Individual Variation in Clutch Size and Clutch Number

Some avian taxa have an invariant clutch size: among seabirds, all albatrosses and petrels lay a single-egg clutch and most penguins (Spheniscidae)

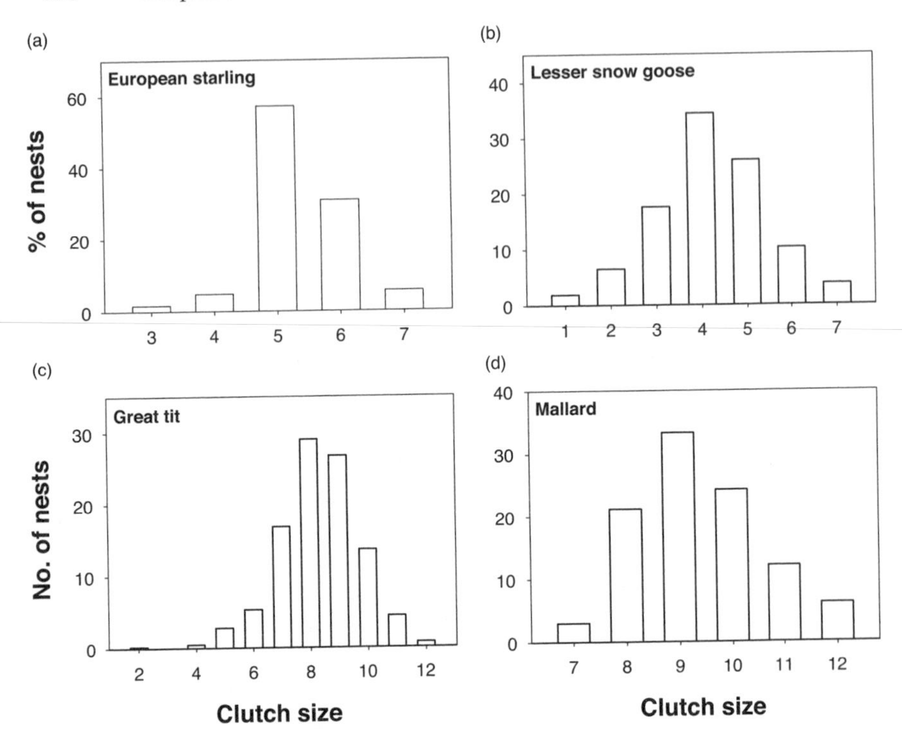

Fig. 5.1. Individual variation in clutch size in species laying relatively small clutches, (a) European starling and (b) lesser snow goose, and those laying relatively large clutches, (c) great tit and (d) mallard. (Based on unpublished data from T. Williams and from data in Cooke et al. 1995; Perrins 1965; and Ball et al. 2002).

lay 2-egg clutches; among the Charadriiformes, most shorebirds (210+ species) lay 4-egg clutches (Lengyel et al. 2009), and most Laridae (gulls) lay 3-egg clutches. Winkler and Walters (1983) actually stated that females lay a characteristic number of eggs "in many other species," but this view masks the marked individual variation in clutch size that exists in many species. For example, in the song sparrow (*Melospiza melodia*), European starling, and lesser snow goose, the modal clutch size is 4, 5, and 4 eggs, respectively, but the clutch size varies from 2 to 5 eggs (Zanette et al. 2006), 3 to 7 eggs (T. Williams, unpublished data), and 1 to 7 eggs (Rockwell et al. 1987) in these three species. Similar, or even larger, individual variation is seen in species laying relatively large modal clutches; for example, in the blue tit, great tit, and mallard duck, the modal clutch size is 10, 8, and 10 eggs, respectively, but the range of individual clutch sizes is 4–17 eggs, 2–12 eggs, and 7–12 eggs (I. Ball et al.

2002; Perrins 1979). Therefore, for both altricial and precocial species, and those laying relatively small and relatively large modal clutch sizes, it is the norm for there to be two- to fourfold variation in potential fecundity among individual females (fig. 5.1), which is as great, or greater, than the differences in mean clutch size between many species.

What factors determine this individual variation in clutch size? In single-brooded species, one of the strongest, most ubiquitous, patterns for any avian reproductive trait is the seasonal decline in clutch size with later laying date (fig. 5.2a). Some multiple-brooded species, such as Anatidae, Galliformes, and Ralliformes and some passerines, also have a linear, seasonal decline in clutch size, with the second clutches being smaller than the first clutches (Feare 1984; Gibb 1950; Lack 1947). However, in other multiple-brooded species, clutch size first increases with date then declines (fig. 5.2b). In an analysis of 66 UK-breeding passerines and non-passerines, Crick et al. (1993) showed that as the number of breeding attempts per year increases, (a) birds lay relatively smaller clutches at the beginning of the breeding season relative to maximum clutch size, (b) clutch size increases more rapidly with date at the beginning of the season, and (c) maximum clutch size occurs later in the breeding season. Although seasonal declines in clutch size occur over relatively protracted laying periods (40–60 days) in some species (fig. 5.2), and this has received the most attention, significant decreases in clutch size can occur over just a few days (see chapter 3). In the greater snow goose, clutch size decreases at a rate of about 0.20 eggs per day (Lepage et al. 2000), and in the marsh tit, clutch size decreases by 0.1–0.2 eggs per day in different years (H. Smith 1993). Even in species with highly synchronous egg-laying, such as the European starling, the mean clutch size can decrease by 1.1–1.5 eggs (20%–30% of modal clutch size) over a 2–3 day period (fig. 5.3). These differences in the rate and pattern of the seasonal variation in clutch size will become important when we consider the fitness consequences of this variation and the potential physiological mechanisms that might determine individual variation in clutch size.

Although egg-laying date is clearly an important determinant of clutch size, in both single- and multiple-brooded species, there can still be marked individual variation in clutch size independent of laying date. For example, in the European starling, clutches initiated on the modal laying date varied from 4 to 7 eggs in the first year (2008) and from 5 to 7 eggs in the second year (2009; T. Williams, unpublished data). Similarly, in a 10-year study of greater scaup in Alaska, although on average the clutch size declined significantly with later laying date in each year, there was a twofold variation in clutch size (5–12 eggs) for a given date in any given year (Flint et al. 2006; fig. 5.4). This date-independent individual variation in clutch size will be another important component

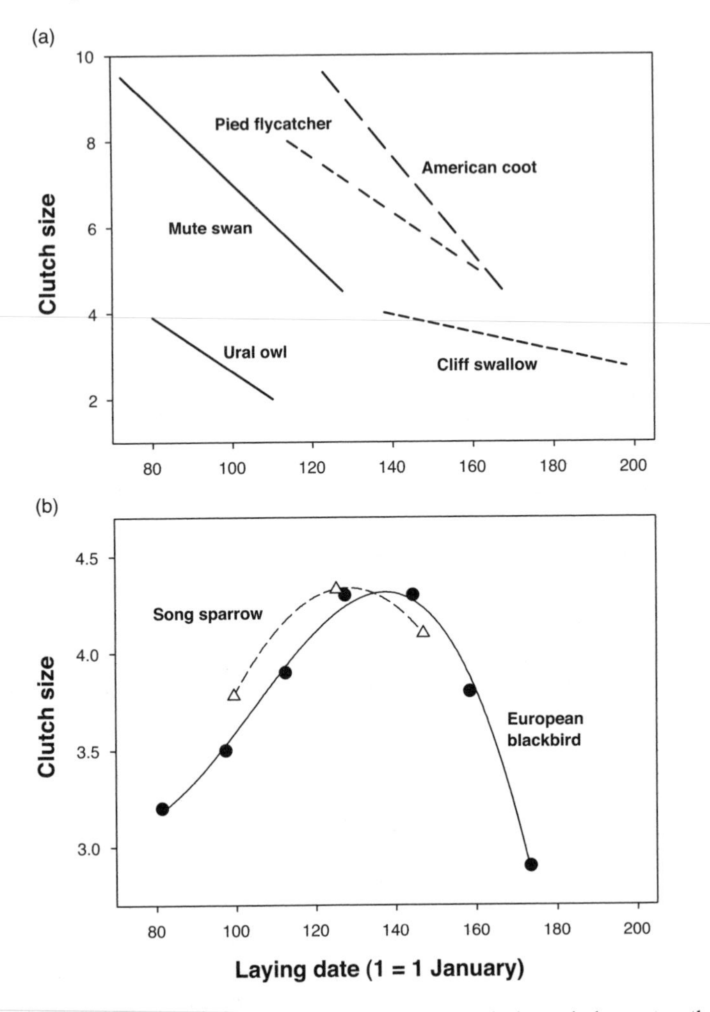

Fig. 5.2. Seasonal variation in clutch size in (a) single-brooded species (based on data in Meijer et al. 1990 and C. Brown and Brown 1999) and (b) multiple-brooded species (based on data in Snow 1958 and Travers et al. 2010).

to consider in relation to underlying physiological mechanisms. Clutch size has also been shown to vary independently of date in relation to nesting-population density (Both et al. 2000), nesting habitat (I. Ball et al. 2002), intensity of nest predation (Travers et al. 2010), and breeding latitude (Jetz et al. 2009). Evans et al. (2005) used the introduction of 11 passerine species from the United Kingdom to New Zealand as a natural experiment to explore geographical variation in clutch size within species. Nine of 11 species had significantly smaller clutches in New Zealand

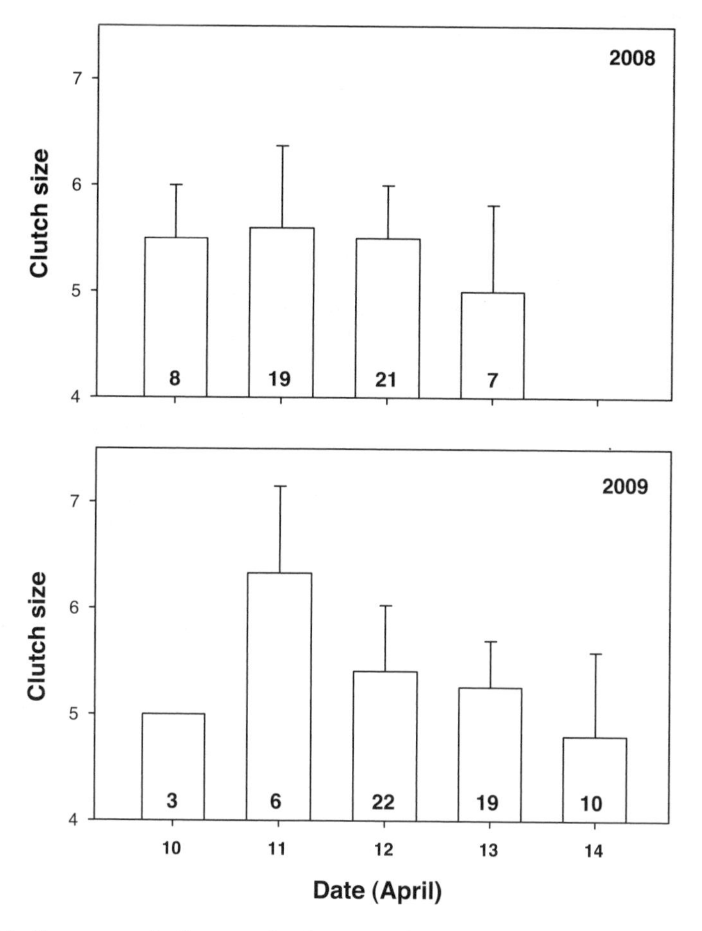

Fig. 5.3. Short-term declines in clutch size with egg-laying date (first egg) in two years in the European starling a species with highly synchronous egg-laying. Numbers in columns indicate sample size.

than in the United Kingdom, but this difference was not explained by the less extreme seasonality or changing day length that each species experienced at the lower latitude. Any putative physiological mechanism(s) accounting for individual variation in clutch size therefore need to explain both date-dependent (seasonal) and date-independent variation in this life-history trait.

In multiple-brooded species there is marked individual variation in the number of breeding attempts made per year, and many species are facultatively multi-brooded. For example, in the black redstart (*Phoenicurus ochruros*), although females can fledge up to three broods per season, the proportion of females initiating only a *single* clutch per season varied from

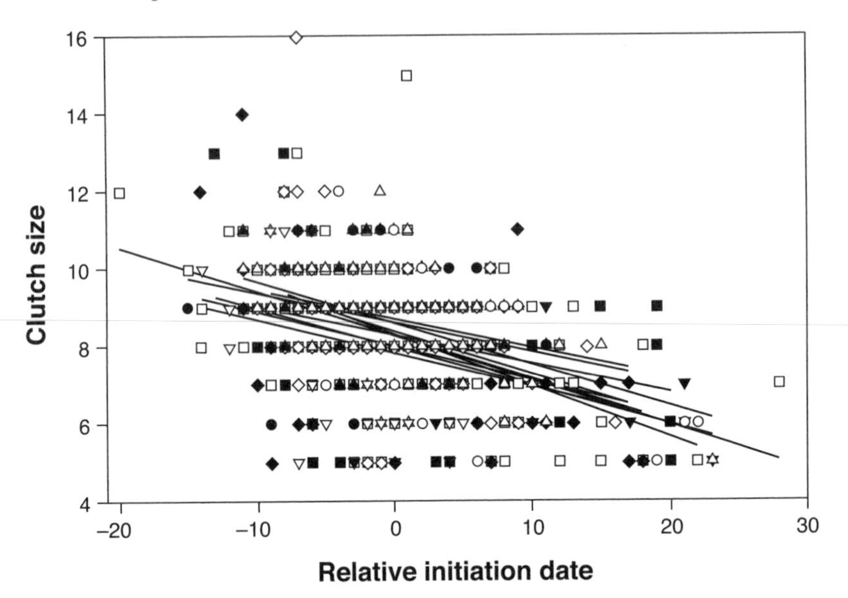

Fig. 5.4. Seasonal and date-independent variation in clutch size in greater scaup. Different symbols indicate different years. (Reproduced from Flint et al. 2006, with permission of the Wildlife Society.)

16% to 58% over 10 years (Weggler 2006). Similarly, the percentage of females that laid a second clutch after successfully fledging a first brood ranged from 0% to 87% (mean 53%, $n = 7$ years) in black-throated blue warblers (*Dendroica caerulescens*), and 0% to 37.8% (mean 9%, $n = 12$ years) in great tits (Nagy and Holmes 2005b; Orell and Ojanen 1983). Individual variation in the number of breeding attempts per season has been variously related to female age, the laying date of the season's first egg, shorter inter-brood intervals (brood overlap), and the occurrence of breeding failures (Nagy and Holmes 2005b; Verhulst et al. 1997; Weggler 2006; Zanette et al. 2006), but the mechanisms, and components of individual phenotype, driving this variation are also poorly understood.

5.2. Why Does Clutch Size Vary among Individuals?

David Lack's seminal paper (1947; expanded and extended in Lack 1966, 1968) set the path for much subsequent work by proposing that clutch size is "ultimately determined by the average maximum number of young for which the parents can find enough food" (a seemingly simple idea, but one described by Ricklefs [2000] as "monolithic"). This hypothesis predicts that, on average, the modal clutch size in a population should be the most productive clutch size. Interestingly, Lack (1947) also argued that the number of eggs a bird lays is "far below the potential [physiological] limit of egg production" and "does not depend on the general bodily condition of the parent," two points I will return to later in this chapter. Lack's ideas remain very influential, and consequently, most studies have focused on the reproductive limitations operating during chick-rearing. Only relatively recently has the role of egg production itself in determining clutch size been revisited (Monaghan and Nager 1997; T. Williams 2005; see section 5.2.4). In the last 50 years Lack's original idea has been extended and modified to incorporate individual variation in parental quality, costs of reproduction, and individual optimization of lifetime fitness (see below), but there remains considerable debate about the adaptive value of individual variation, or plasticity, of clutch size (Pettifor et al. 2001; Tinbergen and Sanz 2004). Other explanations have been proposed to explain variation in clutch size, including nest predation, embryo viability, and limits on incubation capacity. However, in contrast to life-history-based models, few of these ideas provide general explanations for all avian taxa, and many are highly specific to certain taxa (e.g., the *incubation limitation hypothesis* for four-egg clutches in shorebirds). These different hypotheses have different implications for, and impose different requirements on, the putative physiological mechanisms I will go on to consider in section 5.4.

5.2.1. Chick-Rearing Ability and Individual Optimization of Clutch Size

Empirical tests of Lack's hypothesis (1947) have mainly utilized manipulation of brood size to experimentally increase or decrease the number of chicks being reared and have then measured the fitness consequences of this change in reproductive effort. Numerous studies have shown that the observed modal clutch size is actually often *smaller* than the most productive clutch size, at least when estimated using the number of chicks *fledged* in a single breeding attempt. Dijkstra et al.'s review (1990) of experimental brood-enlargement studies showed that parents were able to raise more young to fledging than their original brood size in 29 of 40 (73%) altricial species investigated, and a meta-analysis of the same data by Vander Werf (1992) concluded that the cumulative evidence "did not support Lack's hypothesis." Several explanations were then proposed to reconcile this contradiction, including problems with the use of incomplete measures of fitness, specifically the number of fledglings; the lack of consideration of other potential costs of reproduction associated with rearing more chicks; and the idea that the most productive brood size might be very different for different *individuals* breeding under different conditions (Perrins and Moss 1975). A valid interpretation of Lack's original paper (1947) was that it defined productivity on the basis of the *number* of young *fledged* in a single breeding attempt. In the majority of studies, although parents often rear *more* offspring to fledging in experimentally enlarged broods, these offspring are in poorer condition and have lower body mass and, therefore, potentially *lower* fitness (64.7% of studies, n = 34; Dijkstra et al. 1990). This observation suggests that although parents *can* raise their provisioning rates in response to larger brood size, this increase is not always sufficient to fully compensate for the extra number of young in the nest (Pettifor et al. 2001 and references therein; see chapter 6). Another explanation for the contradictory results of brood manipulation experiments is that the rearing of enlarged broods could also have negative effects on parental survival or future reproduction. So, the short-term benefit of a higher number of fledglings might be balanced by a long-term "cost of reproduction" in adults (e.g., Daan, Deerenberg, and Dijkstra 1996; Gustafsson and Sutherland 1988; W. Reid 1987; H. Smith et al. 1987; Shutler 2006; see chapters 6 and 7). Nevertheless, some studies still appear to contradict Lack's original hypothesis (1947) in that the most productive clutch size is greater than the most common clutch size, sometimes markedly so, even when fitness is measured in terms of post-fledging survival (fig. 5.5a), recruitment (fig. 5.5b), or a composite of parental and offspring survival (Tinbergen and Sanz 2004).

The individual optimization hypothesis reconciles these ideas by suggesting that each *individual* female should be able to adjust her clutch size to her own set of circumstances, such as her condition or quality, laying date, habitat quality, or "potential rearing ability," in order to maximize her individual fitness (Hogstedt 1980; Perrins and Moss 1975; Pettifor et al. 1988, 2001). Thus, there is not one single optimum clutch size, but many individual optima: the optimal clutch size is n for individuals laying n eggs, regardless of the value of n, and individuals laying $n + 1$ egg (or $n - 1$ eggs) will have *lower* fitness then they themselves would have if they had laid n eggs (Hogstedt 1980). The corollary of this is that those birds laying and rearing the largest natural clutches will recruit more offspring because they are "in some way 'better parents than those with smaller clutches" (Pettifor et al. 2001). Several studies have concluded there is unequivocal support for the individual optimization hypothesis (Daan, Dijkstra, and Tinbergen 1990; Gustafsson and Sutherland 1988; Pettifor et al. 1988; Tinbergen and Daan 1990), and Pettifor et al. (2001) suggested "that individual optimization via phenotypic plasticity is the most parsimonious functional explanation for the observed variation in the range in size of clutches laid by individuals." However, other studies have failed to support the predictions from the individual optimization hypothesis (Lessells 1986; Tinbergen and Both 1999), and Tinbergen and

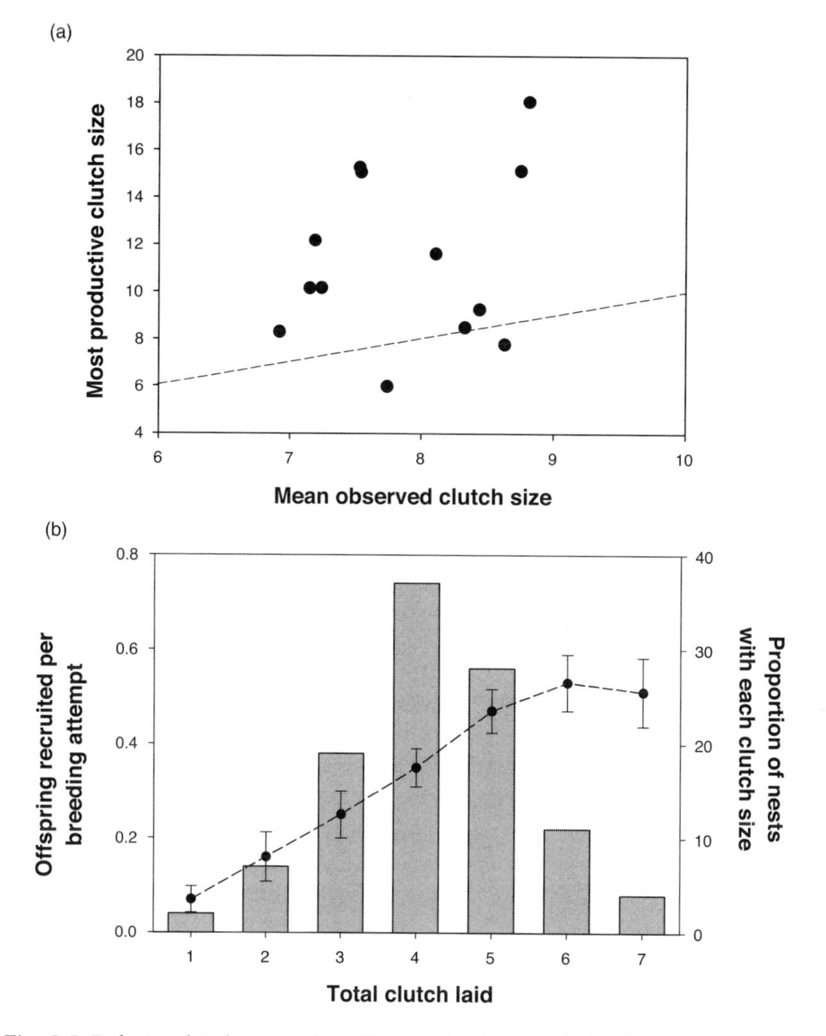

Fig. 5.5. Relationship between breeding productivity and clutch size: (a) the most productive brood size based on post-fledging survival in great tits in different years and mean observed brood size in each year; (b) reproductive fitness measured as offspring recruitment in relation to the distribution of clutch sizes laid in lesser snow geese. (Based on data in Perrins and Moss 1975 and Rockwell et al. 1987.)

Sanz (2004) argued that the conclusion that brood-size experiments are generally in favor of the individual optimization hypothesis "does not hold." Tinbergen and Both (1999) suggested, perhaps most plausibly, that the role of individual optimization differs within and between populations and species, and that it occurs in some years but not in others, although when and why remains poorly understood even in well-studied species such as the great tit (Nicolaus et al. 2009). This important issue therefore remains unresolved, and it is unfortunate that the debate over clutch-size variation appears to have shifted to other questions (Charmantier et al. 2006; see section 5.3).

As with laying date and egg size, we still do not know what components of individual phenotype, including the individual's environment, determines optimal clutch size, but the general concept of individual optimization has been extended to provide an explanation for seasonal variation in clutch size. In species where clutch size declines with date, high-quality females lay relatively early and lay relatively large clutches, and low-quality females lay relatively late and lay relatively small clutches. High-quality females laying small clutches early in the season would clearly not be optimizing their reproductive effort, and low-quality females laying larger clutches late in the season would have a higher cost of reproduction (decreased future fecundity and survival) due to the trade-off between greater reproductive investment and lower reproductive value of late-laid eggs (Daan, Dijkstra, and Tinbergen 1990; Rowe et al. 1994). In multiple-brooded species, fitness will be determined in part by the total number of broods that parents can raise during the entire breeding season, not simply by the success of a single brood. Individuals in multi-brooded species should therefore be selected to start breeding before the date when their optimal clutch size is greatest, and the laying of smaller clutches earlier in the season would be consistent with females making "sub-optimal" investment in their first clutches to maximize total fitness over multiple breeding attempts for the whole breeding season (Crick et al. 1993; Svensson 1995). Even so, in multiple-brooded species, individuals making the most breeding attempts have the highest annual fecundity in terms of fledged chicks. For example, in black redstarts, with each additional successful brood females added on average three fledglings to their seasonal reproductive success (Weggler 2006). Few studies have considered potential trade-offs between number and quality of fledglings, recruitment, or costs of reproduction associated with multiple breeding attempts, but both Weggler (2006) and Parejo and Danchin (2006) failed to find evidence of such costs (but see Verhulst et al. 1997 and Travers et al. 2010). This might reflect the fact that some high-quality individuals can maximize reproductive effort (producing multiple broods) without incurring costs (see chapter 7).

Precocial species do not feed their young, and Lack (1968) eventually supported the view that clutch size in these species has evolved "in relation to the average food available to the female [to support egg production] modified by the size of the egg" (p. 307). However, other aspects of parental care, such as incubation or brooding ability (see section

5.2.3 and chapter 6), vigilance, or anti-predator behavior might limit the number of chicks that can be successfully reared in precocial species (T. Williams et al. 1994; Winkler and Walters 1983). Thus, an optimum clutch size might still be related to individual variation in parental quality or ability. However, these costs might be balanced by other benefits of large brood size in precocial species; for example, larger broods can occupy better feeding territories and consequently fledge more young (Loonen et al. 1999). The idea of nutritional constraints on egg production in precocial species, in relation to clutch size, will be revisited below.

5.2.2. Nest Predation and Clutch Size

Comparative analyses have shown that greater nest predation is associated with a reduced reproductive effort, which is reflected in smaller clutch size and clutch mass (T. Martin 1995; T. Martin et al. 2006; T. Martin et al. 2000). Slagsvold (1984) summarized potential mechanisms by which nest predation and clutch size might be related. First, clutch size might directly influence the risk of nest predation because larger clutches take longer to lay, increasing the duration of exposure to predators, or because larger broods are noisier and more conspicuous. Smaller broods might also be less conspicuous because better-fed nestlings will beg less or because parents can make fewer feeding trips to and from the nest (Eggers et al. 2006). Second, females might lay smaller clutches to reduce the costs associated with egg production, including time delays and physiological costs. This suggests that females adopt a "bet hedging" strategy that increases their ability to renest should nest predation occur, favoring repeated attempts to nest during a single breeding season because less time and energy is invested in each single attempt (Slagsvold 1984). These mechanisms should apply both among and within species, and theoretical models also predict that birds should lay smaller clutches when subject to a high degree of nestling mortality (Lima 1987). However, the effects of nest predation on individual variation in clutch size within species, the ability of birds to use cues to assess variation in predation risk, and the ability of females to adaptively adjust clutch size in relation to nest predation remains largely untested and unknown (Fontaine and Martin 2006).

Doligez and Clobert (2003) artificially increased nest-predation rates via brood removal and also displayed predator models to collared flycatchers and showed that the mean clutch size in the year following the manipulation was smaller in "depredated" than in control areas. However, they were unable to show unequivocally that this was due to phenotypic plasticity of the clutch size in individual females. Eggers et

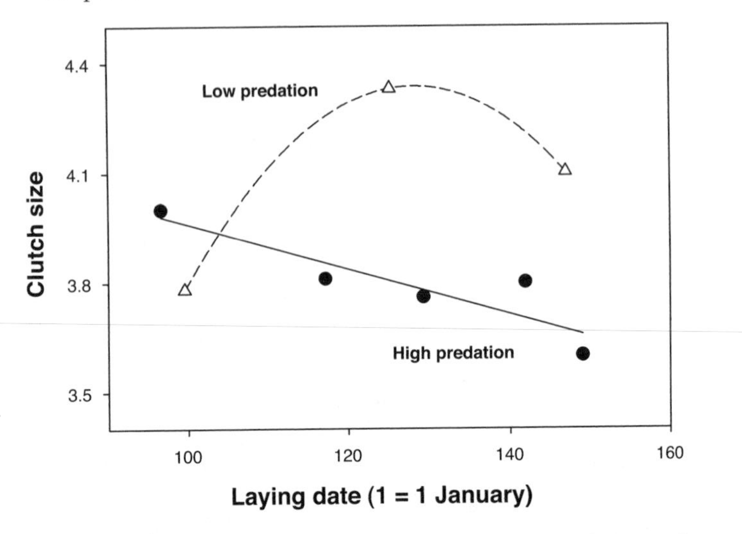

Fig. 5.6. Seasonal variation in clutch size in song sparrows in relation to low and high levels of experimental nest predation. (Based on data in Travers et al. 2010.)

al. (2006) found that Siberian jays (*Perisoreus infaustus*) reduced their clutch size (–1.3 eggs) in the year *after* they were exposed to recordings of nest predator (Corvid) calls presented on their territories, and this reduction *did* involve plastic responses in known individual females. In contrast, in a large-scale natural predator-removal experiment involving eight passerine species, birds in "safer" environments increased their reproductive efforts through increased egg size, clutch mass, and nestling feeding rate but did not adjust the clutch size (Fontaine and Martin 2006). Travers et al. (2010) experimentally manipulated actual nest-predation rates (by clutch removal) in individual female song sparrows and measured the subsequent reproductive effort while controlling for food availability and date and stage of breeding. As expected, females subjected to more frequent experimental nest predation laid more clutches over the season, but they also laid smaller clutches (3.75 vs. 4.21 eggs). In addition, there was a contrasting pattern of seasonal variation in clutch size, with low-predation females showing the pattern typical of multiple-brooded species, but high-predation females showing a linear decline in clutch size with date (fig. 5.6). These results clearly show that the contrasting seasonal pattern in clutch size typical of single- versus multiple-brooded species (fig. 5.2) can also arise from a difference in the probability of nest predation within a single species and that clutch size can clearly vary, and potentially be adjusted, independently of season (date) per se.

5.2.3. Embryo Viability, Incubation Capacity, and Clutch Size

Numerous hypotheses to explain both geographic and seasonal variation in clutch size revolve around the effects of Ta on embryo development (Cooper et al. 2005). Several related hypotheses have proposed that constraints operating during incubation might explain individual variation in clutch size; for example, it has been argued that if the number of eggs being incubated influences incubation energetics or behavior, then the demands of incubation may play a role in determining the optimal clutch size (Monaghan and Nager 1997; J. Reid, Monaghan, and Nager 2002). The general question of individual variation in, and costs of, incubation will be deferred to chapter 6, but this idea is illustrated below by a specific example, the incubation limitation hypothesis, as a potential explanation for the invariant four-egg clutch in shorebirds (Lengyel et al. 2009).

Avian embryonic development does not occur below 24°C–27°C (physiological zero), and therefore in temperate-zone species, many of which delay incubation until the end of laying, cold torpor suspends the development of earlier eggs, allowing embryo development and hatching to be synchronous (although the viability of unincubated eggs maintained in cold torpor can decline slowly; Ewert 1992; Webb 1987). Ambient temperatures between physiological zero and normal incubation temperatures (36°C–38°C) allow some embryonic tissues to develop even in the absence of full incubation and can lead to unsynchronized growth, abnormal development, especially for neurological traits and brain development, and embryo mortality(Deeming and Ferguson 1992; Webb 1987). Embryos might therefore be susceptible to developmental failure under low and moderate temperatures outside of the optimal range once development has been initiated. The *egg viability hypothesis* predicts that warm Ta will select against females laying a large clutch because they might experience a higher probability of hatching failure, due to problems with embryo development at intermediate pre-incubation temperatures (Cooper et al. 2005). Thus, the benefits of laying a larger clutch might be offset by a decline in hatchability with increasing temperatures (T. Arnold 1993; Stoleson and Beissinger 1999). Conversely, selection might favor smaller clutches by maintaining egg viability without imposing the problems of hatching asynchrony that arise if full incubation is initiated early in laying (Sockman et al. 2006). In addition, delayed incubation and exposure of early-laid eggs might decrease embryo viability because of a greater risk of microbial infection (Cook et al. 2005; Cook et al. 2003). Several studies have shown reduced hatchability of eggs exposed to intermediate temperatures between 27°C and 34°C (Stoleson and Beissinger 1999; Webb

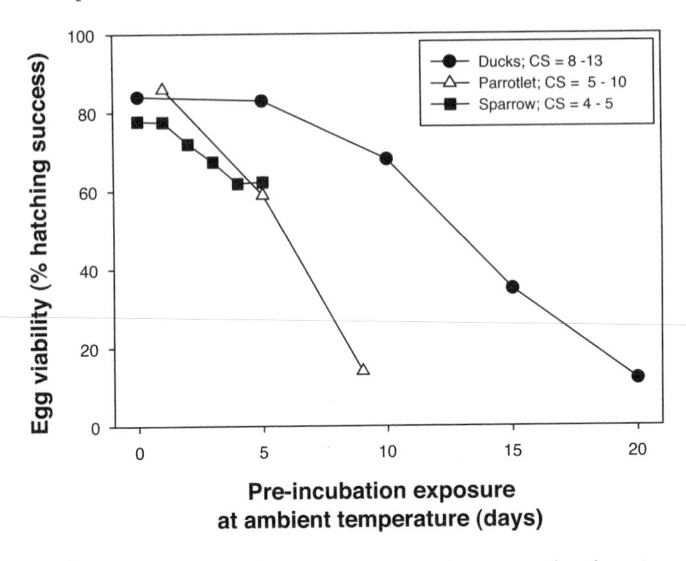

Fig. 5.7. Egg viability (% hatching success) in relation to the duration of pre-incubation exposure to ambient temperature in dabbling ducks, house sparrow, and green-rumped parrotlet. CS, clutch size. (Based on data in Arnold et al. 1993; Stoleson and Beissinger (1999); and Veiga 2002.)

1987) as well as at lower temperatures (T. Arnold 1993; Veiga 1992), even with as little as 3–5 days of pre-incubation delay (fig. 5.7). It has been suggested that the egg viability hypothesis may be broadly applicable to species that lack other mechanisms to synchronize hatching, specifically tropical species (Beissinger et al. 2005; Cooper et al. 2005), but the extent to which embryo viability contributes to individual variation in clutch size remains unclear. Cooper et al. (2005) showed that large-scale nesting patterns of eastern bluebirds (*Sialia sialis*) and red-winged blackbirds (*Agelaius phoeniceus*) are consistent with the egg viability hypothesis in that females appear to initiate incubation before clutch completion when they lay large clutches at lower (warmer) latitudes (see also Badyaev et al. 2003 and Ardia et al. 2006). However, to my knowledge no experimental studies have linked Ta, egg viability, and variation in clutch size, and several studies have failed to support key predictions of the egg viability hypothesis (Beissinger et al. 2005; Wang and Beissinger 2009). This might reflect the fact that complete egg removal, which was used in most of the above studies, does not mimic natural, partial incubation or natural egg "neglect," where there is at least some mother-egg contact (see Wang and Beissinger 2011), but it is also likely that the sensitivity of unincubated eggs to natural variations in incubation behavior and temperature has been overstated (Hebert 2002; see chapter 6).

Rather than focusing on embryo viability, the incubation limitation hypothesis suggests that clutch size is limited by the ability of the *parents* to incubate large clutches (Lack 1947). Most shorebirds (210+ species) have an invariant clutch size of four eggs (Lengyel et al. 2009) even though this taxon displays a wide diversity of mating systems and modes of parental care (G. Thomas et al. 2007). Since most shorebirds do not feed their precocial young and parental care is thought to be less costly than that in altricial birds, most work regarding clutch-size limitations in shorebirds has focused on the costs of incubation. The incubation limitation hypothesis has been tested experimentally through manipulation of clutch sizes (analogous to the manipulations of brood size that are widely used to test optimal clutch size in altricial species), and Lengyel et al. (2009) reviewed eight such studies. As in altricial species, most studies suggest that precocial species can rear more chicks to hatching from enlarged broods with no effect on hatching success, although some other potential costs of incubating five-egg clutches were identified, including an increased duration of incubation and lower chick mass at hatching (Wallander and Andersson 2002; Larsen et al. 2003; Sandercock 1997; but see Székely et al. 1994). T. Arnold (1999) concluded that a combination of "subtle costs," including a higher risk of predation associated with longer incubation and increased frequency of partial clutch loss *were* sufficient to limit the clutch size among most species of shorebirds. However, in common with the problem highlighted in studies of altricial species, none of the above studies quantified the long-term costs of reproduction beyond hatching. Safriel (1975) manipulated brood size in the semipalmated sandpiper (*Calidris pusilla*), adding one chick at hatching, and did show that enlarged broods fledged fewer chicks (1.00 vs. 1.74 chicks per nest). Lengyel et al. (2009) manipulated the clutch size in pied avocets (*Recurvirostra avosetta*), creating five- or six-egg clutches, and also showed that pairs incubating enlarged clutches experienced higher chick mortality and raised fewer young to independence (0.7 vs. 1.2 chicks per nest), even though pairs hatched more chicks from enlarged broods. Lengyel et al. (2009) concluded that the incubation of extra eggs is an energetic cost to parents and that this cost manifests at a later time, during chick-rearing (see chapter 6). These studies suggest that the post-hatching costs of rearing enlarged broods might limit the clutch size in some shorebird species. However, a further criticism of studies involving clutch-size manipulation in precocial species, and with brood- or clutch-size manipulations in general, is that when eggs or chicks are simply added to the nests, females do not incur the physiological costs of producing and laying the additional eggs (Monaghan and Nager 1997). So I will now revisit Lack's (1947) conclusion that the number of eggs a female lays is "far below the potential [*physiological*] limit of egg production."

5.2.4. Constraints on Egg Production

A long-term legacy of Lack's ideas (1947) has been that most avian studies have focused on reproductive limitations operating during chick-rearing. Consequently, until relatively recently the idea that the costs of egg production are minor and unimportant for understanding the evolution of clutch size and the fitness costs of reproduction in birds has continued to dominate the literature (Monaghan and Nager 1997; J. Williams 1996), and these costs are still rarely incorporated into experimental studies (T. Williams 2005). However, there is now substantial evidence for *life-history* costs of egg production, even though the most comprehensive studies have so far been limited mainly to seabirds and some passerines (Nager 2006).

Many birds are "indeterminate layers" (see section 5.4) and will lay additional eggs if eggs are removed from the nest during laying. This observation has been one of the main arguments against there being physiological limitations on egg production (S. Haywood 1993b). However, although birds can lay additional eggs in response to egg removal, the egg quality decreases rapidly with increasing egg number above the normal clutch size. For example, lesser black-backed gulls (*Larus fuscus*) will lay almost three times the normal clutch size of three eggs in response to continuous egg removal, but the eggs laid late in the experimental laying sequence have a 50% lower fledging success (Nager et al. 2000). This decline in egg quality is associated with changes in egg composition that are independent of the changes in egg size (Heaney et al. 1998; Nager et al. 2000; T. Williams and Miller 2003). Egg quality therefore decreases rapidly as additional eggs *above the normal clutch size* are laid, suggesting that the physiological limitations of egg production *do* play a role in the determination of clutch size. The ability of females to lay additional eggs can, in part, be resource dependent, but it is also related to individual female "quality": putative high-quality female zebra finches that had previously laid large eggs produced larger clutches in response to egg removal than females that had previously laid small eggs (T. Williams and Miller 2003). Similarly, in blue tits, late-laying (presumably lower-quality) females had a lower tendency to produce extra eggs in response to egg removal (S. Haywood 1993a). This finding suggests that there is inherent individual or phenotypic variation in *egg-laying ability*, although environment (higher resource levels) can overcome these individual differences to some extent.

Egg-removal experiments make females incur the costs of *producing* additional eggs, that is, it increases both egg-production costs and rearing costs, whereas simply giving the parents extra eggs or chicks only increases incubation and/or rearing costs. Experimental increases

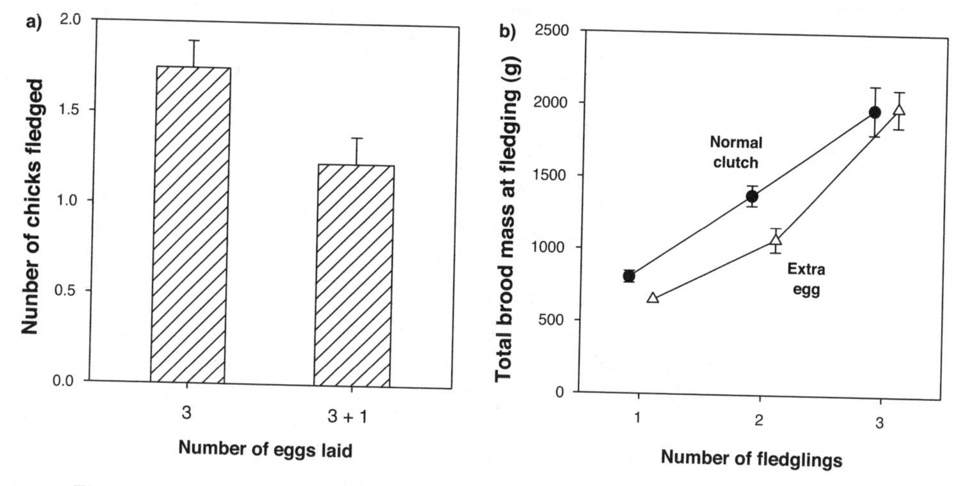

Fig. 5.8. Effect of experimentally increased costs of egg production (+ 1 egg) on (a) number of chicks fledged and (b) total brood mass at fledging in lesser black-backed gulls. (Based on data in Monaghan et al. 1998.)

in egg-production costs have been shown to have long-term fitness consequences in relation to reproductive success in the current breeding attempt, future fecundity, and survival. Females incurring higher egg-production costs are in poorer physiological condition (Kalmbach et al. 2004; Travers et al. 2010; see chapter 7), have a lower chick-hatching mass (which is independent of egg size), reduced chick growth, more early-chick mortality, and decreased chick survival (fig. 5.8) relative to those females that incur only additional rearing costs (Monaghan et al. 1995; Nager et al. 2000). These effects are mediated through the parents' reduced brood-rearing or provisioning capacity (Monaghan et al. 1998). In gulls the negative effect of producing one extra egg on brood mass at fledging was greatest in pairs with lower fledging success, consistent with the idea that lower-quality females have a reduced capacity to produce additional eggs (fig. 5.8b). Although some experiments have involved females laying up to three times the normal clutch size, few studies have reported an incremental effect of the number of additional eggs laid, and negative effects are observed even in females laying only *one* additional egg (Monaghan et al. 1995; Nager et al. 2001). Evidence for effects of the costs of egg production on future fecundity is more limited. In the year following experimentally increased egg production, female lesser black-backed gulls laid on the same date as controls but laid smaller first eggs (Nager et al. 2001), whereas female great skua (*Stercorarius skua*) laid significantly later than the controls but did not lay different-sized eggs (Kalmbach et al. 2004). However, in the lesser black-backed gull (Nager et al. 2001) and great tit (Visser and Lessells 2001), females that were made to lay additional eggs through egg removal had lower return rates the following year, even when laying only one additional egg. Descamps et al. (2009) described the variation in fitness costs of reproduction (large clutch size) as a function of the degree of exposure to avian cholera in common eiders (*Somateria mollisima*). Annual survival was not related to the clutch size before a severe cholera epidemic or during the period of moderate cholera effect but was strongly and negatively associated with clutch size during the period of intense cholera. An increase in clutch size of only one egg was associated with an estimated average decrease in survival of 15%. Thus, the *physiological* costs of egg production should contribute significantly to individual variation in clutch size and should "seriously change our view on selection acting on clutch size," even though that has not yet happened (Tinbergen and Sanz 2004).

5.3. Selection on Clutch Size

Individual optimization, costs of reproduction, and fitness benefits of large clutch size, which were discussed above, make predictions about if, and how, clutch size will respond to selection. If clutch size is individually optimized, then stabilizing selection might maintain the average clutch size. However, while individual optimization might occur, individuals with a higher optimum clutch size might still contribute more genes to the next generation, and thus individuals laying larger clutches should, *all else being equal*, come to predominate in the population (Charmantier et al. 2006; Cooke et al. 1995). If larger clutches are the most productive, then a combination of selection and additive genetic variation should result in an evolutionary response to selection. Clutch size is heritable in birds (h^2 = 0.20–0.40; Christians 2002; Merilä and Sheldon 2000; Sheldon et al. 2003), but despite this additive genetic variation and evidence that clutch size is subject to selection (with larger clutches being the most productive), several early studies found no evidence for an evolutionary response to selection (Boyce and Perrins 1987; Cooke et al. 1990). Numerous explanations have been provided for why clutch size might show a lack of response to selection (Cooke et al. 1990; Merilä et al. 2001; Price and Liou 1989), but the explanation of particular interest in relation to underlying physiological mechanisms is that the evolution of clutch size is constrained by correlations with traits that are primarily environmentally determined, such as the female's nutritional state or her ability to acquire nutrients (see below). Several studies have used animal models (A. Wilson et al. 2010) for the analysis of large sets of pedigreed breeding data to investigate genetic changes in clutch size over time while controlling for potential correlations with other traits (Charmantier et al. 2006; Garant, Kruuk, et al. 2007; Sheldon et al. 2003). A negative genetic correlation between laying date and clutch size, which might constrain the rate of adaptation, has been reported in collared flycatchers (Sheldon et al. 2003) and in great tits in the United Kingdom (Garant, Hadfield, et al. 2007); but in the latter study, only during a period of cold years, not during a more recent warmer period. In contrast, there is no evidence for significant genetic correlations between clutch size and laying date in great tits breeding in the Netherlands (Gienapp et al. 2006; Husby et al. 2010).

Sheldon et al. (2003) found no evidence for fecundity or viability selection acting independently on clutch size in collared flycatchers when controlling for the correlated effect of selection on laying date. Similarly, Garant, Kruuk, et al. (2007) found a negative correlation between the selection gradients for clutch size and laying date in great tits; that is, females that laid early, larger clutches had greater reproductive success, on average, than females laying later, smaller clutches (see fig. 3.7), a finding consistent with explanations for the strong seasonal decline in clutch size. There was positive fecundity selection for clutch size at the population level, which supports the general assumption that, on average, larger clutches are the most productive in terms of offspring produced. However, the estimated selection gradients were smaller than the selection differentials, a result suggesting that the latter were most probably inflated by an indirect effect of selection for laying date (which is consistent with Sheldon et al. 2003). In addition, there was evidence for a non-linear fecundity selection differential for clutch size that was indicative of convex (i.e., stabilizing) selection acting on this trait, although the fitness surfaces for clutch size and laying date were very different even in two local sub-populations of great tits (fig. 5.9). In a third study, Charmantier et al. (2006) investigated the evolutionary response of clutch size to selection in mute swans (*Cygnus olor*), a large-bodied precocial species. In this population there *was* evidence for positive selection on clutch size after controlling for indirect selection acting on laying date. Females were selected for one main adaptive peak with early laying dates and large clutches (fig. 5.10). However, in contrast to the previous studies, the clutch size increased over the study period (1979–2003) both at the phenotypic and genetic levels, and the slope of the phenotypic change over time was not different from that of the predicted response to selection based on the estimated positive selection combined with estimated heritability ($h^2 = 0.19$). Charmantier et al.'s study (2006) did not find constraints on clutch size associated with costs of reproduction because clutch size was positively selected both annually and over a female's lifetime, with no evidence for consistent negative survival selection. Similarly, there was no

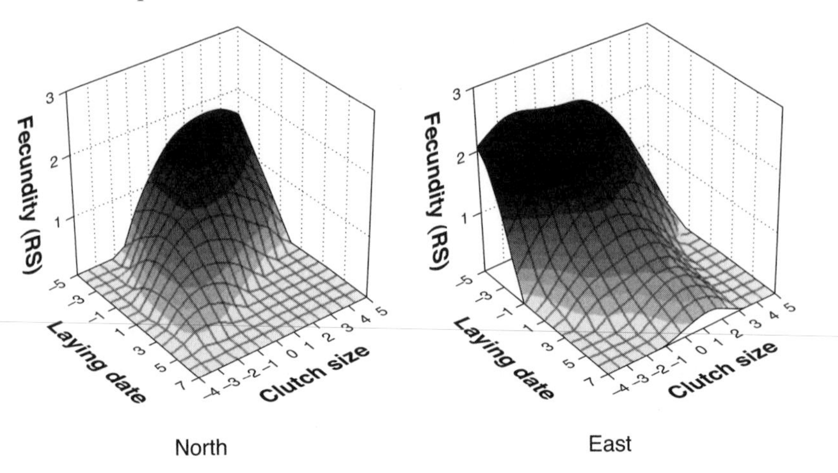

Fig. 5.9. Fecundity fitness surfaces for combinations of laying date and clutch size in two local sub-populations of great tits breeding in Wytham Woods, UK. (Reproduced from Garant, Kruuk, et al. 2007, with permission of John Wiley and Sons.)

evidence to support the individual optimization hypothesis, with a trade-off between clutch size and offspring quality (cf. Sheldon et al. 2003). Rather, in mute swans there was strong positive selection on clutch size both for the number of hatchlings and the number of recruited offspring (as in Garant, Kruuk, et al. 2007). Charmantier et al. (2006) suggested that the evolution of clutch size in this particular population may have been facilitated by a food-supplementation program that relaxed food constraints (the environmental component of variation, sensu Cooke et al. 1990), allowing for a rapid evolution of clutch size.

These complex analyses of large sets of long-term breeding data are potentially powerful but so far have not produced consistent, emergent patterns (Husby et al. 2010). They provide only mixed support for any of the major hypotheses that have been proposed to explain individual variation in clutch size, with the possible exception of the strong, inverse relationship between clutch size and laying date. More studies will hopefully help clarify our understanding, although a concern, as highlighted by Tinbergen and Sanz (2004), is that conclusions from different studies appear to be influenced by the specific analytical approaches taken and/or the specific fitness measures used. These complex theoretical and mathematical approaches do have plenty of "pitfalls and dangers for the researcher wanting to begin using quantitative genetic tools to address ecological and evolutionary questions" (A. Wilson et al. 2010). Furthermore, the promise that these "purely analytical approaches" can

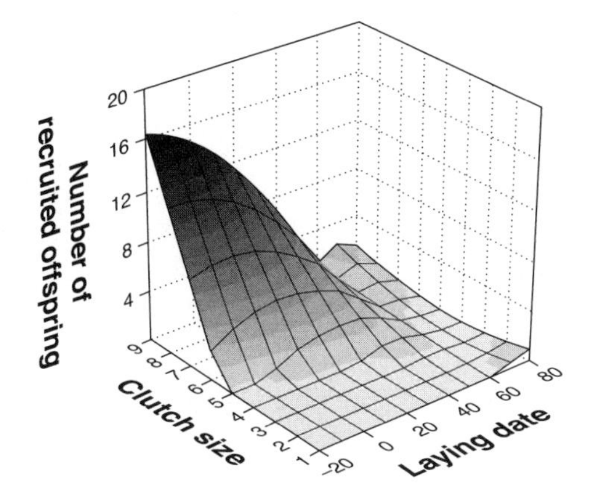

Fig. 5.10. Fitness surface for the relationship between total number of recruited offspring in a female mute swan's lifetime, laying date (1 March = 1), and clutch size. (Reproduced from Charmantier et al. 2006, with permission of the University of Chicago Press.)

be invaluable in identifying key *experimental* manipulations that need to be conducted (Charmantier et al. 2006) has not yet been borne out.

5.4. Physiological Mechanisms of Clutch-Size Determination

In their review, Sockman et al. (2006) commented that, given the central importance of clutch size to reproductive effort in birds, determining the mechanistic basis of clutch-size regulation would be "of particular interest" and noted that "it is surprising that more headway has not been made." This remains a shockingly true statement, given that this was the path Lack (1947, 1968) sent researchers down more than 50 years ago. The strong, even predominant, effect of latitude and season (date) on clutch size suggests that regulation involves some time- or day-length-dependent mechanism, at least for single-brooded species in which clutch size decreases linearly with date. Indeed, Meijer and Drent (1999) concluded that the laying date itself controls the number of eggs laid independently of the food intake and body reserves. In other words, variability in clutch size cannot be understood independently of an understanding of the effect of the timing of breeding, and finding a mechanism for reproductive timing (chapter 4) might well provide the mechanism for clutch-size determination in many species. However, in addition to any mechanism underlying the strong seasonal decline in clutch size, we need to identify additional mechanisms that mediate day-length or date-independent variation in clutch size in order to explain the effects of predation, nesting density, etc., on clutch size (e.g., see fig. 2 in Brommer et al. 2002), and the more complex, convex seasonal pattern of clutch size seen in multiple-brooded species. Finally, we need to revisit the question of whether there are proximate nutritional constraints on egg production in the context of clutch size, as well as the associated hypothesis that food availability or energy balance represents an "environmental" component that explains much of the individual variation in clutch size observed within and between populations.

5.4.1. Determinate versus Indeterminate Laying

Many birds will lay additional eggs if the eggs are removed as they are laid (indeterminate layers), whereas other species will not respond to egg removal by laying additional eggs (determinate layers, Haywood 1993b). Although egg-removal experiments can be informative, the general assumption that the response to egg removal reliably indicates patterns of follicle development in the ovary is not always true, and it is preferable to obtain anatomical or physiological data on ovarian development. At the level of the ovary, clutch size could be determined by mechanisms operating at two distinct stages of follicle development: follicle selection and recruitment, and/or follicle growth and ovulation. The most common pattern of follicle development, which is seen in indeterminate layers, is for females to have the potential to recruit and develop a *greater* number of yolky, vitellogenic follicles than the normal clutch size (Haywood 1993b). Final clutch size is therefore thought to depend on extrinsic cues (e.g., tactile stimuli from eggs in the nest) that interrupt follicle development and inhibit further ovulations (Haywood 1993a, 1993c). In the absence of these extrinsic cues, for example, with egg removal during laying, indeterminate layers will lay supra-normal clutches (Haywood 1993b). This observation suggests that in these species there is no *predetermined* maximum number of follicles that can be recruited and that maximum clutch size is not determined by mechanisms regulating follicle *recruitment*. Rather, clutch size is determined initially by the disruption of follicle growth, which secondarily inhibits further recruitment of follicles (Haywood 1993c). The precise timing of this inhibitory cue relative to egg-laying will be considered further below (section 5.4.2).

It has been suggested that in a very few determinant-laying species, including some New World vultures, petrels, perhaps some penguins (but see Crossin et al. 2011), and pigeons, the number of vitellogenic yolky follicles recruited is equal to the number of eggs laid irrespective of any extrinsic cues (Haywood 1993b). Here all follicles that are recruited develop fully and are ovulated, and clutch size would be determined solely by mechanisms regulating follicle selection and follicle *recruitment* (see chapter 2). The most well-documented case of determinate follicle development occurs in the Columbidae (pigeons), which typically lay a two-egg clutch and in which only two follicles are recruited into RYD, such that only two large yolky follicles are ever present in the ovary before laying (Cuthbert 1945; Goerlich et al. 2009). However, other putative examples of determinant follicle development are less well supported. Haywood (1993b) cited Hector et al. (1986) to suggest that in albatrosses (*Diomedia* spp.), which lay single-egg clutches, only a single yolky follicle is present just before ovulation. However, as figure 2.5 clearly shows, some albatross females recruit and develop more than one yolky vitellogenic follicle. Although albatrosses might not lay additional eggs in response to egg removal (Haywood 1993b), extrapolating these data to specific patterns of follicle development is risky.

To summarize, in most species, the mechanisms underlying follicle selection or follicle recruitment do not appear to set the limit on clutch size. Rather, clutch size appears to be determined by an inhibitory signal (as yet undefined), perhaps linked to tactile stimuli from eggs in the nest, that disrupts follicle growth and subsequently inhibits further ovulations of already developing, vitellogenic follicles (which might *secondarily* inhibit recruitment of additional follicles). This understanding helps to identify a specific component of follicle development that must be part of any physiological mechanism for clutch-size determination.

5.4.2. A General Mechanistic Model for Control of Clutch Size in Birds

In marked contrast to the current situation for timing of breeding, at least in females (chapter 3) and egg size (chapter 4), a mechanistic model for the determination of clutch size has existed for at least 20 years. Meijer et al. (1990) proposed, albeit cautiously, the key elements of a comprehensive model for the "response mechanism" underlying the seasonal decline in clutch size based on their extensive work with the Eurasian kestrel, which built on earlier work (Cole 1930; Haftorn 1985). A similar model for a mechanism underlying individual variation in clutch size, with similar speculation about hormonal regulation, was presented around the same time by Haywood (1993a, 1993c) based on experimental work with small passerines. The key elements of these two models are (a) a close association between the timing of the development of incubation behavior and the termination of egg-laying; (b) a positive feedback loop for egg-contact incubation behavior in which the increased intensity of incubation behavior during the laying period is associated with the presence of eggs in the nest; and (c) the inhibition of follicle growth at some threshold of incubation behavior that occurs sooner in females laying small clutches than in those laying large clutches. In addition, Meijer

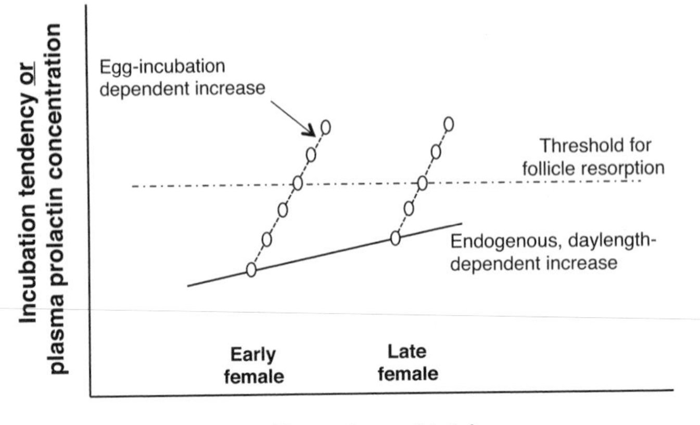

Fig. 5.11. Model for seasonal decline in clutch size in relation to changes in incubation intensity or plasma prolactin levels. (Based on Meijer et al. 1990.)

et al.'s model (1990) proposed an endogenous, pre-programmed decline in clutch size within an annual "reproductive window" in individual females that is synchronized by the annual cycle by day length and is expressed behaviorally as an increasing tendency to incubate earlier during egg-laying with later laying dates. Thus, the threshold incubation level for the disruption of follicle development is reached earlier in late-laying females than in early-laying females, and consequently the final clutch is smaller in late-laying females than in early-laying females (fig. 5.11)

Meijer et al. (1990) and Haywood (1993b, 1993c) both speculated on the physiological nature of the controlling variable underpinning their models for clutch size determination. They suggested that the hormone prolactin plays a key role in mediating both the endogenous seasonal- and egg-contact-related increases in incubation behavior and has antigonadal effects that inhibit further follicle development (see also Badyaev et al 2005 and Meijer et al. 1992; and see Sockman et al. 2006 for a recent review). Thus, these models propose that the seasonal increase in incubation onset and the seasonal decline in clutch size are determined by (a) an endogenous, photoperiod-dependent increase in prolactin release and (b) further elevation of plasma prolactin levels due to stimulation of the brood patch by eggs in nest (fig. 5.11). Meijer et al. (1990) suggested that "some elements in [their] model...may apply fairly generally" and some, but not all, of the components of this model are supported by empirical data that I will now consider.

In many species, incubation behavior and plasma prolactin concentrations both generally increase in females during egg-laying, and there is therefore a broad temporal association between these traits and the

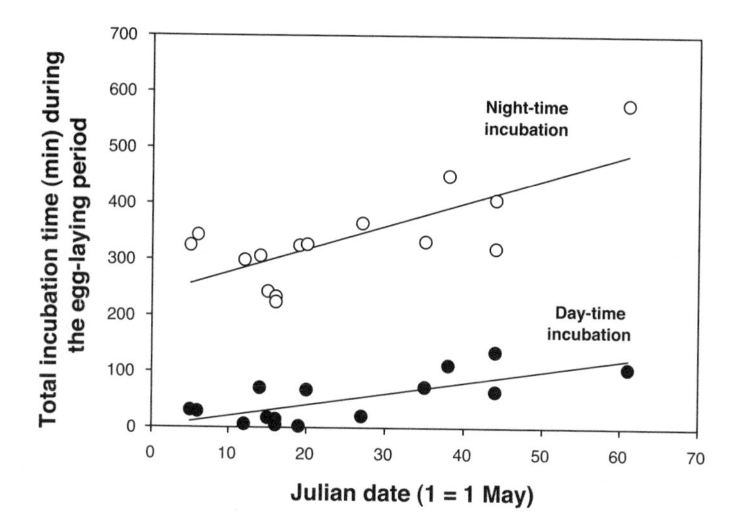

Fig. 5.12. Seasonal increase in incubation intensity for both day-time and night-time incubation during the egg-laying period in great tits. (Based on data in Haftorn 1981.)

cessation of laying (Sockman et al. 2006 and references therein). Furthermore, a seasonal increase in timing of onset of incubation relative to initiation of egg-laying has been documented in a number of species and is evident in both free-living and captive birds (Sockman and Schwabl 2001). Maximal expression of incubation occurs sooner after the initiation of laying in late-season clutches compared with early-season clutches (fig. 5.12; Beukeboom et al. 1988; Haftorn 1981; Müller et al. 2004). However, individual females can show considerable variation in the onset of incubation relative to the initiation of laying and timing of clutch completion (Badyaev et al. 2003; Müller et al. 2004). For example, in canaries, the onset of incubation occurred between 0 and 4 days after the initiation of egg-laying, and this variation was independent of photoperiod or food restriction (Sockman and Schwabl 1999). Thus, there is considerable unexplained individual variation in the timing of incubation onset in the absence of variation in clutch size (Sockman and Schwabl 2001; Sockman et al. 2006).

Early studies by Goldsmith and Williams (1980) and Lea et al. (1981) provided evidence for a positive feedback loop between egg contact and egg incubation involving prolactin, which is a key component of the Meijer and Haywood models. Subsequent work confirmed that this mechanism involves physical (tactile) stimulation of the brood patch by the eggs

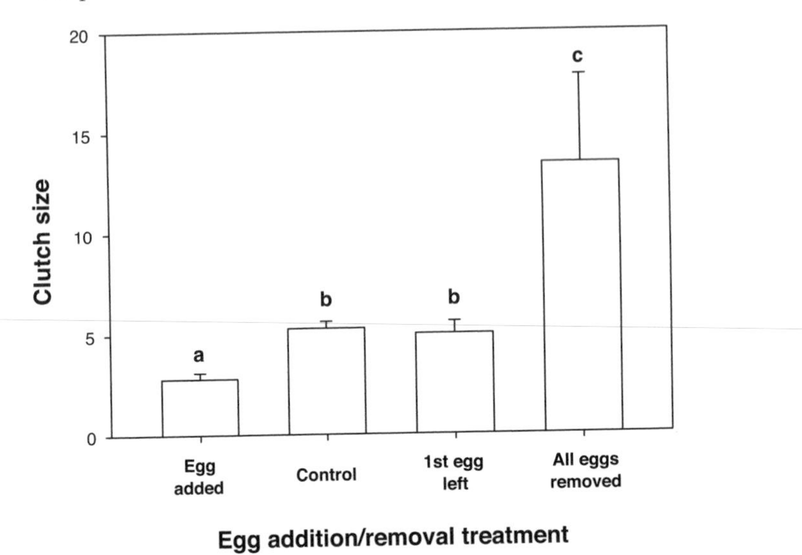

Fig. 5.13. Effect of different egg removal/egg addition treatments on clutch size in the zebra finch. Letters indicate significant differences. (Based on data in Haywood 1993c.)

in the nest. Plasma prolactin levels decrease when the brood patch is desensitized with an anesthetic (M. R. Hall and Goldsmith 1983). Similarly, plasma prolactin levels decrease when the tactile or visual stimuli from the nest or eggs are removed, and increase again if the nest and eggs are returned to incubating females (Goldsmith et al. 1984). Furthermore, the addition of eggs to the nest 3–4 days in advance of laying, thus advancing the onset of tactile stimulation, decreases the clutch size (Haywood 1993b, 1993c). However, as described above, in some species a single egg in the nest, rather than the accumulation of multiple eggs, is sufficient to trigger the cessation of egg-laying (fig. 5.13), so there does not appear to be a simple, linear relationship between the number of eggs and the development of incubation behavior.

There is considerable evidence to support a seasonal, day-length-dependent increase in baseline plasma prolactin levels, such that clutch size is determined in early- and late-laying females when the basal plasma prolactin levels are relatively low and relatively high, respectively, as in figure 5.11 (Sockman et al. 2006). Prolactin secretion is highly responsive to photoperiod, and plasma prolactin levels are stimulated by longer photoperiods (fig. 5.14; Dawson and Goldsmith 1983; Sharp and Sreekumar 2001). Meijer et al (1990) showed that plasma prolactin levels rose from 3.6 ± 2.5 ng/ml in February to 10.3 ± 5.9 ng/ml in early June in female kestrels *before* they initiated egg-laying; that is, the effect was

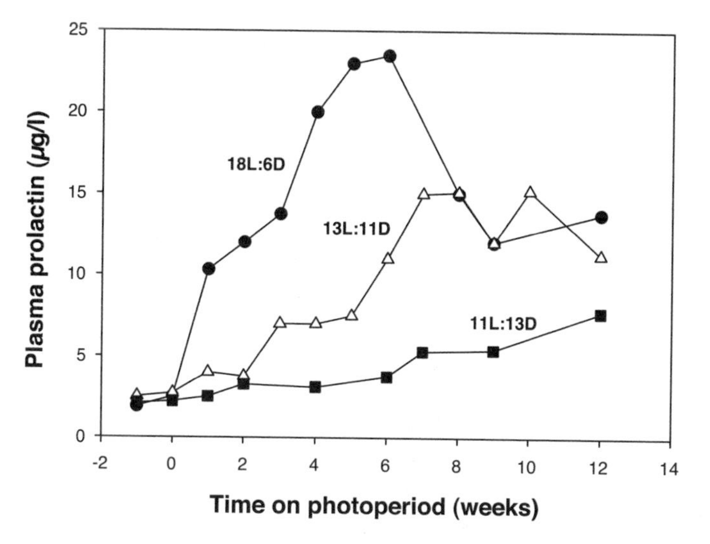

Fig. 5.14. Photoperiod- or day-length-dependent changes in plasma prolactin in captive European starlings. (Based on data in Dawson and Goldsmith 1983.)

independent of any incubation effects. In fact, Meijer et al. (1992) suggested that the seasonal decline in clutch size might more generally be associated with physiological mechanisms underlying the development of photorefractoriness as the season progresses (see chapter 2), which is also associated with higher plasma prolactin levels (Dawson 2002; Dawson and Sharp 2007). However, it seems unlikely that photorefractoriness itself is a critical or general element in any mechanism determining individual variation in clutch size. Many females are clearly making "decisions" about clutch size early in the season but are still capable of relaying at this time; that is, they are not photorefractory. Furthermore, as we have seen, in some populations, the seasonal declines in clutch size are evident well in advance of the onset of photorefractoriness at the population level. Although peak prolactin concentrations do coincide with testis regression, thus marking the onset of photorefractoriness at the end of the season in male birds (Sharp and Sreekumar 2001), it is very different from ovarian "regression" in females at clutch completion, which can occur relatively early in the season and repeatedly (see section 3.5).

What evidence is there for a direct, causal effect of prolactin on incubation behavior and inhibition of gonadal function? In the domestic chicken, turkey, and budgerigar, administration of exogenous prolactin promotes incubation, but in most cases this only occurs in birds in laying condition or following priming with ovarian steroids (El Halawani et al. 1986; Hutchison 1975; Sharp et al. 1989; Youngren et al. 1991).

Immunization with antibodies against prolactin or the avian prolactin-releasing-factor VIP inhibits incubation in chickens and turkeys (Crisóstomo et al. 1998; Crisóstomo et al. 1997; Sharp et al. 1989), although it is not clear if these effects are physiological or pharmacological. Direct evidence for an anti-gonadal effect of prolactin that disrupts the growth of yolky follicles is much more limited (Sockman et al. 2006), although prolactin has long been suggested to have inhibitory effects on the ovary and oviduct (Clarke and Bern 1980). In the turkey hen, prolactin inhibits the release of GnRH in the hypothalamus, reduces the expression of LH-β subunit mRNA and LH release by the pituitary, and reduces the expression of steroidogenic enzyme mRNA, thus inhibiting the production of steroid hormones in the ovary (Rozenboim et al. 2004; Tabibzadeh et al. 1995; You et al. 1995). However, these studies have mainly used in vitro approaches (e.g., cell culture or intra-cranial prolactin perfusion), and it is not always clear that HPG function is inhibited sufficiently to inhibit egg production (Sockman et al. 2006). Inhibition of egg production following prolactin treatment has been reported in laying quail and turkeys (Camper and Burke 1977; Youngren et al. 1991), and in canaries the onset of incubation, which is associated with elevated prolactin levels, is tightly coupled to a marked reduction in plasma E2 levels (Sockman and Schwabl 1999). Treatment of poultry hens with the anti-prolactin dopamine agonist bromocriptine enhances some aspects of egg production (Reddy et al. 2006, 2007), but no similar experimental studies appear to have been conducted in non-domesticated, free-living species. Thus, there is some support for the idea that prolactin has anti-ovarian or anti-estrogenic effects, at least in poultry. However, in the context of the models described above, the mechanism of action remains unclear since Zadworny et al. (1989) suggested that only very small, pre-hierarchical follicles (<1 mm diameter) were the target of prolactin's inhibitory effect.

Both Meijer et al. (1990) and Haywood (1993b, 1993c) proposed that a prolactin-dependent inhibitory mechanism determines clutch size by disrupting follicle growth rather than follicle recruitment, with reabsorption of the yolky follicles that had most recently recruited into the follicle hierarchy. In Eurasian kestrels, this "decision" to reabsorb extra follicles, and hence to terminate laying and fix clutch size, appears to occur at the four-egg stage (day 6 of laying) early in the season (around 10 April, mean clutch size 5.9) and at the two-egg stage (day 2 of laying) later in the season (around 20 May, mean clutch size 4.2 eggs; Beukeboom et al. 1988). Since eggs are laid at 2-day intervals in the kestrel, at least two extra developing follicles would be reabsorbed in this species, but at the time of follicle reabsorption, two further eggs could be laid, one of which was just ovulated and the other which has almost reached its final yolk size (see fig. 5.11). Assuming that, on average, both remaining larger yolky follicles are laid, then the seasonal advance in *sensitivity* to, or timing of, the signal disrupting follicle growth (roughly 0.5 eggs per 10 days; Beukeboom et al. 1988) would be sufficient to explain the seasonal decline in clutch size in this species. In an elegant series of egg-removal and addition experiments, Haywood (1993c) concluded that follicle growth is disrupted on the third or fourth day of laying in zebra finches, which lays 1 egg per day, but that the timing of this event is actually determined on the second day of laying. On average in zebra finches, at the time of follicle disruption two smaller, yolky follicles would be reabsorbed and two larger follicles would be available to complete growth, giving a final clutch size of 5 eggs (see fig. 5.17). However, based on analysis of 12 individual females, Haywood (1993c) also concluded that disruption of follicle growth does *not* occur later in female zebra finches laying larger clutches, which contrasts with Beukeboom et al.'s data (1988) for the kestrel. This seems unlikely, even for captive-breeding zebra finches, in which clutch size can vary from 2 to 8 eggs, unless there are very different fates for the remaining yolky follicles at the time of follicle disruption; that is, either all, or none, go on to be ovulated (see below). Furthermore, Haywood (1993a) did demonstrate variable timing of the disruption of follicle growth in free-living blue tits, which he argued allowed them to lay more variable clutch sizes from 6 to 13 eggs. Haywood (1993c) dismissed the idea that the cessation of egg-laying occurred via the disruption of follicle recruitment in zebra finches since this would produce larger clutches than observed. However, this would be true only if all the remaining vitellogenic follicles were ovulated. Rather it seems plausible, based on currently available data, that the timing of both recruitment and disruption of follicle growth could play a role in determining the final clutch size, especially in light of Zadworny et al.'s conclusions (1989).

Although Meijer et al. (1992) called for "direct experimental tests" of the predictions and proposed physiological mechanisms underlying their model for clutch-size determination, this model, in fact, received scant attention for almost 10 years. Sockman and colleagues (Sockman and Schwabl 2001; Sockman et al. 2000; Sockman et al. 2006) provide the only comprehensive, experimental investigation of the role of prolactin in the regulation of clutch size and onset of incubation behavior in a non-domesticated species. In free-living female American kestrels (*Falco sparverius*), there was a negative correlation between plasma prolactin levels early in the laying period and clutch size (Sockman et al. 2000): individual females with high prolactin concentrations on the day the first egg of a clutch was laid had smaller clutches than those with low concentrations on this day (fig. 5.15a). Millam et al. (1996) also reported a negative correlation between prolactin levels 17 days after the onset of laying and supranormal clutch size in response to egg removal in female

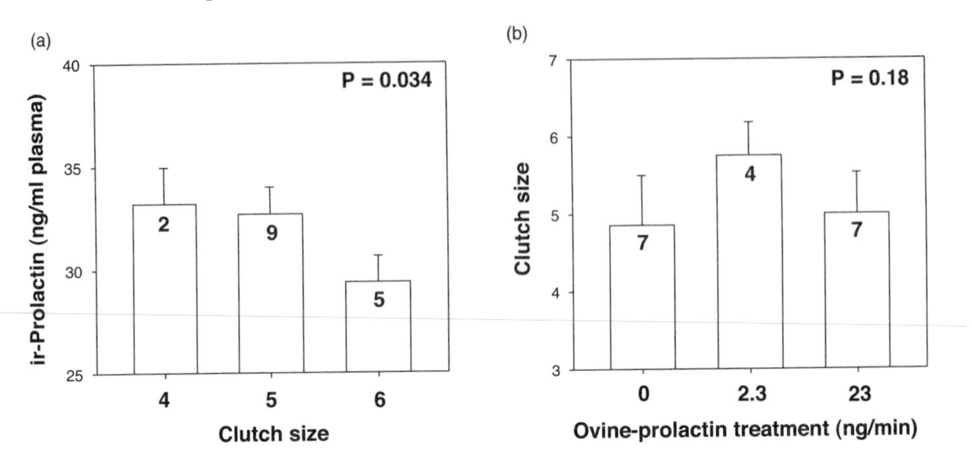

Fig. 5.15. (a) Relationship between ir-prolactin (averaged over days 1, 3, and 4 of laying) and clutch size. (b) Lack of effect of ovine prolactin treatment on clutch size in American kestrels. (Based on data in Sockman et al. 2000.)

cockatiels (*Nymphicus hollandicus*). These results are consistent with the idea that the timing of the rise in prolactin levels relative to some "threshold" during laying might provide the signal for disruption of follicle growth. However, experimental manipulation of plasma prolactin, with osmotic minipumps, on the day the second egg of the clutch was laid had no effect on clutch size (fig. 5.15b) and had no consistent effect on incubation behavior in American kestrels (Sockman et al. 2000). This is contrary to the prediction that higher prolactin levels should reduce clutch size and advance the onset of incubation. Sockman et al. (2000) suggested that their experimental treatment might have elevated prolactin too late in the laying cycle, that is, after the critical period for follicle disruption. In a follow-up experiment, Sockman and Schwabl (2001) initiated exogenous prolactin administration on the day the first egg of a clutch was laid but again observed no effect on clutch size. They suggested that if clutch size is determined by prolactin-induced inhibition of follicular recruitment, this may still have been too late in the follicular cycle. However, if the Meijer and Haywood models are correct, and if clutch size is determined by disruption of follicle *growth* on the second, third, or fourth day of laying (Beukeboom et al. 1988; Haywood 1993c), then it is not clear why the elevation of prolactin prior to this critical period has no effect (albeit with the caveat that these experiments used ovine, not avian, prolactin). Clearly, given the evolutionary importance of clutch size as a life-history trait, much more work on the mechanism of clutch-size determination utilizing the experimental manipulation of prolactin or VIP is warranted.

5.4.3. Potential Mechanisms for Individual and Date-Independent Variation in Clutch Size

Even if the Meijer and Haywood models provide a working hypothesis for physiological mechanisms underlying seasonal, day-length-dependent variation in clutch size, we still require mechanisms for the individual and date-independent variation in clutch size. Given that individual females appear to adjust clutch size in relation to factors such as predation, nesting density, individual quality, and perhaps their nutrient reserve status or energy balance, it seems that there is considerable flexibility, or plasticity, in the physiological response mechanism underlying clutch-size variation. Meijer et al.'s (1990) and Haywood's (1993c) models can be modified fairly easily to generate individual variation in clutch size for a given laying date, although there is currently little empirical data to support these ideas. If individual females have different thresholds for the inhibitory signal that causes follicle resorption (fig. 5.16a) or differences in the slope or rate of the endogenous day-length-dependent increase in plasma prolactin (fig. 5.16b), then females initiating laying on the same date would have different final clutch sizes. This concept is supported by Sockman et al.'s (2000) findings in American kestrels that females with higher prolactin concentrations at the start of laying laid smaller clutches than females with lower prolactin concentrations. As described above, there is only limited evidence so far for individual variation in the timing of the disruption of follicle development (Haywood 1993a). However, clearly for individuals, or species, laying large clutches, the inhibition of follicle growth cannot occur invariantly on the day the third or fourth egg is laid, as suggested for the zebra finch (Haywood 1993c). Rather, if the Meijer and Haywood models are correct, inhibition must occur 2–3 days before the last egg is laid, relative to the final clutch size, for example, on day 11 of egg-laying in an individual laying a 13-egg clutch (see Haywood 1993a); that is, this timing of inhibition must be individually variable.

In addition to individual variation in the sensitivity or threshold for the *timing* of the disruption of follicle growth, individual variation in the fate (resorption, atresia, ovulation) of the remaining developing follicles could provide additional flexibility for females to adjust final clutch size. In the kestrel example cited above, it was assumed that both larger follicles remaining at the time of disruption of follicle growth go on to form eggs. However, if females have the physiological capacity to reabsorb either of these two follicles (one requiring resorption in the oviduct), then this ability would provide additional flexibility to adjust the final clutch size: individual females could go on to lay zero, one, or two additional eggs from their remaining recruited follicles. In fact, using Haywood's model for the zebra finch (fig. 5.17), at the time follicle growth is

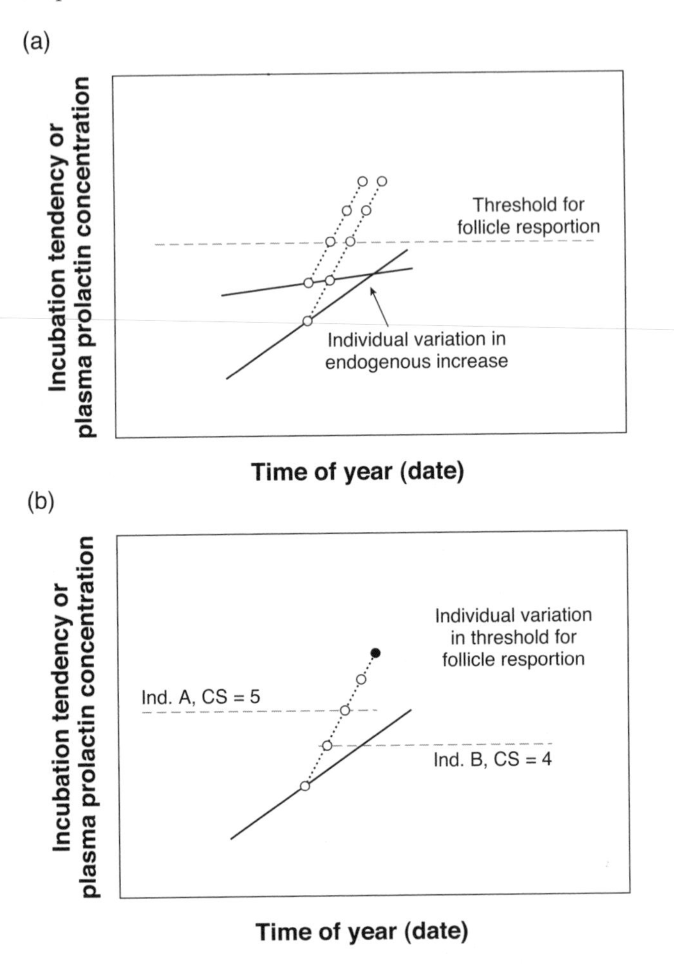

Fig. 5.16. Models for date-independent variation in clutch size involving (a) individual variation in the rate of endogenous increase in incubation intensity or prolactin and (b) individual variation in the threshold for disruption of follicle development. (Based on Meijer et al.'s original model (1990; see fig. 5.11.)

disrupted, between the third and fourth eggs, four follicles would have been recruited. Therefore, individual females could go on to lay zero, one, two, three, or four additional eggs from their remaining recruited follicles, giving a range of clutch sizes of three eggs (if all follicles become atretic) to seven eggs (if all remaining follicles complete development), which is equivalent to the normal range of clutch sizes in this species. Haywood (1993c) recognized that the number of growing follicles that become atretic at the time follicle growth is disrupted might vary and

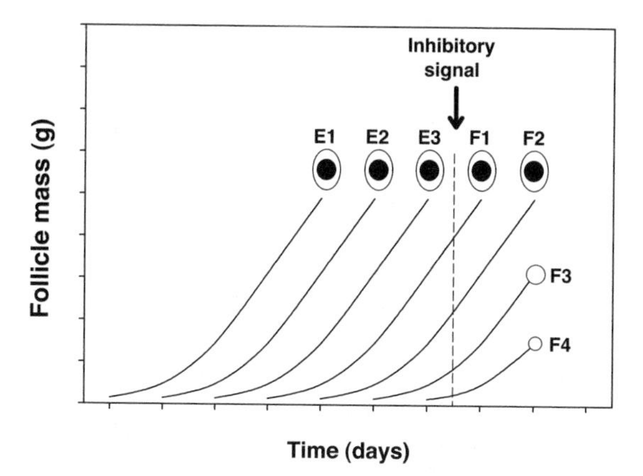

Fig. 5.17. Model for clutch-size determination involving disruption of follicle development and/or follicular atresia derived from studies of small passerines. (Based on Haywood 1993c.)

this number would determine the final clutch size. However, detecting the occurrence and frequency of follicular atresia at the end of egg-laying in order to test these ideas would be difficult, since clutch completion cannot usually be confirmed until 2 days after the last egg is laid, and this might be sufficient time for atretic follicles to be fully reabsorbed.

In conclusion, a two-tier set of mechanisms could allow for individual variation in clutch size—one involving variation in the timing of the inhibitory signal that disrupts follicle development, which sets an individual female's *maximum* possible clutch size, and one involving variation in the fate of the recruited follicles remaining after inhibition, which determines an individual female's final, *actual* clutch size. Presumably one or both of these mechanisms must represent the "target" for the integration of environmental cues that regulate date-independent variation in clutch size, such as predation risk, habitat, or individual quality, although how this is achieved remains completely unknown.

5.4.5. Proximate Constraints on Clutch Size: Food Availability and Nutrient Reserves

Food availability and nutritional constraint are still widely assumed to play a role in determining the variation in clutch size in a number of ways. First, in its simplest form, Lack's hypothesis (1947) requires that at the time of egg-laying, altricial species can predict food availability during the chick-rearing period and adjust their clutch size accordingly. This is the same problem that was discussed in relation to the timing of breeding and to which many of the same points apply (see chapter 3). Second, as we have seen, it has been suggested that the response (or lack of response) of clutch size to selection might be influenced by correlated traits that are mainly environmentally determined, such as the female's nutritional state or her ability to acquire nutrients (Cooke et al. 1995; Merilä et al. 2001; Price and Liou 1989). Third, the individual optimization hypothesis suggests that females make decisions to lay large or small clutches based on an assessment of their own circumstances and that their nutritional state, rate of food intake, or energy balance could be a key component of individual quality on which this decision is based. Finally, following the general assumption that egg production is nutritionally and/or energetically costly, it is widely assumed that (a) large clutches are more costly than small clutches and (b) the additional energetic or nutritional cost of laying large clutches is sufficient to constrain clutch size, most probably through either short- or long-term resource-allocation trade-offs (see chapter 7). Food availability is therefore widely considered to be a major selective force acting on clutch size (Drent and Daan 1980).

The general question of if and how nutritional and energetic costs constrain egg production was considered in chapters 3 and 4, and many of these arguments apply equally to clutch size as to the timing of breeding and egg size. However, there is an additional, and important, logical flaw in the nutrient limitation hypothesis in relation to clutch size that has been recognized by some authors (T. Arnold and Rohwer 1991; Nager 2006) but is still not widely appreciated. Assuming that eggs are produced sequentially at a constant rate, larger clutches do not require a *greater* rate of energy expenditure per unit time; they simply require a longer period of constant, potentially low, or at least sub-maximal, expenditure. Thus, the maximum clutch size is not likely to be dependent on daily *maximum rates* of physiological processes associated with egg formation, but rather on sustaining egg production over multiple days (T. Williams 1996). A female could potentially function at only 50% of maximum physiological or resource-dependent "capacity" producing small eggs, and if she maintained this effort for many days, she could lay a large clutch size (fig. 5.18). This situation contrasts with that of egg size, in which the maximum egg size *is* dependent on the amount of yolk deposited over a very short period (1–2 days) and where the concept of maximum physiological capacity might be relevant in terms of potential constraints (see section 4.5.1; fig. 5.18). With this line of reasoning, there is no a priori reason why increased food (energy) availability would affect clutch size, at least in income breeders that obtain the resources for egg formation from daily dietary intake (Drent and Daan 1980). In capital breeders, if females relied 100% on endogenous nutrient reserves for egg production, then eventually these would constrain clutch size at the point where the female exhausted these stored reserves. However, in most species there is

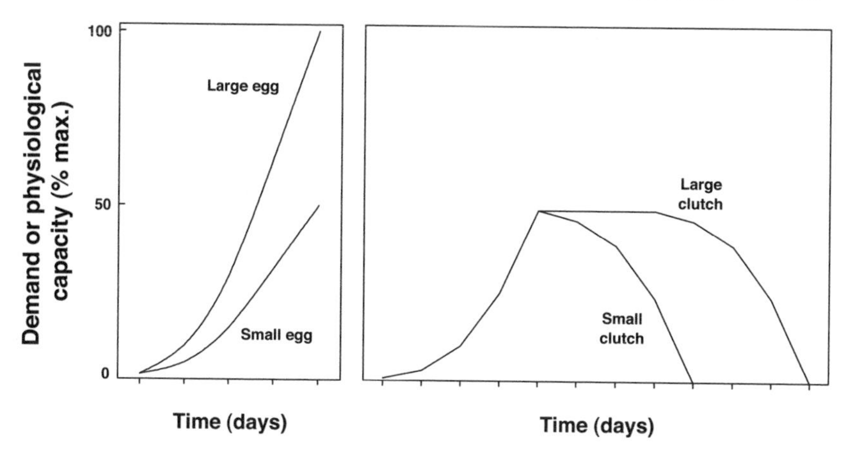

Fig. 5.18. Differential effects of reproductive investment in large versus small eggs, compared with large versus small clutches, on "demand" or physiological capacity during egg production.

little evidence that laying females deplete their nutrient reserves to zero at cessation of laying, in part because following clutch completion, the female has to complete the incubation of the eggs, sometimes with limited feeding opportunities (see below and chapter 6). In support of the prediction that physiological or energy demand is independent of clutch size, the metabolic rate of egg-laying females is not correlated with clutch size (I. Stevenson and Bryant 2000; Vézina and Williams 2005; Ward 1996), or there is only a weak and inconsistent relationship between these traits (Vézina et al. 2006; T. Williams et al. 2009).

As for the timing of breeding and egg size, the standard experimental approach to the question of food availability and clutch size has been to provide egg-laying females with supplemental food, with the prediction that the clutch size will increase if this supplementation releases birds from some nutritional or energetic constraint. These studies have been reviewed numerous times (Meijer and Drent 1999; Nager et al. 1997), and most recently Nager (2006) summarized the results from 51 studies on 32 species in which birds were provided with supplemental food during the egg-laying period (his table 1). Less than half of all the studies (20 of 51, or 39%) found an increase in the clutch size of food-supplemented females compared with those of control birds, and only a third of all the studies (29%) found an effect of supplemental food on clutch size that was independent of advance in laying date. Furthermore, even in studies that showed a significant effect of supplemental food, the increase in clutch size was relatively small (mean 0.7 eggs, range 0.2–1.8, n = 14 studies; Nager et al. 1997) compared with individual variation in clutch

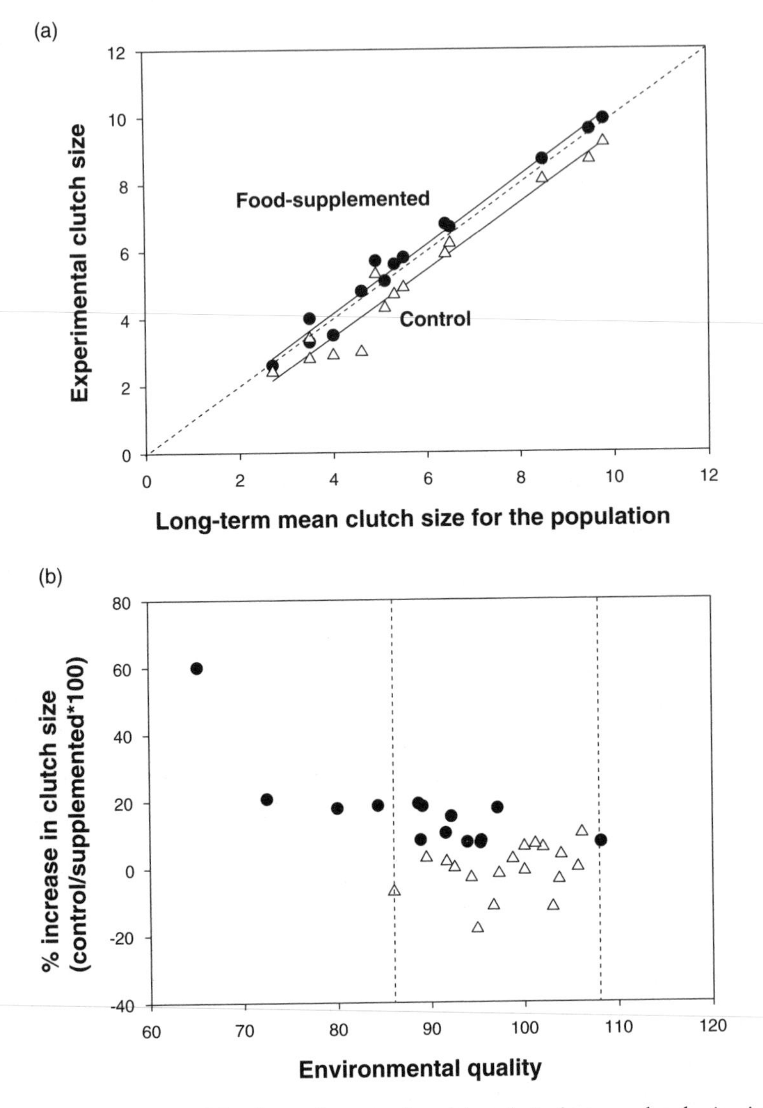

Fig. 5.19. Effects of food supplementation (closed circles) on clutch size in relation to (a) the long-term mean clutch size for each study population and (b) environmental quality, which is defined as ([mean clutch size of controls – long-term mean clutch size]*100). (Based on data in Nager et al. (1997).

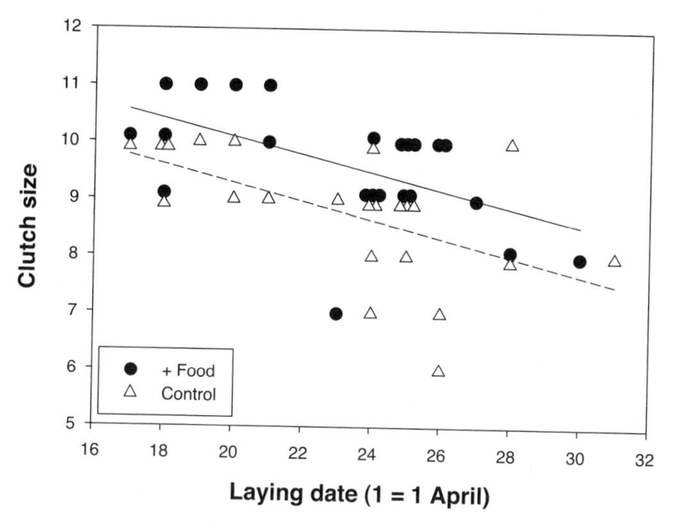

Fig. 5.20. Seasonal decline in clutch size in both food-supplemented (circles) and control (triangles) great tits. (Based on data in Nilsson 1991.)

sizes among food-supplemented females. The effect of supplemental food is independent of clutch size among species (fig. 5.19a) even though it is widely assumed that the nutritional constraint would be greater in species laying larger clutches (but see section 4.5.1). More importantly, food supplementation does not decrease individual variation (variance) in clutch size, and the clutch size of food-supplemented birds is increased only to the long-term mean clutch size for that population, not to larger clutch size (fig 5.19a). In other words, supplemental food does not allow individuals with "small clutch" phenotypes to express "large clutch" phenotypes. In great tits, food supplementation increased the clutch size independently of date by 0.55 eggs, but clutch size varied between 6 and 12 eggs within treatments (Nager et al. 1997). Similarly, in the study reporting the largest effect of food supplementation, the clutch size of supplemented birds (4.8 eggs) was not higher than the average clutch size for the general population (4.6 eggs), and the clutch size still varied between 4 and 7 eggs in food-supplemented birds (Newton and Marquiss 1981). Finally, food-supplemented females show the same seasonal decline in clutch size as control birds (T. Arnold 1994; Dijkstra et al. 1982; Nilsson 1991; Ramsay and Houston 1997); fig. 5.20). Similarly, clutch size showed the same seasonal decline (0.05 egg per day) in relation to natural annual fluctuations in food availability (phase of the vole cycle) in Eurasian kestrels (Korpimaki and Wiehn 1998).

Nager (2006) provides numerous explanations for why food-supplemented birds might not lay larger clutches. A common argument is that clutch size should only increase with food addition when natural food supply is low (Boutin 1990). Nager et al. (1997) tested this idea and concluded that changes in clutch size due to supplementary feeding *did* depend on environmental conditions during the experiment (where environmental quality was defined as the clutch size of control birds divided by the long-term average clutch size reported for that population; see their fig. 3). However, an alternative view of these data is that for 10 of 14 studies that reported a significant effect of food supplementation, "environmental quality" was within the range of that for the 18 studies reporting no effect (fig. 5.19b; and note the strong influence of the extreme left data point, which came from Newton and Marquiss's study, 1981). Another commonly proposed explanation is that total food availability represents the wrong "currency" and that specific nutrients, rather than energy, might limit egg formation (Nager 2006; Patten 2007). However, as described in chapter 4, four out of five studies that specifically manipulated protein during the laying period found no effect on clutch size (see table 2 in Nager 2006). Therefore, this idea does not provide unequivocal support for the idea of nutritional constraints on clutch size.

Many large-bodied, precocial species (e.g., waterfowl) produce large clutches of energy-rich eggs (Carey 1996; Lack 1967), and it has been suggested that the energetic costs of clutch formation are higher, and thus the importance of food availability or nutrient reserves will be greater, in these species. Lack (1967, 1968) specifically proposed that clutch size in waterfowl has evolved in relation to the amount of food the female can store for the young in the egg, and therefore in relation to the food available to the laying females, an idea termed the *egg production hypothesis* (Ankney et al. 1991). Consequently, there has been long and intense debate about the importance of nutrient reserves during reproduction in waterfowl but, even here, there is little unequivocal evidence that nutritional constraints determine individual variation in clutch size (Ankney et al. 1991; T. Arnold 1999; T. Arnold and Ankney 1997; T. Arnold and Rohwer 1991; Drobney 1991; Winkler and Walters 1983). T. Arnold and Rohwer (1991) argued, correctly in my view, that even though it is clear that endogenous nutrient reserves can be used during laying, this use does not demonstrate that they are *required* for laying or that they *limit* clutch size. I echo their comment that the "plausibility of the nutrient limitation hypothesis has led to its uncritical acceptance … an indication of how entranced we are with the idea that energetics dictate life history." They concluded that although most female waterfowl use endogenous nutrient reserves during laying, such reserves have only a small effect on *variation* in clutch size (except perhaps for arctic-breeding waterfowl). This conclusion was based on two key results: (a) individual variation in nutrient reserves greatly exceeds individual variation in clutch size, and

(b) variability in nutrient reserves remains high in post-laying females, whereas it should decrease if females with larger reserves use these additional nutrients to lay more eggs. T. Arnold and Rohwer (1991) concluded that only 2 of 24 (8%) studies showed a significant reduction in the variation of endogenous reserves among post-laying females (these two populations were large arctic-nesting waterfowl). Meijer and Drent (1999) also concluded that the idea that nutrient reserves have only a small effect on clutch size "seems to be more generally true" for a wider range of altricial and precocial species.

If a female with access to more resources during egg-laying is less "nutritionally constrained" or is more capable, in Perrins' (1970) words, of forming eggs "without risk to herself," we might expect higher breeding productivity in food-supplemented females laying larger clutches. In fact, of the 15 studies reviewed by Nager (2006) that reported a significant effect of supplemental food on clutch size (independent of laying date), only 7 considered the short-term fitness consequences of clutch-size manipulation, and none measured the long-term effects on future fecundity or survival in females laying experimentally enlarged clutches. Furthermore, 4 of these 7 studies provided supplemental food from laying through chick-rearing. Although in each of these studies, food-supplemented birds had higher breeding productivity, this observations is most probably the result of greater food availability during chick-rearing rather than increased clutch size per se (Arcese and Smith 1988; V. Gill and Hatch 2002; Högstedt 1981; Soler and Soler 1996). In the 3 studies where birds received supplemental food only prior to or during laying, supplementarily fed birds either did not have higher breeding success than controls (Nager et al. 1997; Newton and Marquiss 1981), or they had significantly lower breeding success (Sanz and Moreno 1995), despite laying larger clutches. This result suggests that food supplementation might cause birds to lay at a less favorable time or produce clutches that are no longer "individually optimised" (Nager et al. 1997; Sanz and Moreno 1995) or both. Consistent with this idea of a maladaptive response, in a recent study T. Harrison et al. (2010) showed that food supplementation between laying and hatching *decreased* clutch size and brood size in blue and great tits despite its advancing the laying date, leading these authors to suggest that "[food] supplementation can disrupt normal reproduction." In any case, these results are more consistent with food availability acting as an environmental cue for early reproductive decisions in females, rather than having a direct effect via nutritional constraint.

Far fewer experimental studies have investigated the role of food availability in determining the *number* of breeding attempts per year in facultative multiple-brooded species. In general there is a positive correlation between food availability and the number of nesting attempts under natural conditions (Verboven et al. 2001). However, in many species the frequency of multiple brooding decreases through the season, and therefore the timing of the initial breeding event (the first clutch) is an important determinant, with early-breeding birds more likely to be double brooded (Ogden and Stutchbury 1996; Orell and Ojanen 1983). In great tits the occurrence of multiple breedings is related to the timing of breeding relative to peak food availability: when the first clutch is laid early relative to peak caterpillar abundance, then a greater proportion of birds produce second broods, but caterpillar biomass per se has no effect on the occurrence of multiple breeding (Husby et al. 2009; Verboven et al. 2001). Chronic food supplementation, throughout breeding, had no effect on the probability of second broods in *Parus* spp. (von Brömssen and Jansson 1980), and it increased clutch number in only one of two years in the dunnock, *Prunella modularis* (Davies and Lundberg 1985), with the latter result confounded by the fact that fed birds laid up to 20 days earlier. Rodenhouse and Holmes (1992) experimentally reduced caterpillar availability for black-throated blue warblers, but the birds made significantly fewer nesting attempts in only one of three years, and food reduction did not significantly lower the annual production of young per pair. When food was supplemented to females only during the nestling stage of the first broods, the number of subsequent nesting attempts increased in two studies (Nagy and Holmes 2005a; Simons and Martin 1990) but not in a third study (Verboven et al. 2001). These first two studies might suggest that food limitation during the later stages of breeding can be a constraint on multiple breeding (Nagy and Holmes 2005a), but in general, most other studies are equally consistent with the idea that food provides a cue for breeding decisions, and the data are equivocal that there are nutritional constraints on multiple breeding.

5.5. Conclusion

Life-history-based models, which incorporate the ideas of individual optimization of life-time fitness, costs of reproduction, and variation in individual quality, clearly provide the most satisfactory and compelling explanations for the adaptive significance of individual variation in clutch size. These models provide general explanations for most systematic patterns of variation in clutch size for both single- and multiple-brooded species. Nevertheless, despite more than sixty years of research in this area, there remains considerable debate about the adaptive value of individual variation, or plasticity, in clutch size (e.g., compare Pettifor et al. 2001 and Tinbergen and Sanz 2004). As Ricklefs (2000) pointed out, "understanding variation in life-history attributes among populations is an inherently difficult problem," and perhaps the conclusion of Tinbergen and Both (1999) that individual optimization of clutch size would be *expected* to differ within and between populations and species, and that it might occur in some years but not in others is the best we can do by way of explanation. However, given the importance of variation in clutch size to avian life-histories, I think that this question still represents one of the major, unresolved challenges in the field. Given this, it would be troubling if the focus of the debate has indeed shifted to other, albeit related, questions (Charmantier et al. 2006). If Tinbergen and Both (1999) are correct, then it is important to determine those components of individual phenotype (including individual "quality," habitat quality, potential rearing ability) that dictate individual clutch-size decisions and under what circumstances individuals do or do not optimize clutch size. Thus, the focus going forward should be on individual-level analyses and experimentation. In this context, brood-size manipulation (giving birds "free" chicks) will be of limited utility for understanding individual variation in clutch size, unless this is coupled with simultaneous manipulation of the costs of incubation ("free eggs") and the costs of egg production ("full costs" sensu Visser and Lessells 2001). The fact that there is strong evidence for the costs of egg production (decreased future fecundity and survival), and that individual females appear to have a limited ability to produce additional high-quality eggs beyond their normal clutch size (section 5.2.4; cf. Lack 1947), means that experimental manipulations must be conducted within this complete life-history context. Further quantitative studies of selection using long-term breeding data sets will be valuable to clarify when and under what circumstances clutch size does respond to selection, although it will also be useful if these studies can carry through with the promise of identifying key *experimental* manipulations to further our understanding of clutch-size variation.

More than sixty years on, it seems somewhat of an oversight that we have not identified a *physiological mechanism* for clutch-size determination. Mea culpa! Again, this would appear to represent a worthy and important goal for future research that will benefit from the integration of physiology and evolutionary biology with a focus on individual variation. The Meijer and Haywood models (Haywood 1993c; Meijer et al. 1990) appear to provide a working hypothesis for the mechanism of clutch-size determination and makes a set of clear, testable predictions. Although there are data supporting some aspects of this model, these are largely correlational and rely heavily on studies of poultry. In contrast, there are currently few data supporting other key predictions of this model. The most important of these is that the hormone prolactin plays a central, integrating role in both development of incubation and termination of egg-laying via anti-gonadal effects on the ovary. I echo Sockman et al.'s conclusion (2006) that "a role for prolactin in regulating clutch size in any species is not firmly established, and evidence for some species indicates that clutch size might not be coupled to the timing of incubation onset" (see chapter 6 for more on this subject). More experimental studies of the sort described in Sockman et al. (2000) on a wide range of species, should be a research priority, but we should also investigate other hormonal signals that might be inhibitory to ovarian function (e.g., GnIH, inhibin, and intra-ovarian factors). Finally, any mechanism of clutch-size determination must accommodate individual variation in clutch size in relation to individual and environmental quality. If inhibition of vitellogenic follicle development is a key determinant of clutch size, then we need to identify how environmental cues (e.g., food availability, physiological "condition") are integrated into this process; something that should have broad parallels with environmental control of timing of breeding.

5.6. Future Research Questions

- Are current life-history-based models sufficient to explain the diversity of patterns of variation in clutch size with laying date in single- and multiple-brooding species? Can these models explain the very rapid changes in clutch size that occur over just a few days in highly synchronous breeders, and the considerable date-independent individual variation in clutch size seen in many single-brooded species?
- If the role of individual optimization of clutch size differs within and between populations and species and occurs in some years but not in others, when and why does this occur?
- How widespread, and how important, are effects of nest predation on *individual* variation in mean clutch size and seasonal variation in clutch size?
- How important is variation in incubation behavior and nest temperature for embryo viability and development and offspring phenotype and fitness? Can experimental studies mimicking more natural "partial incubation" confirm that egg or embryo viability and onset of incubation are significant factors affecting individual variation in clutch size?
- Are the "costs of egg production" evident in precocial species when females are made to incur the costs of producing additional eggs by egg removal during laying (rather than simply being given "free" eggs as with clutch-size manipulation)?

- Do the Meijer and Haywood models of clutch-size determination provide a robust, general physiological model for seasonal declines in clutch size?
- Is prolactin the "master hormone" regulating onset and development of incubation and clutch size? What role do ovarian steroids play in this process? Is prolactin an anti-gonadal or anti-estrogenic hormone? Does variation in plasma prolactin levels inhibit ovarian function and vitellogenesis in free-living birds, thus directly causing inhibition of egg-laying? Does experimental manipulation of prolactin modulate incubation behavior and clutch size in free-living birds?
- Can the Meijer and Haywood models be modified to provide an explanation for other patterns of variation in clutch size, such as the seasonal pattern in multi-brooded species, and date-independent variation in clutch size among individuals? Is there individual variation in thresholds for the inhibitory signal that causes disruption of follicle growth, the fate of vitellogenic follicles at the time of inhibition, or differences in the slope or rate of the endogenous day-length-dependent increase in plasma prolactin?

CHAPTER 6

Parental Care

INCUBATION AND CHICK-REARING

It is axiomatic that both incubation and chick-rearing are required for successful reproduction; these are important traits, or composite functions of traits, that contribute to fitness. If parents do not incubate, then eggs will not hatch, and if chicks are not brooded and fed, they will die. It is not surprising, therefore, that the lifetime number of fledglings produced is one of the main determinants of individual variation in fitness (see fig. 1.1). Many aspects of avian incubation have been extensively reviewed (Deeming 2002a), including the energetics of incubation (Thomson et al. 1998; Tinbergen and Williams 2002; J. Williams 1996), hormonal control (Buntin 1996; Goldsmith 1991; Vleck 2002), and the costs of reproduction associated with incubation effort (J. Reid, Monaghan, and Nager 2002). Similarly, chick-rearing has been reviewed previously, with a focus on the energetic costs of parental care, optimal foraging, chick provisioning, offspring growth and development, and costs of reproduction (Clutton-Brock 1991; Maurer 1996; Starck and Ricklefs 1998; T. Williams and Vézina 2001). Here I will not duplicate these reviews, but rather, I will focus on individual variation in parental care, the relationship between individual variation and maternal fitness, and the physiological mechanisms underlying this variation. For example, does individual variation in incubation behavior (e.g., the timing of onset, intensity, constancy) determine the number and quality of chicks at hatching? Or, does individual variation in chick-provisioning or other components of parental investment (e.g., brooding, guarding, feeding) explain the number and quality of chicks that fledge and ultimately are recruited to the breeding population? If there are fitness consequences of variation in parental care, then what are the physiological mechanisms that underpin individual variation in parental effort?

6.1. Comparative Aspects of Variation in Parental Care

Elsewhere in this book I have largely avoided any consideration of *interspecific* variation in life histories, but it is a necessary starting point

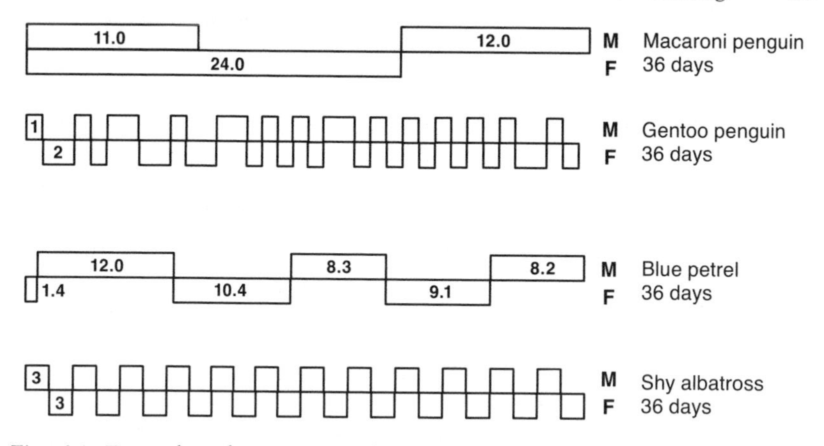

Fig. 6.1. Examples of species-specific variation in incubation routines in large, long-lived species with short (1–3 days) or long (10–20 day) alternating incubation bouts.

in this chapter since the broad patterns of variation in parental care, and especially incubation routines, in some senses set the bounds for potential physiological costs and should help identify the most likely mechanisms through which the costs of parental care might operate. In approximately half of all bird species, both parents contribute to incubation—coordinating and alternating incubation bouts, and thus allowing for a high level (>90%) of nest attendance (Deeming 2002a). With this type of *continuous incubation* (J. Williams 1996), eggs can be maintained more or less constantly at the temperatures required for normal embryo development (36°C–38°C). In those species with frequent, short incubation bouts, feeding can occur frequently, albeit intermittently, so that incubation is unlikely to be associated with prolonged fasting, extreme mass loss, or depletion of endogenous nutrient reserves. Other species, however, alternate feeding bouts with more prolonged incubation shifts, which can range from 1–3 days to 10–20 days, during which the incubating birds must fast (fig. 6.1). Longer incubation bouts make it more likely that the potential physiological or nutritional demands of incubation, the adaptations of fasting physiology, and the ability of individuals to extend incubation if their partner does not return for nest relief will be important determinants of successful incubation.

Only two avian orders (Passeriformes and Apodiformes) show lower levels (<75%) of nest attentiveness, and this is often associated with uniparental, female-only incubation, in which incubating females have to leave the nest intermittently to self-feed (Deeming 2002a). Pied flycatcher females spend about 75% of their approximately 17-hour "active day" incubating, with periods on the nest averaging 15.8 minutes, and 25% off the nest, with periods of feeding averaging 4.8 minutes (Lundberg and Alatalo 1992). This is an example of *intermittent incubation* (J. Williams 1996), and in this case eggs can potentially experience frequent temperature fluctuations during incubation as the female balances her energy needs with the thermal requirements of the eggs. Egg temperatures can drop to around 26°C–30°C (close to "physiological zero") when females are off the nest, although there is, in fact, little evidence that this natural variation, with periodic cooling, affects hatching success (J. Williams 1996 and references therein). Indeed some female-only, intermittent incubators are able to maintain constant, high mean egg temperatures; for example, for the goldcrest (*Regulus regulus*), overall mean egg temperature was 36.5°C, and females regulated egg temperature by varying the duration of off-nest periods, which were short (<10 min.) and inversely correlated with air temperature (Haftorn 1978). This appears to a common phenomenon since mean egg temperatures for 38 species of passerines with intermittent incubation (34.6°C ± 2.1°C) did not differ from those of other species showing continuous incubation (Tinbergen and Williams 2002; J. Williams 1996). Therefore, for uniparental, female-only incubators with frequent self-feeding, physiological adaptations for fasting and "nutritional constraint" are unlikely to be important in explaining the costs of incubation. In many passerines with intermittent incubation, the female's physiological demands of incubation can be further reduced through incubation feeding or courtship feeding by the male partner (Hatchwell et al. 1999; Klatt et al. 2008; Lifjeld and Slagsvold 1986). However, there is little evidence that variation in the male partner's provisioning rate affects the duration of a female's incubation bouts, the total incubation period, or the hatching success of her eggs (Hatchwell et al. 1999; Lifjeld and Slagsvold 1986).

In contrast to passerines, some uniparental, female-only incubators have very high levels of nest attendance and, consequently, very short and infrequent nest reliefs for feeding, with no male courtship feeding. In the female common eider, nest attentiveness is 99.6% (range 99.0%–99.9%) over the 26-day incubation period, with an average of 11 minutes of "recess" every 56 hours (Parker and Holm 1990). Similarly, female Canada geese (*Branta canadensis*) spend an average of 97.5% (95.3%–98.6%) of the 28-day incubation period on their nests, with a total recess (off-nest) time of only 22 minutes per day early in incubation (Aldrich and Raveling 1983). In these species, incubating females can lose up to 30%-40% of their initial body mass (fig. 6.2), which suggests that nutritional constraints and physiological adaptations for fasting *might* be more important in determining incubation behavior and hatching success (but see below).

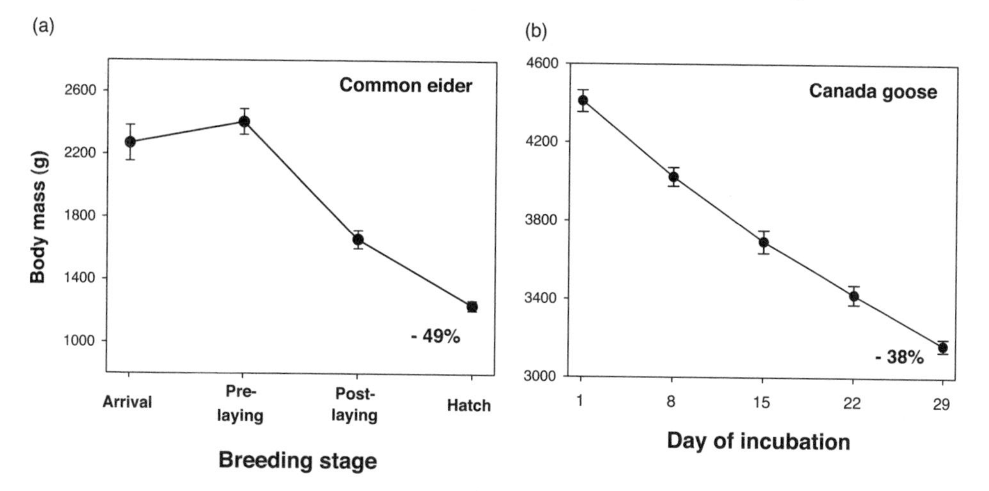

Fig. 6.2. Change in body mass during incubation in (a) the common eider and (b) Canada goose. (Based on data in Aldrich and Raveling 1983 and Parker and Holm 1990.)

In altricial species, where young are hatched at a relatively early stage of development, so that they are often blind, relatively unfeathered, and unable to thermoregulate themselves, there is a variable post-hatching period of obligate parental nest attendance and brooding, which is often a continuation of the species-specific incubation routine (with the same implications for physiological demands). Brooding can be uniparental and continuous, involving prolonged fasting (some seabirds, sea ducks); uniparental and intermittent, with short but relatively frequent off-nest periods for self-feeding (e.g., many passerines); and biparental, allowing continuous brooding with more or less prolonged periods for feeding and self-maintenance.

6.2. Individual Variation in Parental Care

6.2.1. Incubation

There are no simple, easy-to-obtain metrics for incubation effort, unlike for laying date, egg size, or egg number (described in chapters 3, 4, and 5). Detailed information on variability in incubation effort can be obtained only by very frequent nest visits (which is often impracticable) or by the use of temperature data loggers, passive integrated transponder tags, or video camera technology, which are logistically complex and relatively expensive and which, in turn, limit the sample size and the completeness of data for known individuals over long incubation periods. The development of increasingly sophisticated, low-cost data loggers should facilitate this work in the future. What data exist suggest that there is marked individual variation in all components of incubation behavior, which is superimposed on the general taxon-specific pattern of incubation and brooding behavior described above. However, this variability has not been as well characterized as it has for some other female reproductive traits. In particular, the extent to which this particular kind of variation reflects repeatable, "consistent individual differences" in behavior (Biro and Stamps 2010), behavioral flexibility within or between individuals, or environmentally determined variation (e.g., the effects of predators, Conway and Martin 2000) remains largely unknown.

As we saw in chapter 5, there is individual variation in the timing of the onset of incubation relative to clutch completion. Wang and Beissinger (2009) used a continuous record of nocturnal and diurnal incubation (obtained using HOBO data loggers) to reveal marked diversity in the onset of incubation, which they classified into 11 potential patterns across five species of North American passerines. Within species, two main patterns dominated for development of *diurnal* incubation, with either a monotonic or irregular increase in incubation intensity to full incubation (≥80% of nests). However, the development of full *nocturnal* incubation was more variable, with some individuals incubating fully from the first egg, and 50% or more of nests showing either a monotonic or irregular increase in incubation intensity; that is, nocturnal incubation was more likely to start earlier and with a higher duration than diurnal incubation. Potti (1998) reported similar individual variability in female pied flycatchers, with most birds starting incubation on the day of clutch completion (47.6%) or 1 day before clutch completion (33.9%) but with some females initiating incubation 2 or 3 days before clutch completion. The onset of incubation occurs earlier relative to clutch completion in larger clutches in some species (Lessells and Avery 1989; Wang and Beissinger 2009) but not others (Ardia and Clotfelter 2007).

The total duration of incubation, from laying to hatching, can vary markedly among individuals, even in those species with very high levels of nest attentiveness (>95%), such as eiders and geese, in which the incubation period can vary between 28 and 30 days and between 22 and 26 days, respectively. In species with highly coordinated, alternating incubation bouts, such as macaroni and gentoo penguins (fig. 6.1), incubation periods can vary from 28 to 40 and from 32 to 42 days, respectively (T. Williams 1995). In extreme cases, such as in the fork-tailed storm petrel (*Oceanodroma furcata*), incubation can vary from 37 to 68 days (mean 49.8 days), but this is associated with periods of prolonged "egg

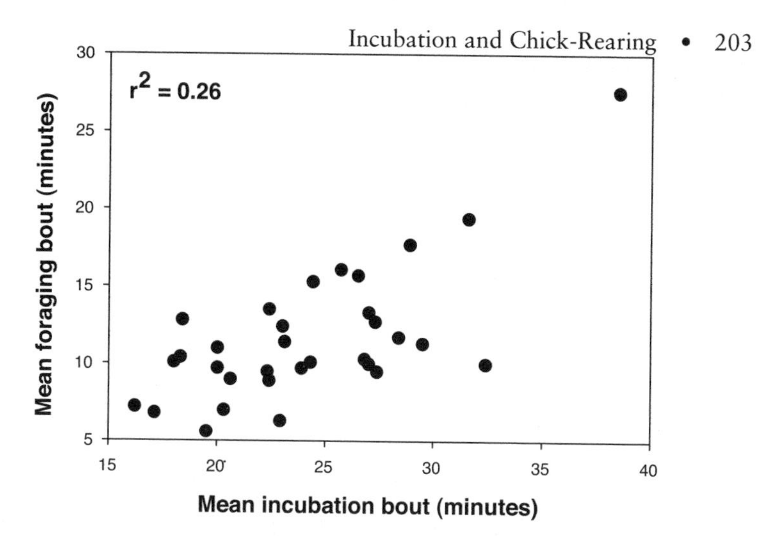

Fig. 6.3. Individual variation in the mean duration of foraging bouts and the mean duration of incubation bouts in female long-tailed tits. (Drawn from data in Hatchwell et al. 1999.)

neglect," when the parents do not incubate the eggs for several days at a time (Boersma et al. 1980). In intermittent-incubating passerines, incubation periods vary from 14 to 21 days among individual females in long-tailed tits (*Aegithalos caudatus*; Hatchwell et al. 1999), and from 12 to 18 days in pied flycatchers (Lundberg and Alatalo 1992). The total duration of incubation is inversely related to nest attentiveness in some species; for example, female Canada geese that are the most attentive have the shortest incubation periods (Aldrich and Raveling 1983). However, the variation in the duration of incubation among females (2 days) is more than the difference in the total amount of recess time taken among females (28 hours), so nest attentiveness does not fully explain individual variation in incubation period.

Finally, there is marked individual variation in the frequency and duration of incubation bouts and off-nest bouts (fig. 6.3), although, in general, these behavioral differences do not explain variation in the total duration of incubation (Hatchwell et al. 1999; Lundberg and Alatalo 1992; Robinson et al. 2008). This lack of relationship between the time spent incubating and the duration of incubation suggests that there might be individual variation in the *effectiveness* of incubation, in terms of embryo development times, even when the birds are sitting on the eggs and "fully" incubating. This variability in effectiveness might occur, for example, because of variation in treatment of individual eggs (egg turning, number of eggs incubated at any one time in large clutches) or because of characteristics of the female's brood patch (size, stage of development, blood flow, etc.), although few if any data are currently available to address

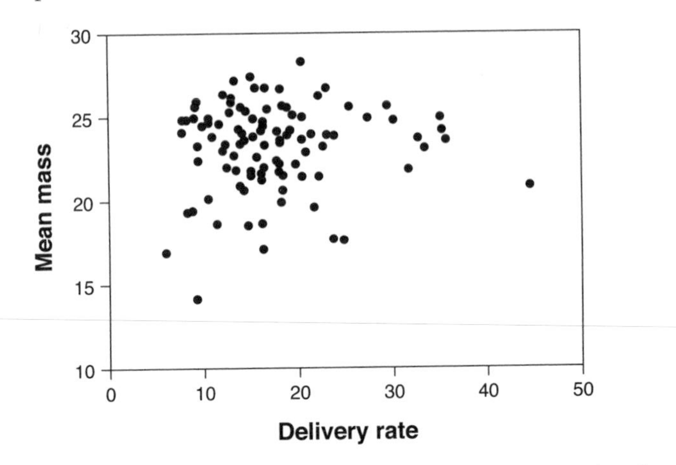

Fig. 6.4. Individual variation in parental care in house sparrows, showing the lack of relationship between mean offspring mass per brood and total number of food deliveries per hour. (Reproduced from Schwagmeyer and Mock 2008, with permission of Elsevier.)

these possibilities (but see Massaro et al. 2006). In addition, other factors intrinsic to the egg or embryo might determine variation in development times and incubation periods (Robinson et al. 2008).

6.2.2. Chick-Rearing

Anyone who has ever sat and watched chick rearing parents returning to a nest to feed chicks will need no convincing of the huge individual variation that characterizes provisioning rates. In small-bodied birds, such as passerines, 5- to 10-fold variation in hourly or daily provisioning rates have been reported (fig. 6.4), for example, from 300 to 1,500 visits per nest per 24 hours by both parents in blue tits (Nur 1984c), and from 5 to 45 deliveries per hour in house sparrows (Schwagmeyer and Mock 2008). In the European starling, the provisioning rate of females feeding 6–8-day-old nestlings varies from 1.3 to 16.0 visits per hour, and the overall provisioning rate of both parents, expressed relative to brood size, varies almost 8-fold (0.58–4.00 visits per chick per hour).

Brood size itself does not explain much of the individual variation in provisioning rate: for a given brood size in European starlings, the number of visits per chick per hour still varies 3–4-fold (fig. 6.5a). However, individual birds can adjust their provisioning effort in a number of other ways in addition to simply varying the number and frequency of nest visits, for example, by changing load size and the type and size of prey delivered to the nest per visit, or by changing the foraging distance

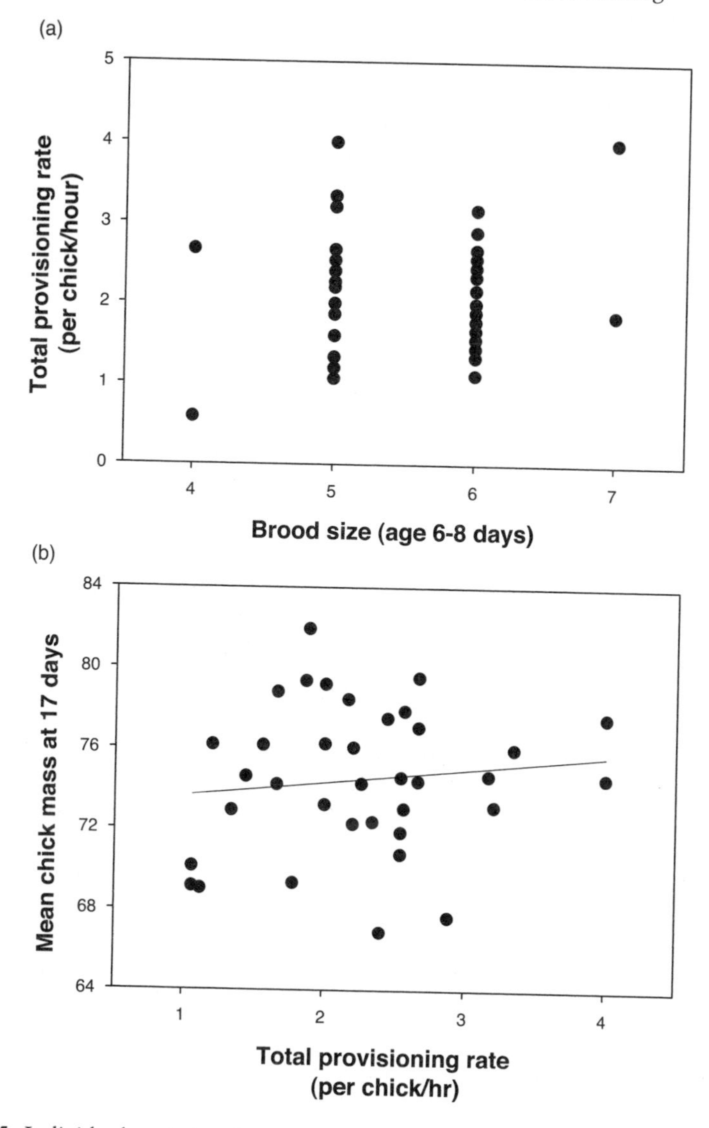

Fig. 6.5. Individual variation in provisioning rates in European starlings measured 6–8 days post-hatching in relation to (a) brood size at 6–8 days of age and (b) mean chick mass per brood at 17 days of age.

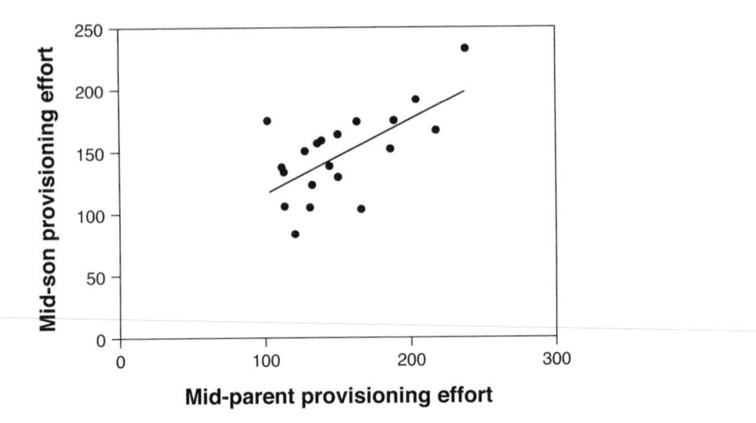

Fig. 6.6. Heritability of provisioning rate, based on mid-son to mid-parent regression, in long-tailed tits. (Reproduced from MacColl and Hatchwell 2003, with permission of John Wiley and Sons.

and travel time to and from the nest (Stodola et al. 2010; Wright et al. 1998). It is generally more difficult to collect large amounts of data on these other components of provisioning, but this potential complexity of foraging behavior does highlight the possibility that, although relatively easy to obtain, the number of visits to the nest might be a poor indicator of individual variation in parental provisioning effort or parental "quality." In species such as seabirds, which make long-distance foraging trips at sea, chick provisioning is much less frequent (once per day or less) and therefore is inherently less variable, and load (meal) size and prey quality are likely to be much more important indicators of provisioning effort. It is also clear that for individual chicks, nestling "state" and behavior can affect food allocation within the nest, either through sibling competition (Kacelnik et al. 1995; Kolliker and Richner 2004), sibling cooperation (Forbes 2007), or skewed within-brood food allocation by the parents (Mock et al. 2009), and how food is allocated within-broods might also have implications for the parent's inclusive fitness.

There is some limited evidence for repeatability of individual provisioning rates during parental care, suggesting that individual differences do not simply reflect labile, behavioral, and environmentally determined variation. With long-tailed tits for which individual variation in feeding visits was measured in at least two years, parental care was highly repeatable ($r = 0.70$ for males, $n = 16$; $r = 0.37$ for females, $n = 10$; MacColl and Hatchwell 2003). In house sparrows there was also significant repeatability of parental effort in both males ($r = 0.51$) and females ($r = 0.57$; Dor and Lotem 2010). The heritability of provisioning effort based on mid-offspring mid-parent regression was $h^2 = 0.59$ in long-tailed tits (excluding helpers, fig. 6.6) and $h^2 = 0.57$ in house sparrows, and the estimated heritability in long-tailed tits using an animal model including parents of both sexes was $h^2 = 0.42$.

6.3. Fitness Consequences of Individual Variation in Parental Care

The most direct effect of variation in parental care on fitness will be between "successful" and "unsuccessful" parents; that is, birds that care for chicks sufficiently well to produce one or more fledglings versus those that abandon the breeding attempt either during incubation or rearing and have zero breeding productivity. However, given that there is marked individual variation in numerous components of incubation and chick-rearing behavior among *successful* birds (as described above), a more interesting question is, Does individual variability in parental care contribute to individual variation in relative fitness among these "successful" birds? For example, it is widely assumed that variation in the timing of onset of incubation will affect fitness via the effects of hatching asynchrony, and this could, in theory, provide some degree of parental "control" over embryo development times and hatching spread, affecting breeding productivity (Sockman et al. 2006; Stoleson and Beissinger 1997). Similarly, parents that provision their chicks at a higher rate might be expected to produce more or larger chicks with higher recruitment probability. It would therefore be predicted that individual variation in parental care *should* be closely (positively?) related to hatching success, chick growth, survival or recruitment, and ultimately, to variation in a female's lifetime fitness. But is this the case?

In fact, predicting the direction and shape of the relationship between variation in parental care and fitness is not straightforward, and for some traits, counterintuitively, negative relationships are plausible (see fig. 1 in T. Williams and Vézina 2001). Most obviously, "better" parents or higher-quality individuals might work more efficiently, rather than simply working harder, to rear chicks (Daunt et al. 2006; Lescroël et al. 2010). An additional problem is the issue of "constraint" versus "restraint," which we have encountered with other reproductive traits, such as clutch size. For example, if parents are maximizing fitness over multiple breeding attempts, then low productivity in any one breeding attempt might not reflect a simple, direct constraint of parental ability; rather, it might reflect a "strategic" decision by the parents to trade-off current and future offspring, adaptively reducing investment in a given brood as a means of increasing residual reproductive value and *lifetime* fitness (Drent and Daan 1980; Mock et al. 2009). Because of the problems in obtaining detailed incubation data for a large number of individuals, there has so far been no comprehensive analysis of variation in individual lifetime fitness or quantitative selection studies (e.g., fecundity or viability selection) in relation to variation in parental behavior (of the sort described for primary reproductive traits in chapters 3, 4, and 5). Indeed, relatively few studies have considered the long-term effects of variation in parental care on female survival and future fecundity (but see de Heij et al. 2006, below).

6.3.1. Incubation Effort, Constancy, and Duration

The main consequence of a female's decision to start incubation before clutch completion is that it will cause asynchronous embryo development in earlier-laid eggs relative to later-laid eggs leading to hatching asynchrony. Hatching asynchrony can set up size hierarchies among chicks in a brood, with larger, earlier-hatched chicks having a competitive advantage over their later-hatched siblings (Clark and Wilson 1981; Magrath 1990; Mock and Parker 1997). A common assumption is that from the female's perspective, asynchronous hatching allows for rapid brood reduction in unpredictable environments, matching reproductive effort to available resources. Thus, hatching asynchrony creates "a caste of low-cost, disposable offspring" (Forbes et al. 1997) that allows females to secondarily adjust clutch or brood size by tracking resources that might be unpredictable at the time of clutch initiation (Forbes et al. 2001). The adaptive significance of hatching asynchrony and sibling developmental hierarchies has been extensively reviewed (Forbes 2007, 2009; Sockman et al. 2006) and will not be considered further here. However, in reality, few studies have unequivocally shown that individual variation in the onset of incubation has significant long-term effects on maternal *fitness* (but see Badyaev et al. 2003 and Forbes and Wiebe 2010).

It is difficult to manipulate the components of incubation behavior, such as frequency or duration of incubation bouts and off-nest feeding periods, directly (but see Davis et al. 1984), so the standard experimental approach has been to manipulate overall incubation effort by varying the number of eggs being incubated or by changing the thermal environment of the nest. Researchers then investigate the immediate effects of this manipulation on incubation behavior, as well as the effects on subsequent breeding productivity that might be causally related to the changes in parental behavior. This approach generally assumes that such experimental manipulations modify the energetic and/or nutritional costs of incubation (see section 6.4) that could provide insight into the physiological mechanisms underlying any observed changes in fitness.

6.3.2. *Short-Term Effects of Incubation Effort*

Clutch-size manipulations of incubation effort have typically involved females incubating 30% extra or fewer eggs, for example, manipulation of one egg in two-egg clutches (Heaney and Monaghan 1996), two eggs in normal six-egg clutches (Sanz 1997), and three eggs in modal nine-egg clutches (de Heij et al. 2006). In most cases these studies have shown that incubating enlarged or reduced clutches has no *short-term* effect on the duration of the incubation period, the amount of time females spend on the nest, hatching or fledging success, or the size and probability of second clutches. One exception is the study of J. Reid et al. (2000a), which used a within-clutch design, increasing the clutch size by two eggs (modal clutch size = 4) in European starlings to investigate the effects on "sibling" eggs reared in control or enlarged clutches. Eggs incubated in enlarged clutches had a lower hatching success, but Reid et al. concluded that hatching success was determined largely by incubation conditions, not by energetic or nutritional constraints on the incubating female per se. Eggs in enlarged clutches lost more mass during incubation, perhaps because clutch enlargement altered the nest microclimate, causing greater water loss during the incubation period (Reid et al. 2000a).

In studies that *have* reported significant effects of manipulated clutch size on nestling growth and survival, these results are typically confounded by rearing costs, since in most cases birds also reared the enlarged or reduced broods (Sanz 1997). To isolate the effects of variation in incubation behavior, it is necessary to restore the original clutch size at the end of incubation, allowing for the manipulation of incubation effort while keeping the subsequent rearing costs constant. Using this approach in common terns (*Sterna hirundo*), Heaney and Monaghan (1996) showed that second hatched chicks in broods of two, but where 3-eggs had been incubated, had reduced growth rate and fledging mass compared with control pairs which incubated only 2 eggs It was suggested that this was due to reduced adult provisioning capacity associated with greater incubation effort, though this was not demonstrated directly (Heaney and Monaghan 1996). In a study of house wrens, where the clutch-size manipulation was reversed at hatching, allowing the parents to provision their natural brood sizes, females that incubated enlarged clutches had longer incubation periods than control females, but the increased incubation effort did not adversely affect the allocation of subsequent parental effort to nestling provisioning (Dobbs et al. 2006).

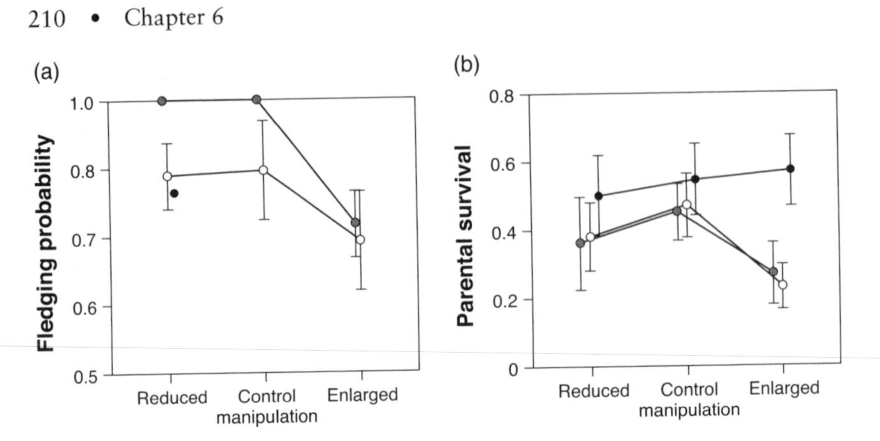

Fig. 6.7. Effect of clutch-size manipulation of incubation effort on (a) fledging probability of offspring from second (non-manipulated) clutches and (b) average local survival of parents in great tits. Different symbols represent three different years. (Reproduced from de Heij et al. 2006, with permission of the Royal Society.)

6.3.3. *Long-Term Effects of Incubation Effort*

Studies on clutch-size manipulation therefore provide little evidence to support the idea that incubation effort (the number of eggs being incubated) causes immediate or short-term variation in incubation behavior that then influences hatching success. However, several studies have reported significant *longer-term* effects of incubating more or fewer eggs. De Heij et al. (2006) reduced and enlarged clutches by three eggs in three different years (mean pre-treatment clutch size = 9.3 eggs) in great tits, restoring the original clutch size before hatching. The fledging probability was lower for second (non-manipulated) clutches of enlarged-clutch females (fig. 6.7a), and the local survival (return rate to the breeding area) of female parents was lower for enlarged-clutch females in the two years when females laid second clutches (40%–46%; only 3% of birds laid second clutches in 1 year; fig. 6.7b). In other words, effects of the increased incubation effort were not manifest in the first, manipulated, breeding attempt, but long-term costs were evident in terms of future fecundity and survival. In the same population, there was evidence for positive selection on clutch size during the nestling phase, as revealed by brood-size manipulations (Tinbergen and Sanz 2004). De Heij et al. (2006) therefore estimated the overall fitness of birds rearing manipulated clutches from incubation onward (fig.6.8) and concluded that the costs of incubation were strong enough to change positive selection on clutch size during the nestling phase into stabilizing selection in two of three years (in the third year selection remained directional and positive); that is, selection

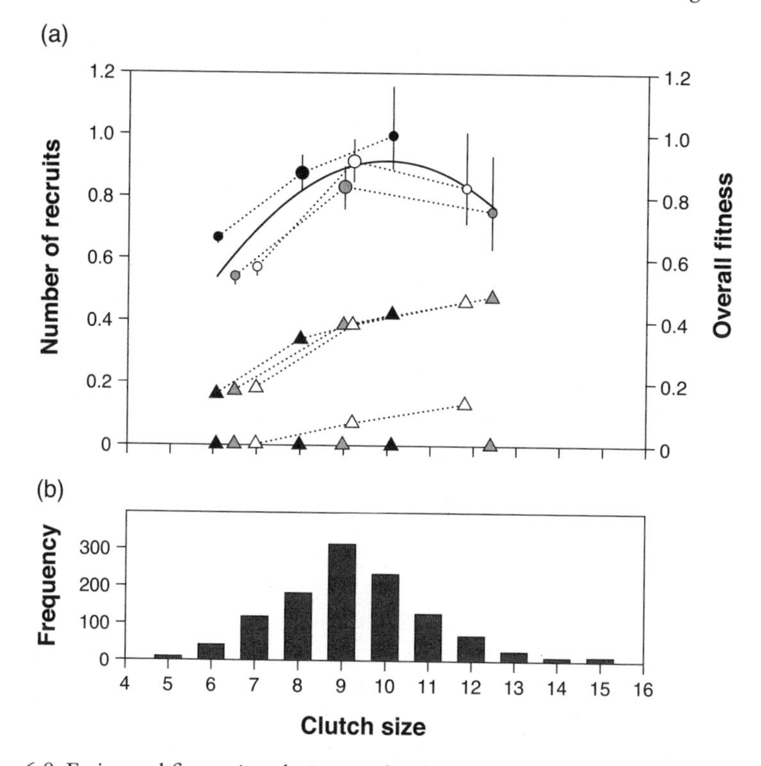

Fig. 6.8. Estimated fitness in relation to clutch-size manipulation of incubation effort in three years (indicated by different symbols) in great tits. Triangles indicate fitness estimated from brood-size manipulations (in Tinbergen and Sanz 2004) excluding costs of incubation, and circles indicate overall fitness from incubation onward, including costs of incubation. bottom panel gives the frequency distribution of observed clutch sizes for this population. (Reproduced from de Heij et al. 2006, with permission of the Royal Society.)

on clutch size can act indirectly through the effects of incubation effort. Visser and Lessells (2001) also demonstrated a cost of incubation in blue tits in terms of decreased offspring recruitment and female survival, although overall this appeared to be less significant than the fitness costs associated with increased egg production.

As we have seen already on numerous occasions in this book, "individual quality," albeit poorly defined, appears to be a key determinant of variation in timing of breeding, egg size, and clutch size. Hanssen et al. (2003) similarly suggested that the relative long-term costs of incubation revealed by clutch-size manipulation were dependent on individual quality in female common eiders. Putative low-quality females that laid four-egg clutches did not respond to clutch enlargement (addition of one egg) by modifying their incubation effort, as measured by relative loss of body mass, but they consequently had lower hatching success than controls. High-quality females that laid five-egg clutches were apparently able to incubate an extra egg successfully by facultatively increasing their incubation effort, as reflected in higher relative mass loss, thereby hatching and rearing more young. However, there was no effect of clutch-size manipulation on local survival (return rate) in either low- or high-quality females. Hanssen et al. (2003) interpreted this as indicating that incubation costs limit the clutch size for low-quality females (since these females were not able, or willing, to increase their incubation effort when clutch size was increased) so that they could *maintain* their survival (return rate) and future reproductive value.

In summary, although there is marked individual variation in most components of incubation behavior (timing of onset, bout length, duration and frequency of off-nest periods), there is little evidence that these determine variation in the duration of the incubation period, or hatching success. Furthermore, variation in incubation effort (at least as determined by number of eggs incubated) does not have consistent short-term effects on incubation behavior or hatching, suggesting that the immediate nutritional or energetic costs of incubation do not mediate or constrain immediate behavioral responses (see below). In contrast, robust data are accumulating for the long-term effects of variation in incubation effort in terms of future fecundity and survival. These data are most consistent with the idea that incubation effort, and perhaps individual variation in specific incubation behaviors, are determined within a long-term life-history framework (as with the timing of breeding and clutch size), where individual decisions are part of an overall strategy to maximize a female's *lifetime* fitness. In relation to "costs of incubation," the putative physiological mechanisms underlying the variation in incubation effort therefore need to be able to link the short-term components of incubation to these long-term effects (see chapter 7). Once again, the optimization of incubation effort within the females' overall life-history context appears to be dependent on "individual quality," although, as with other traits, it remains unknown which intrinsic or extrinsic components of phenotype determine variation in quality with regard to incubation.

6.3.4. Chick-Rearing, Provisioning Effort, and Nestling Productivity

Parental provisioning of nestlings is widely assumed to be positively correlated with fitness because higher provisioning rates should lead to higher fledging mass, which in turn might predict higher post-fledging survival and recruitment. Significant positive relationships between late nestling size and recruitment have been shown for at least 22 species in a diverse array of taxa, although a robust lack of a relationship between fledging size and recruitment has been demonstrated in at least 12 other species (Magrath 1991; Schwagmeyer and Mock 2008). Despite this predicted, intuitively logical, relationship, relatively few behavioral studies have demonstrated strong, positive correlations between provisioning rate and short-term components of fitness such as chick growth or fledging success, perhaps because parents can adjust provisioning effort in many different ways (as described above). Maigret and Murphy (1997) found a positive, but weak, correlation ($r^2 = 0.18$) between fledgling productivity and provisioning rate in eastern kingbirds (*Tyrannus tyrannus*) when controlling for brood size. However, in house sparrows, mean offspring mass was independent of the provisioning rate despite almost 10-fold variation in the rate of food delivery per nest (fig. 6.4). Rather, in this study, offspring mass was positively correlated to the delivery rate of the largest prey items, which were rare in diets and actually negatively correlated with total deliveries (Schwagmeyer and Mock 2008). In European starlings, chick mass just prior to fledging, at 17 days of age, is independent of variation in the provisioning rate during the most rapid growth phase (days 6–8, fig. 6.5b). In black-throated blue warblers, the nestling mass at day 6 was positively related to the male's provisioning rate but was independent of the female's provisioning rate, and nestling mass was also independent of the estimated biomass provided (Stodola et al. 2010). Finally, Nur (1984c) actually found in blue tits that parental visits were higher at nests with lower chick weights, independent of brood size, and suggested that breeding birds try to "compensate" for low nestling weights by increasing the feeding rates. So, in summary, surprisingly few studies provide support for the predicted strong, positive relationship between provisioning rate and chick growth or fledging success.

6.3.5. Short-Term Effects of Variation in Chick-Rearing Effort

Many studies have used brood-size manipulations to test the capacity of parent birds to increase their provisioning efforts when faced with feeding more chicks (Martins and Wright 1993; Nur 1984b, 1984c; Wright et al. 1998). Typically, the average feeding rate per nest either increases with increasing brood size or there is a non-linear relationship in which very small broods have lower provisioning rates, but a general finding is that the feeding rate per chick *decreases* with increasing brood size. In blue tits the feeding rate decreased from 140 to 110 feeds per nestling per 24 hours as the brood size increased from 3 to 6 chicks, but it then levelled off at approximately 75 visits per nestling per 24 hours as the brood size increased from 8 to 15 chicks (Nur 1984c). The decelerating increase in feeds per chick with increasing brood size can be associated with larger loads per visit and changes in the type of prey delivered to the nest (Wright et al. 1998). As a consequence of this flexibility in provisioning effort, chicks in large and small broods can actually have similar rates of food intake, although they might experience differences in the nutritional quality of their food (Wright et al. 1998). A further general finding from experiments using brood-size manipulation is that chicks in very large broods are lighter at fledging (though not always structurally smaller), either directly due to differences in feeding rate or because smaller meal sizes or lower-quality food is brought back to the nest in larger versus smaller broods (Maigret and Murphy 1997; Wright et al. 1998). This lower growth is consistent with the idea that limits to parental effort in large broods generate short-term costs in terms of the decreased quality of offspring. However, it is important to realize that in such experiments parents are attempting to rear more chicks than they would have reared based on their original clutch size; that is, they might be being forced to rear individually "sub-optimal" brood sizes.

Alternative approaches to manipulation of parental effort during chick-rearing have included clipping flight or tail feathers, to decrease flight performance, and adding weights to provisioning birds, but these have also provided inconsistent results in terms of fitness costs (see Wright and Cuthill 1989 and references therein). Although provisioning rates in manipulated birds are often reduced, this has relatively little, or no, effect on hatching or fledging success, chick growth, or fledging mass (Lifjeld and Slagsvold 1988; Love and Williams 2008a; Verbeek and Morgan 1980; Winkler and Allen 1995), at least in the first (manipulated) brood. In one of the few experimental studies to consider offspring sex, Rowland et al. (2007) found that feather clipping of provisioning mothers significantly reduced chick growth in daughters (~ 6%) but not in sons. One problem with all of these experimental approaches is that unless both parents are manipulated, the non-manipulated parent can increase its provisioning rate, which can compensate for the decreased contribution of its partner (Wright and Cuthill 1989), and manipulated birds might also make other provisioning adjustments, for example, greater load size per nest visit (Lifjeld and Slagsvold 1988). Future experiments using these experimental techniques therefore need to account for this more complex, but more realistic, individual flexibility in provisioning behavior.

6.3.6. Long-Term Effects of Variation in Chick Rearing Effort

As with incubation, there is marked individual variation in chick-rearing effort as measured by provisioning rates, but there is surprisingly little consistent evidence that it is positively correlated with short-term components of fitness such as chick growth, fledging mass, or fledging success. Individual birds appear to have a considerable capacity to increase provisioning rates, or to make other adjustments, when given larger broods to rear; that is, they can adopt a "flexible investment" strategy (Erikstad et al. 2009). However, do individual parents pay a long-term cost for the decision to provision chicks at a higher rate? Certainly it is a widely held assumption that any increase in the parental provisioning effort should have energetic costs (Wright et al. 1998), as well as potentially raising the risk of predation, which should have consequences for the survival and future reproduction of the parents. Some brood-manipulation studies have shown that the body mass of female parents near the end of rearing is negatively correlated with feeding frequency, and Nur (1984c) argued that this relationship was a direct effect of females losing more weight because they fed larger broods more often (but see Maigret and Murphy 1997). But in a subsequent paper, Nur (1984a) argued that the hypothesis that a loss in mass primarily reflects the *physiological* cost, or "stress," of feeding nestlings could be rejected based on the time course of mass loss and on sex-specific differences in mass loss. Moreover, any effect of provisioning on parental mass loss seems relatively short lived; for example, in the swift (*Apus apus*), although both parents lost body mass during the breeding season, and mass loss was higher for birds rearing larger broods, the adults regained all lost body mass by the end of the nestling period as chick-feeding rates slowed before fledging (Martins and Wright 1993). Although many studies have reported a significant correlation between brood size and female mass, or mass change, female body mass is poorly predicted by brood size alone (Nur 1984a). Therefore, the evidence is less than compelling that high provisioning rates, which are associated with large brood sizes, might cause long-term costs through decreased body condition or mass of the female parent (see below for more on this point). In addition, there is always the possibility that loss of body mass might not be a "cost" but, rather, could reflect adaptive mass change (see section 6.4.4).

A few studies have provided evidence for the long-term effects of increased parental effort on adult survival, even if these are unlikely to be mediated by effects on body mass, while other studies have failed to find such effects. Daan, Deerenberg, and Dijkstra's (1996) work on Eurasian kestrels stands out as one of the few robust studies showing the effects of the work load associated with parental care on true survival (cf. "local" survival, or return rates). In an analysis of the known time of death of 63 kestrels that had raised broods of experimentally manipulated size, Daan and coworkers showed that 60% of the parents raising two extra nestlings were reported dead before the end of the first winter, compared with only 29% of those raising either normal-size or reduced broods

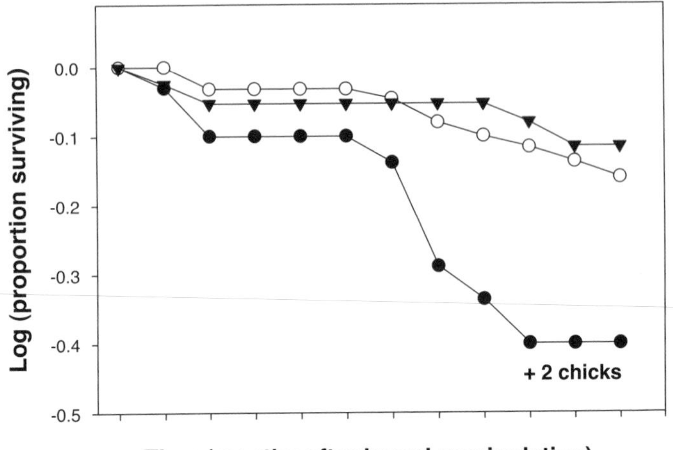

Fig.6.9. Natural mortality in Eurasian kestrels as a function of time of year for parents after having raised experimentally manipulated or control broods. Closed circles indicate enlarged brood (+2 chicks), and open circles and triangles indicate reduced (+2 chicks) and control broods, respectively. (Reproduced from Daan, Deerenberg, et al. 1996, with permission of John Wiley and Sons.)

(fig. 6.9). Since the higher mortality occurred in the winter following the brood enlargement, and previous work had confirmed that kestrels adjusted their DEE to the modified brood size (see below), Daan et al. concluded that increased parental effort directly entailed a higher risk of mortality. Glaucous-winged gulls raising experimentally enlarged broods (4–7 chicks) also had significantly lower survival rates than adults raising normal broods (1–3 chicks), although there was no effect of brood size on future fecundity and no differences in survival or fecundity over the range of natural brood sizes (W. Reid 1987). Reid similarly concluded that these data provided evidence for an "incremental cost of reproduction" involving *physiological* costs related directly to increased adult foraging time and decreased adult body mass in birds rearing larger broods. Nur (1984a) showed that adult female local survival declined with increasing brood size (3–15 chicks) in blue tits, but not in a linear fashion. The recapture rate was highest for females rearing broods of 3 (41%), intermediate for those rearing 6 and 9 young (24%–25%), and lowest for those rearing 12 and 15 young (19% for both groups). However, in contrast with W. Reid's suggestion (1987), survival in this case was independent of female mass at day 10 of chick-rearing (except for the very lightest birds) and independent of mass loss during rearing. However, in contrast to the three studies cited above, numerous studies have *failed* to find any effect of experimentally manipulated brood size on local survival (DeSteven 1980; Gustafsson and Sutherland 1988; Maigret and Murphy 1997; Roskaft 1985).

Does investment in parental care during chick-rearing affect future fecundity or future reproductive value of adults? Slagsvold (1984) showed in great tits that reproductive effort in re-nesting was inversely related to the parental effort expended in rearing the first brood. First broods were experimentally manipulated to contain either fewer (3–4) or more (10–11) young, and at 15 days of age, both nests and young were removed to mimic nest predation. Parents that reared fewer nestlings in their first brood re-nested sooner (albeit only slightly: 5.6 vs. 6.3 days) and also reared a greater number of fledglings in the re-nesting attempt (6.4 vs. 3.8 chicks) compared with those which had reared an increased number of young. In Roskaft's study (1985), rooks rearing enlarged broods in the preceding year laid a week later and reared fewer young in the subsequent year, mainly due to lower hatching success (92% in year 1 vs. 23% in year 2), suggesting that the effects of parental care on future fecundity can be prolonged. In contrast, in collared flycatchers, manipulated brood size had no effect on adult survival (Gustafsson and Sutherland 1988). However, enlarged broods produced more fledged young that were lighter and smaller at fledging and were less likely to recruit, and female offspring that did recruit from enlarged broods had lower fecundity as measured by recruitment of their offspring. In the cooperatively breeding long-tailed tit, the total provisioning effort per nest predicted the survival of male fledglings (MacColl and Hatchwell 2003), although parental care included provisioning by multiple caregivers; in this species up to six adults (parents and helpers) may invest care in a single brood. Female (but not male) long-tailed tits with higher mean provisioning effort had a greater chance of at least one fledgling recruiting into the local breeding population, and there was a positive relationship between individual lifetime fitness and variation in provisioning effort in female, but not male, parents (fig. 6.10; MacColl and Hatchwell 2004; this appears to be the only study to have related actual measures of provisioning effort to lifetime fitness).

6.4. Physiological Mechanisms Underlying Individual Variation in Parental Care

Studies of variation in parental care, in particular those that posit limits to parental care or costs of high levels of parental care (as described above), are replete with comments that the "work load" associated with rearing chicks (and to a lesser extent with incubating eggs) entails a high

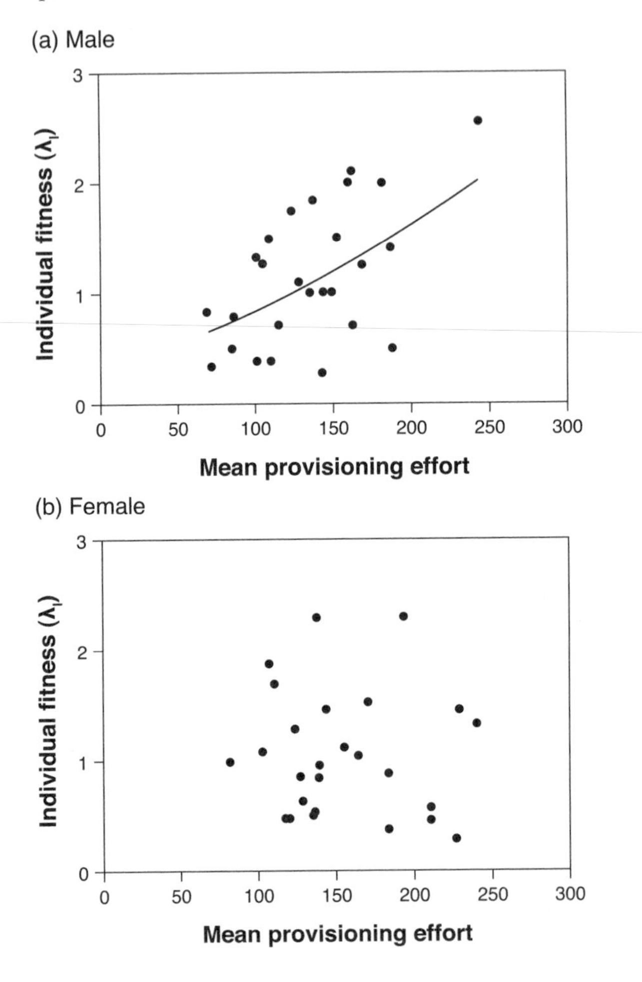

Fig. 6.10. Individual variation in lifetime fitness in relation to mean provisioning effort in (a) male and (b) female long-tailed tits. (Reproduced from MacColl and Hatchwell 2004, with permission of John Wiley and Sons .)

physiological demand, and that the latter requirement *directly* generates the physiological costs of parental effort. The idea that parents with large broods feed their chicks less often than do parents with fewer young, because they are somehow physiologically "*fatigued*" or *unable* to feed their chicks more, has a long history (Gibb 1950), although alternative explanations were also advanced relatively early on (Royama 1966). W. Reid (1987) concluded that the decreased survival of breeding adult glaucous-winged gulls rearing enlarged broods resulted from the *physiological* costs of reproduction. Nur (1984c) also concluded that higher feeding frequency was costly "in the physiological sense" to the female. Daan, Deerenberg, and Dijkstra (1996) speculated that the increased mortality of kestrels rearing enlarged broods might be related to a "temporary suppression of vitality" leading to a greater "risk of death by exhaustion," although they acknowledged that "the mechanisms mediating these effects of the rate of parental work on mortality remain to be unraveled." Thus, the *concept* of physiological costs has been central to studies of variation in parental care, but our understanding of the physiological mechanisms mediating these costs remains rudimentary to say the least. Indeed, the physiological basis of variation in avian foraging in general is very poorly known (Maurer 1996) despite the fact that this has, without doubt, been the most intensively studied breeding stage over the last 60–70 years.

If the work load associated with parental care causes birds to become "*exhausted*" or "*fatigued*," some individuals must have an inadequate capacity to sustain a high rate of physiological work. It is widely assumed that energy or nutrients are the limiting resources that constrain the physiological capacity for work in this sense. A corollary of this argument is that individuals with high levels of parental care should have a higher demand for resources and might be more likely to incur a short-term energy or nutrient deficit or have to reallocate resources from other functions (see section 6.4.4). Here I will focus on three questions. First, are incubation and chick-rearing energetically and nutritionally costly? Second, and more important, does individual variation in parental care (number of eggs incubated, timing, duration and constancy of incubation, provisioning rate) determine nutritional or energetic demand such that some individuals are better able to meet these demands or some individuals incur costs due to increased demands of higher levels of parental care? Third, are there other resource-independent regulatory mechanisms (sensu T. Williams 2005) that might help explain individual variation in parental care?

Physiological mechanisms underlying the long-term costs of reproduction are considered more generally in chapter 7, so here I will mainly focus on how parental care might generate physiological costs that lead to short-term fitness costs such as breeding failure or differential breeding success within a single breeding attempt. As I have pointed out previously (T. Williams 2005), transient, short-term energy or nutrient deficits could conceivably cause short-term fitness costs due to resource-allocation decisions *within* breeding stages or among different stages of a *single* breeding attempt (perhaps explaining the "temporary suppression of vitality" of Daan, Deerenberg, and Dijkstra 1996). However, it seems far less likely that transient shortfalls, or energy or nutrient debts, would generate negative effects for the relatively long time frames over which costs of reproduction, reduced survival, and future fecundity are mediated (months or years; see chapter 7).

6.4.1. Energetic Costs of Incubation

Early studies suggested that the additive energetic costs of incubation might be small relative to the basal, or resting, metabolic rate (BMR; Walsberg 1978, 1983), although this view was based primarily on indirect or theoretical estimates of energy expenditure. As a consequence, it has been commonly thought that energy demand during incubation is low relative to that in other stages of breeding. In contrast, more recent reviews of studies in which the energy expenditure of incubating birds has been measured directly have concluded that parents spend as much energy during incubation as when feeding nestlings or producing eggs (Tinbergen and Williams 2002; J. Williams 1996; T. Williams and Vézina 2001). Metabolic rates during incubation are generally two to three times the BMR, with the metabolic cost of incubation strongly dependent on Ta (Tinbergen and Williams 2002). One reason for this disparity of views might be that under thermoneutral conditions there is, on average, only a small elevation of the incubation metabolic rate above the BMR: none of 19 studies found incubation metabolic rates in excess of the BMR (Thomson et al. 1998). However, as Thomson et al. (1998) and others have pointed out, most birds incubate at temperatures well below their thermoneutral zone. For example, many small passerines have body temperatures of 42°C–44°C, but at temperate latitudes, incubation occurs when Ta are 5°C–15°C or lower (even though nests provide favorable microclimates for incubation, nest temperatures are still considerably below body temperatures even in cavity-nesting birds; fig. 6.11).

This question of the metabolic costs of incubation is therefore far from resolved, and some studies continue to support the idea of relatively low energetic costs of incubation in some species (e.g., see Spaans et al. 1999). In contrast, arctic-breeding sandpipers have metabolic rates equivalent to four to six times the BMR during the incubation phase (Cresswell et al. 2004). The emerging consensus appears to be that incubation behavior entails significant energetic costs even though these can vary considerably depending on nest microclimate, parental behavior, and species (J. Reid, Monaghan, and Nager 2002; Tinbergen and Williams 2002). Energetic costs would be predicted to be higher, and energetic constraints more likely, with uniparental incubators, for which intermittent incubation is characterized by a variable temperature regime and for which parental investment might be higher due to the high costs of egg rewarming to maintain egg temperature (J. Reid, Ruxton, et al. 2002). In biparental species, coordination of incubation bouts, and thus

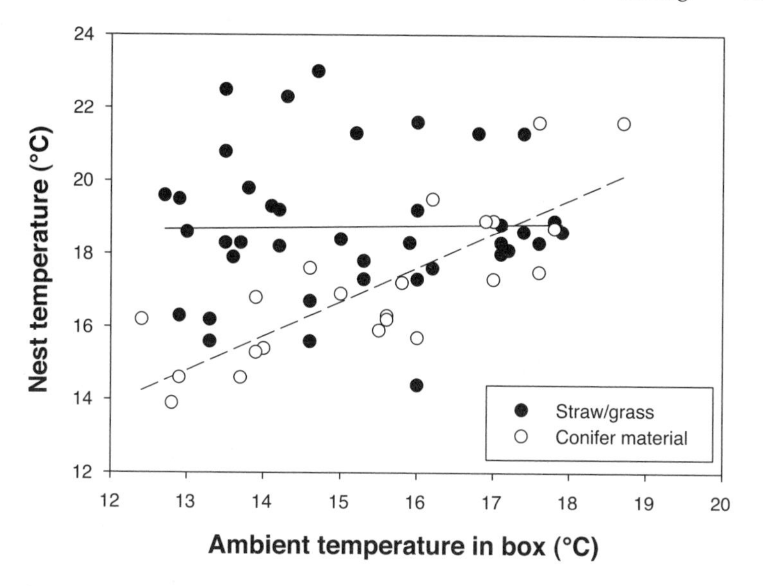

Fig. 6.11. Nest temperature in relation to variation in ambient box temperature for "decomposing" (solid line, straw/grass material) and "non-decomposing" (dashed line, conifer material) nests of European starlings. (Based on unpublished data from O. Love and T. Williams.)

the ability to extend incubation bouts if the partner does not return, are probably more important and could also be energy limited.

Overall, therefore, in terms of DEE, birds seem to be working as hard during incubation as they are during egg production or chick-rearing (Tinbergen and Williams 2002; T. Williams and Vézina 2001). However, does *individual* variation in incubation behavior or incubation effort explain, or directly determine, energetic demand? Are some individuals better able to meet the energetic demands of incubation, or do some individuals incur greater costs due to increased energetic demands of greater investment in incubation? Do the higher energy demands of incubating large clutches constrain clutch size (chapter 5)? Do the levels of energy reserves, or energy expenditure, of the sitting parent determine the lengths of incubation bouts (the parental energy threshold model; see Chaurand and Weimerskirch 1994), or other specific components of incubation effort, and consequently breeding success, at the *individual* level? Most studies have addressed these questions by manipulating the putative energetic demands of incubation, either by varying the number of eggs birds have to incubate or by changing the thermal environment of incubation (e.g., increasing or decreasing nest temperature), and then assessing the changes in incubation behavior and breeding productivity (see above).

However, relatively few studies have combined these experimental manipulations with direct measurements of energy expenditure to confirm that behavioral changes in incubation effort involve associated changes in energy demand.

Experimental manipulation of clutch size has few consistent effects on incubation behavior or hatching success, even when birds incubate considerably more eggs than their normal clutch size (section 6.3.2). It is perhaps not surprising, therefore, that relatively few studies have reported significant increases in energy expenditure for birds incubating larger clutch sizes. Some studies have suggested that larger clutches are energetically more demanding and more costly to incubate, at least at low Ta (Biebach 1981; Haftorn and Reinertsen 1985; Moreno and Sanz 1994), but other studies have found little or no support for a positive effect of clutch size on DEE when eggs are maintained at typical incubation temperatures (de Heij et al. 2008; J. Reid et al. 2000a, 2002b). De Heij et al. (2007) found only a moderate (6%–10%) increase in metabolic rate during *nocturnal* incubation under field conditions in free-living female great tits incubating three additional eggs, and females did not have lower metabolic rates when incubating *reduced* clutch sizes. Furthermore, in a follow-up study, de Heij et al. (2008) showed that incubating enlarged clutches in female great tits had no effect on DEE or any aspect of nest attendance behavior, although the DEE was related to Ta (fig. 6.12). They suggested that the lack of effect on mean DEE was the result of individual variation in the way females responded to enlarged clutches: some spent more time foraging and gained mass, whereas others spent more time on the nest and lost mass. In other words, individual females adopted different self-maintenance or offspring investment "strategies." This idea is very similar to that of individually variable energy management during egg production, which was proposed by Vézina et al. (2006) and T. Williams et al. (2009), and highlights the importance of considering *individual variation* in parental care strategies.

Several studies have attempted to directly manipulate energy expenditure during incubation by modifying the thermal environment of the nest. J. Reid et al. (2000b) placed heating pads under the nests of European starlings and found that the pads retarded the cooling of the eggs when the females were absent from the nest, thus potentially decreasing the females' energy investments in rewarming the eggs and, therefore, reducing total incubation costs. There was no effect of the treatment on the duration of the incubation period, mean incubation temperature, hatching success or chick mass soon after hatching (day 3), but the proportion of first-clutch eggs from which young fledged was significantly higher in the experimental nests than in the control nests (74.9% vs. 51.0%). Ardia et al. (2009) also used experimental warming of nest boxes during

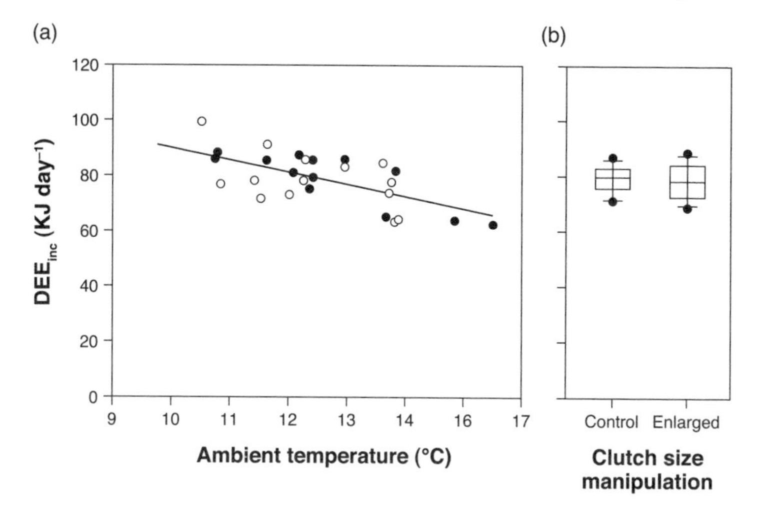

Fig. 6.12. Variation in daily energy expenditure during incubation (DEEinc in female great tits (a) in relation to mean ambient temperature measured over a 24-h period for control (closed symbols) and experimentally enlarged (open symbols) clutches, and (b) residual DEEinc for females incubating either enlarged or control clutches. Reproduced from de Heijj et al. 2008, with permission of John Wiley and Sons.)

incubation in female tree swallows (*Tachycineta bicolor*) and showed that females in heated nest boxes spent more time incubating and maintained higher on-bout and off-bout egg temperatures (i.e., they invested more in embryo development). Although both Reid et al. and Ardia et al. concluded that incubation was energetically constrained, that their manipulations *reduced* the energy demands of incubation, and that resources saved during reduced-demand incubation were reallocated to other aspects of reproduction, neither study directly measured energy expenditure. When experimental manipulation of nest temperatures has been coupled with direct measurement of energy expenditure, there has been little evidence for a strong effect of thermal incubation environment on energy demand. Bryan and Bryant (1999) heated the nest boxes of great tits (a female-only incubator) at night to an average of 6.2°C above Ta, compared with 2.8°C in the control boxes. Although females in heated boxes increased their incubation constancy, from 86.3% in the controls to 90.1% in birds in heated nests (which is equivalent to an additional 55 minutes of incubation per day on average), there was only a weak effect on DEE in terms of a statistically significant treatment×ambient temperature interaction ($P = 0.047$); that is, DEE increased relatively more at lower temperatures in control nest boxes relative to heated boxes. Cresswell (2003, 2004) manipulated nest temperatures in semipalmated and pectoral sandpipers

(*Calidris melanotos*) and showed that birds in warmed nests increased the mean length of incubation bouts length by about 10% (mean 11.1 hours) and increased mean nest attendance by 3.6% (mean 81.5 ± 0.6%). However, experimental heating had no effect on DEE when controlling for differences in nest attendance. Cresswell (2004) argued that the predicted negative effects of increased nest temperature on DEE were not observed because DEE was primarily determined by the energy expended off the nest, which would support the idea that individual variation in incubation behavior, or clutch size, per se has little effect on the individual variation in DEE.

6.4.2. Energetic Demands of Chick-Rearing

Energy expenditure has been measured directly, using the doubly labelled water method (Speakman 1997), in relation to variation in brood size or provisioning effort in more studies than for any other fitness-related component of reproduction. These data were reviewed by T. Williams and Vézina (2001), and no additional studies have been published more recently to change the main conclusions of their review: that evidence for a systematic relationship between individual variation in DEE and current reproduction (parental care) or future reproduction and survival is "less than compelling." Most work has failed to find any relationship between DEE and brood size (14 of 20 studies [70%] not significant) or between DEE and provisioning rate (10 of 15 studies [67%] not significant). Where significant positive relationships between parental care and DEE have been reported, they are often inconsistent within studies or within species. For example, in the pied flycatcher, DEE was positively correlated with feeding rate and brood size in males, but not females, in one study (Moreno et al. 1995), whereas DEE was independent of either variable in both sexes in a second study (Moreno et al. 1997). Similarly, in the great tit, DEE has been reported to be independent of brood size in both females and males (Tinbergen and Dietz 1994; Verhulst and Tinbergen 1997) and positively related to brood size in females only in other studies (Sanz and Tinbergen 1999; Sanz et al. 1998). D. Thomas et al. (2001) reported that DEE was lower in individual blue tits that were feeding young closer to the peak of the caterpillar food supply and suggested that those individuals that timed breeding to better match prey availability had a lower energy expenditure; however, Verhulst and Tinbergen (2001) found no evidence for this relationship in great tits (fig. 6.13). There appears to be only a single experimental study that has shown that greater energy expenditure during chick-rearing in relation to larger brood size has long-term fitness consequences. This singular exception is the work of Daan, Deerenberg, and Dijkstra (1996; described above), in which

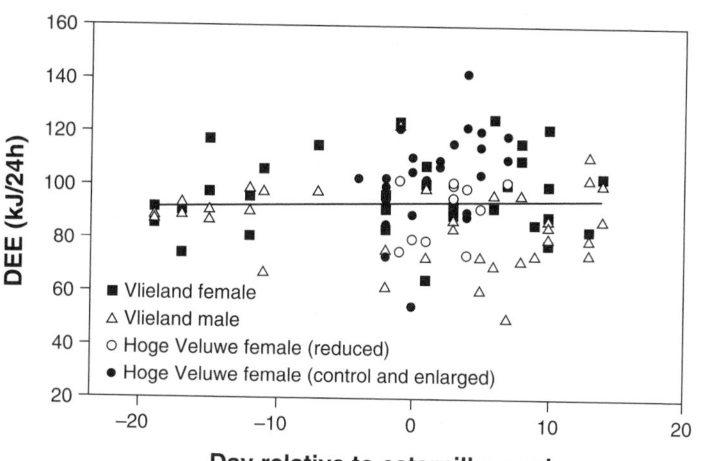

Fig. 6.13. Variation in parental daily energy expenditure (DEE) during chick-rearing relative to the peak in caterpillar abundance in two populations of great tits in the Netherlands. (Reproduced from Verhulst and Tinbergen 2001, with permission of the American Association for the Advancement of Science.)

kestrels rearing experimentally enlarged broods had lower adult survival rates. Previous work by the same researchers had shown that DEE was positively related to brood size for this experimental manipulation and that individuals with higher DEE had reduced local survival, leading them to conclude "that DEE affects local survival directly" (Deerenberg et al. 1995). T. Williams and Vézina (2001) recommended that future studies linking energy expenditure and parental effort should focus on individual variability (cf. comparative studies), and also that future studies should reevaluate some of the widely held assumptions that have formed the basis of this work to date. However, they also highlighted several biological, analytical, and technical problems with the doubly labelled water method that must be adequately dealt with if robust, meaningful analysis of *individual* variation in energy expenditure is to be achieved (see also Vézina et al. 2006).

6.4.3. Nutritional Demands of Incubation and Chick-Rearing

The energy demands of incubation and chick-rearing must be met either from stored nutrient reserves or from daily food intake, and a common alternative approach to assessing the resource demands of parental care has therefore been to look at changes in body mass and body composition. If birds are in negative energy balance when incubating large clutches or when rearing large broods, they should lose mass in general or, perhaps, just specific endogenous nutrient stores. It is widely assumed that the endogenous nutrient reserves available for incubation are a key factor in determining incubation constancy (Aldrich and Raveling 1983; Chastel et al. 1995; Chaurand and Weimerskirch 1994) and that the lack of sufficient nutrient reserves, or poor body condition, explains breeding failure (nest or brood desertion).

The nutritional demands of incubation would be expected to be highest in species with prolonged incubation fasts (e.g., some penguins, seabirds; see fig. 6.1) or in uniparental incubators with very high levels of nest attentiveness (e.g., some ducks and geese). The physiological changes during long-term fasting associated with prolonged incubation shifts have been well characterized in penguins, and nest desertion has been shown to occur just prior to, or at the time of, a switch from phase II fasting—a period when metabolic rates and activity decrease and use of lipid stores as the major fuel source is prolonged—to stage III, when fat stores are largely depleted and protein catabolism increases, as indicated by a rapid increase in plasma uric acid (Cherel et al. 1988; Groscolas 1986; Vleck and Vleck 2002). Thus, at some threshold body mass, metabolic and hormonal changes, possibly related to the limited availability of lipid stores, are thought to provide a "refeeding signal." This stimulus would, in turn, lead to behavioral changes, such as decreased incubation intensity, increased locomotor activity (fig. 6.14), or motivation to find food, that ultimately would be followed by nest desertion (Groscolas et al. 2000; Robin et al. 1998). In blue petrels (*Halobaena caerulea*) and king penguins (*Aptenodytes patagonicus*), the body mass of the leanest failed breeders leaving the breeding colony to refeed at sea is close to, or only slightly less, than the body mass at the transition between phases II and III of fasting (Ancel et al. 1998; Olsson 1997), suggesting that this refeeding mechanism allows individuals in long-lived species to trade off survival versus short-term breeding success.

The extent to which metabolic mechanisms dependent on condition or nutrient reserves underlie breeding failure in other species is less clear. Metabolic and hormonal changes associated with nutrient utilization during incubation have been less well studied in smaller species in which females can feed regularly during incubation and have less extreme incubation routines. Some work has found that failed breeders (deserters) are in worse body condition than non-deserters (Merilä and Wiggins 1997; Wiggins et al. 1994), an observation that would be consistent with the association of nest failure and nutritional constraint. In Wiggins et al.'s study (1994), collared flycatcher females that deserted during incubation laid later and smaller clutch sizes (independent of date) than successful birds and had lower return rates the following year, suggesting an interaction of individual quality, nutritional constraint, and breeding failure.

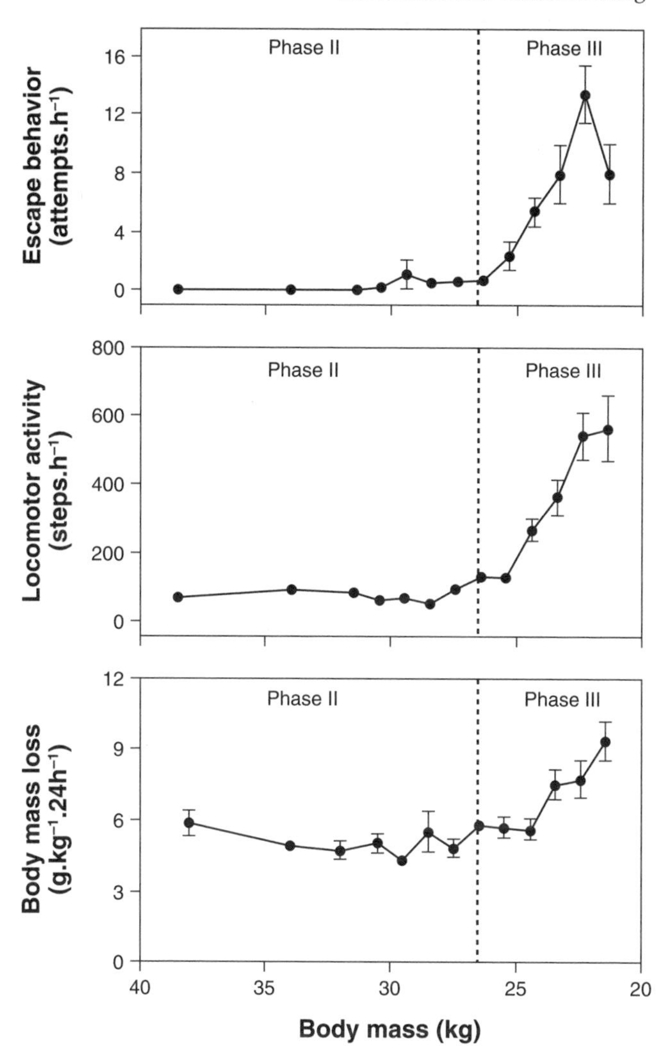

Fig. 6.14. Changes in specific daily body mass loss, locomotor activity, and escape behavior in relation to body mass in captive male emperor penguins during phase-II and phase-III fasting. (Reproduced from Groscolas and Robin 2001, with permission of Elsevier.)

However, many other studies on passerines have shown no change in body mass during incubation (Moreno et al. 1991; Woodburn and Perrins 1997). Furthermore, body mass, or "condition," of birds that desert their nest or brood is often *higher* than that of non-deserters (Bleeker et al. 2005; Hõrak et al. 1999), suggesting that breeding failure might involve reproductive restraint. Kern et al. (2005) showed in breeding female pied flycatchers that plasma triglyceride and glucose levels were not different between incubating and chick-rearing birds, but that incubating birds had higher uric acid and lower β-hydroxybutyrate levels despite also having higher free-fatty acid levels and higher body mass. Kern et al. (2005) concluded that females relied on lipids as an energy source during incubation (i.e., they avoided protein catabolism) and that the lipids were obtained from dietary intake rather than from endogenous adipose stores. Although female pied flycatchers had lower body masses during chick-rearing compared with earlier stages of breeding, there was little evidence that mass loss was a consequence of a physiological response to intermittent fasting between refeeding bouts (Kern et al. 2005). Hõrak et al. (1999) also used analysis of plasma metabolites to assess parental health in relation to brood desertion in great tits by comparing the plasma metabolite profiles on the 8th day of the nestling period of individuals with broods in which all chicks subsequently died and those with broods in which one or more chicks survived to fledging. Females that deserted their broods were on average 1g (>5%) heavier than non-deserters, and Hõrak et al. (1999) concluded that these birds had been in a *better* nutritional state prior to desertion than successful birds, with a plasma metabolite profile indicative of fat deposition (higher triglyceride, lower β-hydroxybutyrate), not fat metabolism. Norte et al. (2010) also showed that physiological indices of female body condition tended to be positively, not negatively, correlated with breeding performance, including chick fledging mass. Thus, data from physiological analyses of plasma metabolites are more consistent with the idea that breeding failure reflects reproductive restraint rather than nutritional constraint and that deserting females making a "decision" to direct resources to self-maintenance and future fecundity or survival.

6.4.4. Are There Resource-Allocation Trade-Offs during Parental Care?

If incubation or chick-rearing requires a high level of energy expenditure or nutrient utilization, then another way that individuals could meet this demand would be to divert resources away from other physiological processes during parental care; that is, there could be immediate or short-term resource-allocation trade-offs. What is the physiological evidence for these? Kullberg and coworkers (Kullberg, Houston, and Metcalfe 2002; Kullberg, Metcalfe, and Houston 2002) suggested that in small passerines, females might have reduced flight ability during incubation, which they linked to a loss of protein from the pectoral muscle, to meet the demands of egg production (Kullberg, Houston, and Metcalfe2002; but see chapter 7). For example, female pied flycatchers flew on average

10% slower during incubation compared with during early chick-rearing, although this was probably compounded by the fact that females maintain an *elevated* body mass throughout incubation (Kullberg, Metcalfe, and Houston 2002). However, this loss appears to represent an adaptive physiological adjustment (adaptive mass loss) rather than a "cost" of incubation because mass loss continued throughout incubation and chick-rearing and involved a further 12% decrease in the flight muscle index from incubation to early chick-rearing (see also Freed 1981; Vézina and Williams 2003). Thus, these changes not only potentially reduce wing loading and flight costs during chick provisioning, but also *increase* escape-flight ability when females are rearing chicks (Kullberg, Metcalfe, and Houston 2002).

Resource reallocation associated with high energy demands of parental care has been proposed as an explanation for the suppression of immune function that has been observed during incubation and chick-rearing, which might then lead to a greater susceptibility to parasitism and reduced future fecundity or survival. A recent meta-analysis of 26 studies showed that there was a "moderate" negative effect ($r = 0.297$) of reproductive effort on immune responsiveness and that experimental manipulation of brood size during chick-rearing produced stronger effects than manipulation of clutch size during incubation (Knowles et al. 2009). There was also a relatively weak but "well-supported" positive effect ($r = 0.09$) of reproductive effort on the levels of blood-parasite infections across all studies, with larger effect sizes for infection intensity (proportion of parasitized cells within infected hosts, $r = 0.155$) than for infection prevalence (presence vs. absence of infection among hosts, $r = 0.058$). However, Knowles et al. (2009) highlighted the fact that not only were the *physiological* mechanisms underlying these effects unresolved, but the significance of these effects for long-term reproductive costs "remains questionable" at present.

In collared flycatchers, females rearing experimentally enlarged broods had a reduced ability to mount an immune response to the Newcastle disease virus (Nordling et al. 1998). This appears to represent a true cost, since females rearing enlarged broods had a higher intensity of infection by naturally occurring *Haemoproteus* blood parasites at the end of the nestling period, and infected birds had lower survival (fig. 6.15). However, it has also been suggested that immunosuppression might be adaptive if females avoid potentially harmful autoimmune responses (Råberg et al. 1998) or the potential costs of mounting a full immune response while facing resource limitations during incubation. Hanssen et al. (2004) induced immune responses against artificial infections in incubating female common eider by injecting three different non-pathogenic antigens—sheep red blood cells (RBCs), diphtheria, and tetanus—early

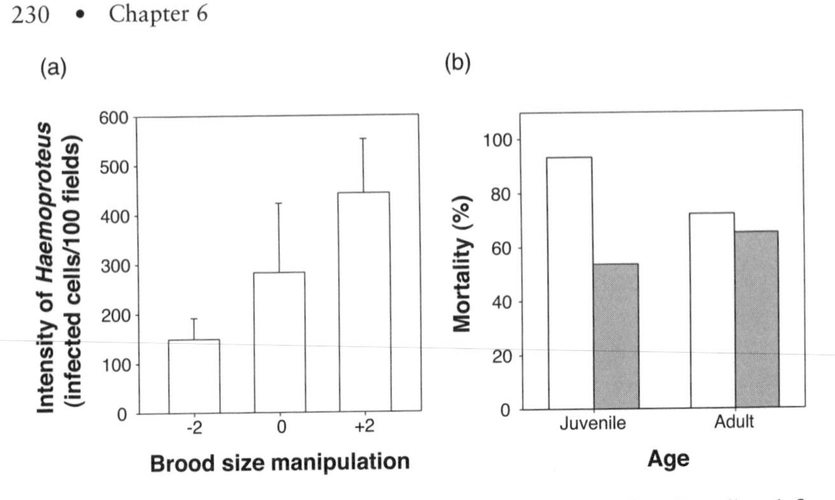

Fig. 6.15. Reproductive effort and immunosuppression in female collared flycatchers rearing reduced (−2), control (0), and increased (+2) brood sizes: (a) intensity of *Haemoproteus* parasites in peripheral blood; (b) differences in annual mortality rates between females infected by *Haemoproteus* and uninfected controls in relation to age. (Redrawn from Nordling et al. 1998, with permission of the Royal Society.)

in the incubation period. In females that mounted a humoral immune response against sheep RBCs, the return rate was only 27%, compared with 72% in females that did not mount a response against them. Moreover, responding against diphtheria toxoid when also responding against sheep RBCs led to a further reduction in return rate (<5%). Hanssen et al. (2005) experimentally manipulated incubation "demand" by creating three- and six-egg clutches (natural range three to six eggs) and confirmed that females incubating large clutches had decreased immune function (lymphocyte levels and specific antibody response to the non-pathogenic antigens diphtheria and tetanus). Although there was no detectable effect of clutch size on current reproduction or return rate in the next breeding season, females with high incubation demand had a delayed date of clutch initiation in the year after manipulation and laid a lower clutch size that same year (year $n+1$). These data are consistent with the idea that mounting a full immune response during incubation can have important fitness consequences, although they do not directly confirm that it is mediated by energy or nutrient constraints or resource reallocation. Thus, there is only weak evidence at present (in one or two species) that physiological mechanisms of resource allocation mediate short-term trade-offs associated with energy or nutrient demands of incubation or chick-rearing (see chapter 7 for further discussion).

6.5. Hormonal Mechanisms Underlying Individual Variation in Parental Care

As discussed at the beginning of section 6.2, in terms of underlying physiological mechanisms, it is important to distinguish the all-or-none effects of parental care that determine breeding success or complete breeding failure from the continuous effects of individual variation in parental care among *successful* breeders that might give rise to differential fitness in each breeding attempt. Hormones have been implicated in both of these components of breeding productivity. For example, egg or chick abandonment might occur because individuals reach some threshold level of energy depletion that induces metabolic and behavioral shifts (e.g., refeeding) that are hormonally regulated (Groscolas et al. 2008). However, the extent to which hormones might regulate individual variation in parental care in *successful* breeders has been less well studied. Early studies established a role for prolactin in the regulation of parental behavior in birds (Eisner 1960; Lehrman and Brody 1961), and this idea has dominated physiological studies of parental care until very recently. In some altricial species, prolactin can be elevated both during incubation and throughout chick-rearing, so prolactin has therefore been considered to be involved in the physiological control of incubation, the brooding of post-hatch chicks, and chick provisioning (Angelier and Chastel 2009). More recently, corticosterone has been implicated in the regulation of parental care, mainly during chick-rearing or in species with long incubation fasts, but in two seemingly contradictory ways that set up a potential corticosterone-mediated "reproductive conflict" (Love et al. 2004). Corticosterone regulates adaptive physiological and behavioral responses to stressful events, and elevated corticosterone levels have been associated with abandonment of reproduction. It has therefore been suggested that birds would adaptively *down-regulate* corticosterone secretion to reduce the chance of nest abandonment when their investment in offspring is substantial (e.g., during chick-rearing) or when the probability of renesting is low (see below). Conversely, corticosterone's role in regulating energy balance and food acquisition, via increased locomotor activity or foraging behavior, predicts that breeding birds would *elevate* corticosterone to meet the energy and behavioral demands of raising offspring. Prolactin and corticosterone might also interact in the regulation of parental care, and Angelier and Chastel (2009) recently concluded that "both corticosterone and prolactin are very likely to mediate parental effort and parental investment in birds." Indeed, a recent study by Ouyang et al. (2011) has suggested that pre-breeding corticosterone and prolactin levels might predict individual differences in reproductive success. Some work has also been conducted on the possible role of thyroid hormones in the context of thermogenesis and the expenditure of energy during incubation. However, plasma T3 is not correlated with natural variation in clutch size, and elevated T3 occurs in females incubating both experimentally enlarged and reduced clutch sizes (Criscuolo et al. 2003), so there are few consistent data to support a key role for this hormone at present.

6.5.1. Prolactin and Parental Care

Prolactin's role in incubation behavior is largely predicated on the broad temporal association between increases in plasma prolactin levels at the onset of incubation and decreases in plasma prolactin levels at the end of the incubation or the brooding period. However, as we saw in chapter 5, there is in fact relatively little experimental evidence that circulating prolactin levels are "causally or permissively" related to the onset of incubation (Sockman et al. 2006). Treatment of females with exogenous prolactin induces development of incubation behavior only in combination with ovarian steroids, and prolactin has no effect in ovariectomized females (El Halawani et al. 1986; Hutchison 1975; Lea et al. 1986). However, this ovarian-steroid dependence of incubation does not persist *after* initiation of this behavior (Sharp et al. 1988; Wood-Gush and Gilbert 1973); for example, ovariectomy of incubating hens does not inhibit incubation, suggesting that prolactin might play a role in the *maintenance* of incubation. The pattern and duration of elevated plasma prolactin levels during breeding varies among species in relation to their specific patterns of incubation and/or chick-rearing (fig. 6.16). In many waterfowl, which have precocial young, prolactin levels increase during egg-laying and then decrease rapidly at hatching, but in some seabirds with prolonged periods of brooding of altricial young, and in some passerines, elevated prolactin levels can be maintained through chick-rearing (Criscuolo et al. 2002; Goldsmith 1991; Hector and Goldsmith 1985). In some multiple-brooded passerines (e.g., great tit, Europeans starling; fig. 6.16), plasma prolactin rises rapidly in females during egg-laying to a peak in incubation but then declines during the inter-brood interval before rising again during incubation of the second brood (Silverin (1991). However, in other multiple-brooded species, such as the white-crowned sparrow and song sparrow (fig. 6.16), plasma prolactin remains elevated during the feeding of first-brood nestlings and through the second breeding cycle, not decreasing until the molt following the fledging of the second brood (Hiatt et al. 1987; Wingfield and Goldsmith 1990). Thus, plasma prolactin levels can remain high even during periods when individuals are not expressing incubation or chick-rearing behavior, and there are numerous other examples in which elevated plasma prolactin levels and parental care can be dissociated (see below). The most noticeable case is in brood parasitic brown-headed cowbirds (*Molothrus ater*), in which plasma prolactin levels increase in females between mid-May (<20 ng/ml) and late June (>40 ng/ml), even though females neither incubate eggs nor rear chicks. Similar changes in plasma prolactin occur in male cowbirds, even though they never visit the nests in which eggs are laid and are therefore not exposed to any visual stimuli from eggs or chicks (Dufty et al. 1987).

It has been suggested that the maintenance of elevated plasma prolactin, which in turn might maintain incubation behavior and parental care, involves continued tactile stimuli from eggs in the nest or other stimuli from young chicks in the nest, although evidence to support this causal relationship is limited (Buntin 1996; Goldsmith 1991). In some species, removal of eggs or the nest from incubating birds, denervation of the brood patch, or treatment with anti-prolactin antibodies leads to a rapid decline in circulating prolactin levels. However, incubation behavior can be maintained for up to 5 days even though prolactin decreases within 4–24 hours of manipulation, and females will readily resume incubation if the eggs or nest are returned even in the absence of high circulating prolactin levels (Sharp et al. 1988; Sharp et al. 1989). In species such as seabirds, which make prolonged foraging trips during incubation or chick-rearing, elevated plasma prolactin levels are thought to be maintained even when birds are at sea and away from the eggs or nests. In both male and female king penguins, plasma prolactin rose from 15 ng/ml around the time of copulation to more than 40 ng/ml in incubating birds, and these high levels were maintained throughout the first chick-feeding period (until mid-April). After an absence of about 5 months, birds returning to the colony for the second chick-feeding period still exhibited high prolactin levels (32–34 ng/ml), and prolactin did not decline to baseline (5 ng/ml) until molt (Jouventin and Mauget 1996). Similarly, in other penguins (Lormee et al. 1999; Vleck et al. 2000) and albatrosses (Hector and Goldsmith 1985), birds returning to breeding colonies to relieve their incubating partner after extended foraging bouts at sea had high plasma prolactin levels similar to those of their incubating partner, and prolactin did not increase further when the birds assumed incubation. This observation suggests that elevated prolactin levels are either maintained when birds are off nest, independent of tactile or visual stimuli from eggs or chicks, or that prolactin secretion is endogenously timed, perhaps increasing after some set duration of foraging in anticipation of return to the nest. Sustained prolactin secretion in birds foraging at sea could still reflect that prolactin is mediating and maintaining components of parental behavior but is doing so through different behavioral responses (e.g., foraging effort) and possibly very different target tissues

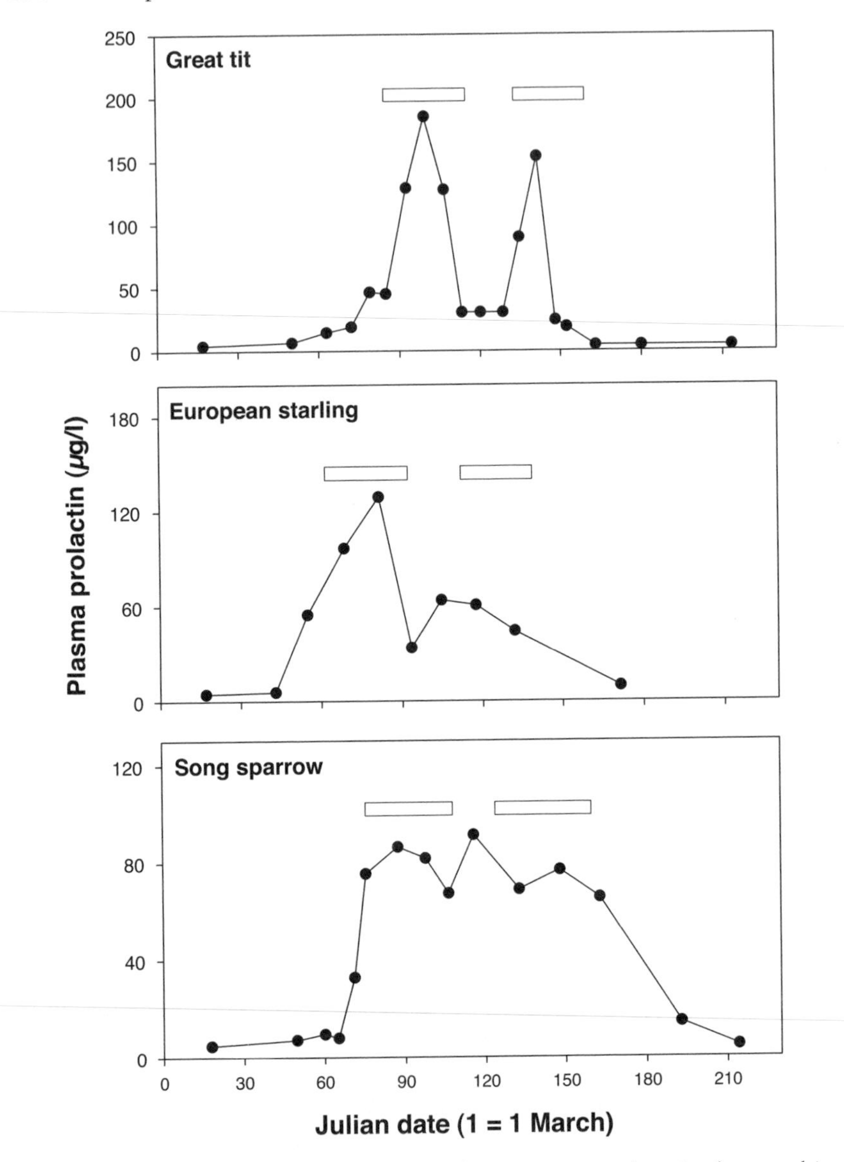

Fig. 6.16. Seasonal variation in plasma prolactin concentrations in three multi-brooded species. Open bars indicate periods of parental care for first and second breeding attempts. (Redrawn from data in Goldsmith 1991.)

(e.g., balancing self-feeding with retention of undigested food for offspring; Vleck 2000).

Some studies have artificially extended or shortened the incubation period—for example, by replacing more developed eggs with less developed ones, and vice versa—and investigated associated changes in plasma prolactin. In free-living pied flycatchers, when incubation was extended from 16 to 19 days, elevated prolactin levels were *not* maintained to the end of the extended incubation (Silverin and Goldsmith 1984). Similarly, in *Diomedea* albatrosses, the timing of the decline in prolactin levels at the end of normal incubation remained constant, even when the incubation period was artificially lengthened or shortened (Hector and Goldsmith 1985). This result suggests that there is an endogenously determined "window" for incubation-stimulated prolactin secretion (Goldsmith 1991), but this in itself does not determine the duration of incubation behavior (see also Ketterson et al. 1990). In king penguins, plasma prolactin concentrations were lower at egg and chick abandonment compared with successful birds; however, at chick abandonment, plasma prolactin levels remained elevated (fig. 6.17a) perhaps due to a higher stimulatory effect of chicks than eggs (Groscolas et al. 2008). In red-footed boobies (*Sula sula*), failed breeders had plasma prolactin levels close to basal, prebreeding levels within 6–24 hours of nest failure, but some individuals had elevated prolactin levels up to 4 days before nest failure (Criscuolo et al. 2002). Unless birds are sampled just prior to breeding failure, it is difficult to determine cause and effect: Do prolactin levels decrease because of breeding failure or in advance of breeding failure potentially *causing* failure? However, these studies suggest that breeding failure is not always closely associated with rapid changes in plasma prolactin levels.

6.5.2. Prolactin and Individual Variation in Parental Care

Many of the above studies simply confirm that there are broad, temporal correlations between elevated plasma prolactin levels and incubation behavior, but they also provide many examples in which prolactin and behavior can be dissociated. None of these studies unequivocally demonstrate cause and effect. What evidence is there for a systematic, quantitative relationship between *individual* variation in plasma prolactin levels and individual variation in parental care? In other words, is prolactin correlated with, and potentially causally related to, individual variation in the amount of parental behavior (Goldsmith 1991)? Despite a recent conclusion that "a positive correlation between the intensity/quality of parental care and plasma prolactin has been observed in many parental birds" (Angelier and Chastel 2009), I would argue that few studies have addressed this question appropriately, and most studies have been

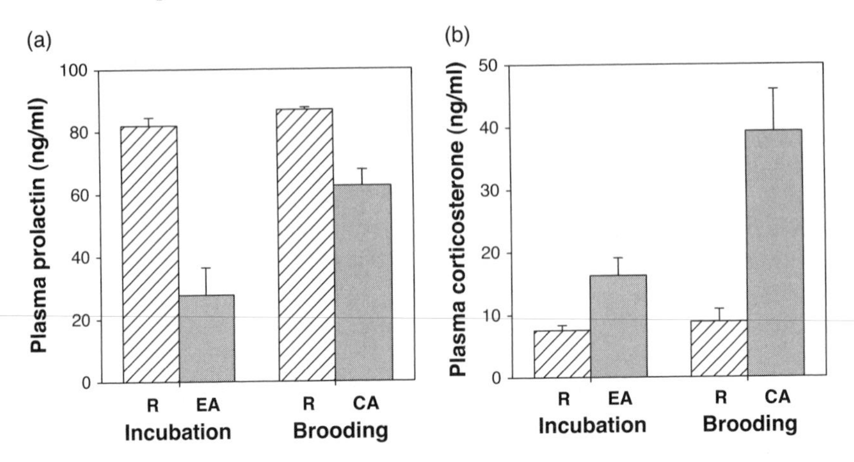

Fig.6.17. Plasma levels of (a) prolactin and (b) corticosterone in relation to nest and chick abandonment in free-living king penguins. R, nest relief at end of incubation or brooding shift; EA ,egg abandonment; CA, chick abandonment. (Based on data in Groscolas et al. 2008.)

correlational. In ring doves, at the end of incubation (day 14 or 15), the average amount of time spent on the nest each day was independent of prolactin-like activity (measured with a cell-based assay), but there was a weak negative correlation ($P = 0.06$) with day-to-day variability in nest occupation; that is, females with lower incubation constancy had lower prolactin-like activity (Buntin et al. 1996). During chick-rearing in ring doves,VIP-like immunoreactivity was higher in birds rearing two chicks rather than one chick, but this might simply reflect the effect of greater stimulation by the larger number of young (Cloues et al. 1990). Miller et al. (2009) analyzed individual variation in plasma prolactin in mourning doves (*Zenaida macroura*) in relation to adult weight, as a measure of current adult state, and to nestling weight, as a measure of parental effort by the adults. Variation in parental prolactin was negatively correlated with nestling age (as would be expected) and positively correlated with nestling mass in young nestlings (2 days of age) but not older nestlings (4 or 6 days of age). Miller et al. (2009) argued that these results indicate that prolactin levels *positively* affect the feeding rates of young, leading to faster growth of nestlings in parents that have higher prolactin levels, although this was not directly demonstrated. In cooperatively breeding Florida scrub jays (*Aphelocoma coerulescens*), prolactin levels were significantly correlated with the number of visits to the nest and the amount of food delivered to the young when non-breeding helpers and the breeding pair were analyzed together, but among breeding birds only parental effort was independent of plasma prolactin levels (Schoech et al. 1996).

Similarly, in dark-eyed juncos (*Junco hyemalis*), the frequency and duration of female parental behaviors (time spent brooding, nestling feeding) were unrelated to the circulating levels of prolactin (Ketterson et al. 1990). Finally, in polygamous species some females rear chicks without help from a male partner, spend a much greater amount of time brooding chicks, and have much higher feeding rates than females that are assisted in parental care by a male. However, there is no difference in the plasma prolactin levels of aided and unaided females despite this greater level of parental effort, a result indicating that variation in parental behavior is unrelated to circulating levels of prolactin (Ketterson et al. 1990; Silverin and Goldsmith 1984).

Experimental studies that have manipulated plasma prolactin levels in relation to post-hatching parental care are surprisingly rare. In willow ptarmigan, exogenous prolactin treatment did not affect incubation constancy, although there was a positive effect of treatment on nest defense or distraction displays (Pedersen 1989). The only experimental study of prolactin in relation to chick provisioning in a non-domesticated bird appears to be that of Badyaev and Duckworth (2005). In house finches there was a positive relationship between feeding visits to the nest and plasma prolactin in *males*, confirming the result based on very small sample sizes reported in the widely cited paper by Duckworth et al. (2003). Experimental elevation of prolactin with VIP-implants increased feeding rate in "non-parental," red-morph males, which have naturally lower prolactin levels and lower provisioning rates. Conversely, the inhibition of prolactin with bromocriptine caused "parental," yellow-morph males, with naturally higher prolactin levels and higher provisioning rates, to stop feeding their nestlings (Badyaev and Duckworth 2005).

In summary there is, rather surprisingly, only weak and inconsistent support for the hypothesis that *individual* variation in parental care is related to *individual* variation in plasma prolactin levels, and these data come largely from correlational studies, which simply show broad temporal associations between prolactin and behavior. Clearly, additional experimental studies of the sort described by Badyaev and Duckworth (2005) are required to further support the widely held assumption that prolactin *causally* affects individual variation in parental care in birds.

6.5.3. Corticosterone and Breeding Failure

Does corticosterone play a key role in regulating parental care through its roles in mediating stress (acute or chronic), metabolic and energetic homeostasis, or foraging behavior (Angelier and Chastel 2009; Angelier, Clément-Chastel, et al. 2009; Angelier, Shaffer, et al. 2007)? Specifically, is corticosterone causally related to the abandonment of reproduction, perhaps redirecting resources to parental self-maintenance and survival? Some studies provide compelling evidence for a temporal association between elevated plasma corticosterone levels and abandonment of parental care. King penguins undergo prolonged fasts during incubation and in every case of abandonment (14 spontaneous abandonments of eggs or young chicks), birds were in phase III fasting, which is characterized by accelerated protein catabolism (Groscolas et al. 2008). For freely incubating birds, plasma corticosterone was two- to fourfold higher at egg and chick abandonment compared with birds that were successfully relieved by their partner (fig. 6.17b). These data are consistent with corticosterone having both metabolic effects (stimulating increased protein breakdown at onset of phase III fasting) and behavioral effects (increasing locomotor activity and stimulating parental refeeding). In captive, fasting female eiders, experimental elevation of plasma corticosterone increased the loss in mass, supporting the idea of a role for corticosterone in mediating metabolic changes during incubation (Bourgeon and Raclot 2006). However, Bourgeon et al. (2006) found no change in endogenous plasma corticosterone levels over the incubation period in *non-manipulated* female eiders, despite the prolonged fasting and mass loss that occur in this species. Few studies have investigated the role of corticosterone in breeding failure in species that have intermittent incubation, in which nutritional constraints would be predicted to be less severe (and in which, as we have seen, there are no consistent relationships between breeding failure, mass change, and physiological condition). Silverin (1986) elevated plasma corticosterone to *stress-induced* levels in pied flycatchers using silastic implants at the end of incubation, and breeding was not abandoned (in a second experiment, involving only males, corticosterone-treated birds did abandon their nests when corticosterone was elevated to supra-physiological levels: 50–130 ng/ml, cf. 10 ng/ml in non-manipulated males and 30 ng/ml in males exposed to handling stress, Silverin 1998). However, corticosterone-treated birds of both sexes did reduce parental behavior: nestlings were fed less, fewer fledglings resulted, and young that did fledge weighed less than fledglings from control implanted birds. Furthermore, corticosterone-treated birds had lower mass loss during chick-rearing, a finding consistent with corticosterone re-directing resources toward self-maintenance or self-feeding (Silverin 1986).

6.5.4. Corticosterone and Individual Variation in Parental Care

Does corticosterone play a more general role in regulating *individual* variation in parental care, in particular the foraging or provisioning effort, among successfully breeding birds? When birds are not acutely stressed (probably most of the time), *baseline* corticosterone levels will regulate day-to-day variation in the energy balance (metabolic homeostasis) and foraging behavior. This would predict (a) that breeding birds might *elevate* corticosterone to meet the energy and behavioral demands of raising offspring, and (b) that baseline plasma corticosterone levels should be *positively* correlated with variation in parental care (e.g., feeding rates). This *corticosterone-adaptation hypothesis* predicts that better parents should have higher baseline corticosterone levels, in contrast to the *corticosterone-fitness hypothesis*, which predicts that more stressed, lower-quality individuals (poorer parents) will have elevated corticosterone (Bonier et al. 2009). Love et al. (2004) provided an explanation for how parents can physiologically modulate the effects of elevated corticosterone to resolve this potential "reproductive conflict" whereby corticosterone can have a positive function during offspring rearing (e.g., increasing foraging effort) yet also potentially play a role in brood desertion and nest abandonment.

Some studies support a role for corticosterone in regulating food intake in adult birds, but generally only under conditions of nutritional stress (Astheimer et al. 1992; Gray et al. 1990), during migratory hyperphagia (Landys et al. 2004), or in relation to food caching in wintering birds (Pravosudov 2003). In captive white-crowned sparrows and song sparrows, corticosterone-treated birds fed for longer periods and with greater intensity when provided with food after a 24-hour fast, but foraging behavior and feeding rates were unaffected by exogenous corticosterone in fed birds (Astheimer et al. 1992). Does corticosterone play a similar role in regulating provisioning effort in parental birds during chick-rearing? There is only limited and contradictory data to support this idea. Miller et al. (2009) showed that individual variation in baseline corticosterone, and baseline prolactin, levels in adult mourning doves was positively related to nestling weight at early nestling ages (<6 days), a finding consistent with the prediction of a *positive* relationship of hormone levels to current parental effort of adults. In the polymorphic white-throated sparrow (*Zonotrichia albicollis*), tan-striped males provision nestlings at higher rates than do white-striped males. Horton and Holberton (2009) were able to reverse this morph-specific behavior in males by experimentally manipulating baseline corticosterone levels. Tan-striped males with experimentally elevated baseline corticosterone levels fed nestlings at lower rates, which were similar to the provisioning rates in white-striped males. Conversely, white-striped males treated with a glucocorticoid receptor antagonist (RU486) fed nestlings at higher rates than white-striped controls and at similar rates as tan-striped control males. These results suggest that increases in baseline corticosterone directly or indirectly *inhibit* nestling provisioning behavior, in contrast to the conclusions reached by Miller et al. (2009) and Love et al. (2004), and clearly, more studies of this kind are required.

There is some evidence that corticosterone is involved in parental care in large, long-lived species of birds, some of which make extended foraging trips associated with prolonged incubation bouts (fig. 6.1). In the black-browed albatross (*Thallasarche melanophris*), baseline corticosterone and prolactin levels were positively and negatively correlated, respectively, with the time spent incubating and fasting on the nest. In Adelie penguins, birds with higher corticosterone levels before their foraging trips spent less time at sea, stayed closer to the colony, had greater foraging effort in terms of diving activity, and lower mass gain than birds with low pre-trip corticosterone levels (Angelier et al. 2008). These results suggest that pre-trip baseline corticosterone was *positively* related to maximization of the rate of energy delivery to the chicks, perhaps at the expense of the parent's body reserves (Angelier et al. 2008). However, in black-browed albatrosses, the probability of successfully fledging a chick was *negatively* related to baseline corticosterone levels for both sexes combined (fig. 6.18a; Angelier, Weimerskirch, et al. 2007). Furthermore, in a second study of black-browed albatrosses (Angelier et al. 2010), baseline corticosterone levels were unrelated to individual female quality when estimated as number of fledged chick produced over a 5-year period (fig. 6.18b).

Few studies to date have confirmed these correlational results with experimental manipulation of corticosterone levels and direct assessment of foraging behavior in free-ranging birds during chick-rearing. In black-legged kittiwakes, corticosterone manipulation during chick-rearing affected time-activity budgets in a context-dependent manner through an interaction with body condition: males in good body condition increased their time spent flying/foraging in response to increased corticosterone levels, whereas males in poor body condition did not (Angelier, Clément-Chastel, et al. 2007). These preliminary studies suggest, at best, a very complex, context-dependent relationship between corticosterone and foraging or provisioning effort during chick-rearing, but clearly, further experimental work is warranted, in particular in non-seabird species.

If corticosterone does regulate parental care, is there evidence that these corticosterone-mediated mechanisms are adaptive in the long-term, that is, that they explain individual variation in fitness? Few studies have linked variation in glucocorticoids to variation in fitness, either for baseline corticosterone (Bonier et al. 2009) or stress-induced corticosterone levels (Breuner et al. 2008). Bonier et al. (2009) reviewed the evidence for fitness correlates of *baseline* plasma corticosterone levels in eight studies in which plasma corticosterone was experimentally manipulated; of these, only three found a significant negative relationship

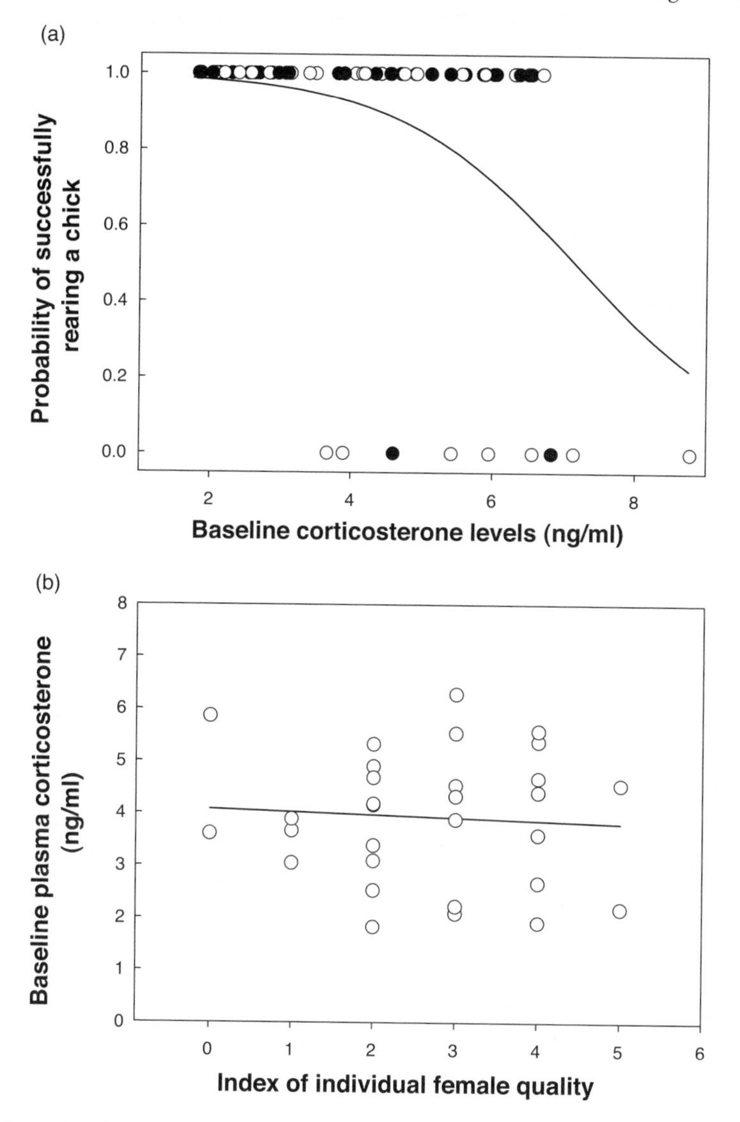

Fig. 6.18. Baseline plasma corticosterone levels in relation to (a) fledging success in a single year for male (open circles) and female (closed circles) parents combined, and (b) individual female quality (number of fledged chicks across 5 years) in black-browed albatrosses. (Redrawn using data in Angelier, Weimerskirch et al. 2007 and Angelier et al. 2010, with permission of the Royal Society and Springer Science and Business Media, respectively.)

between fitness and baseline corticosterone, and only two of these studies included measurements of parental care. Only one experimental study to date (Love and Williams 2008a) has manipulated both environmental conditions and plasma corticosterone, and it found a positive relationship between corticosterone and female fitness, evidence that supports the corticosterone-adaptation hypothesis (see below). Breuner et al. (2008) concluded that most data on corticosterone and parental care support the *maternal/offspring match hypothesis*, which proposes that while condition-dependent maternal glucocorticoids may have negative effects on offspring growth and survival, this might better match maternal provisioning ability ("quality") to offspring demand in stressful environments, and so maximize parental fitness over the long term.

Studies by Love and colleagues (Chin et al. 2009; Love et al. 2005; Love and Williams 2008a, 2008b; Love, Wynne-Edwards, et al. 2008) link condition-dependent variation in corticosterone levels to offspring and maternal fitness through a combination of early effects (egg production and hormonally mediated maternal effects) and late effects (chick provisioning). In European starlings, low-quality females in poor body condition early in the breeding season have higher plasma corticosterone levels than mothers in good condition, and yolk corticosterone levels in eggs vary in proportion to those in the maternal plasma. Experimental manipulation of maternal corticosterone levels confirmed that females with higher plasma corticosterone levels produced eggs with higher yolk corticosterone (Love et al. 2005). Furthermore, corticosterone-treated male offspring were smaller at hatching, grew more slowly, and had higher mortality than males in control nests (Love et al. 2005), thus potentially decreasing "demand" on low-quality mothers during chick-rearing. To confirm that low-quality mothers benefited from corticosterone-mediated changes in offspring phenotype, matching brood demand to their own parental quality, Love, Salvante, et al. (2008) set up "matched" and "mismatched" females by combining (a) experimental manipulation of offspring phenotype and brood demand using corticosterone injections of eggs, and (b) experimental manipulation of post-hatching female quality using feather clipping. All experimental females were subsequently followed across multiple breeding attempts and years to measure long-term reproductive success and (local) survival. Females that raised offspring that were "matched" to their rearing ability (low-quality, feather-clipped females, and corticosterone-treated offspring) were (a) in better condition at the start of their second clutch, (b) raised more and higher-quality offspring in their first year and the subsequent breeding season (fig. 6.19a), and (c) had higher survival during the two years post-manipulation (fig. 6.19b), compared with mismatched females (low-quality female and non-corticosterone-treated offspring). Overall, therefore, females

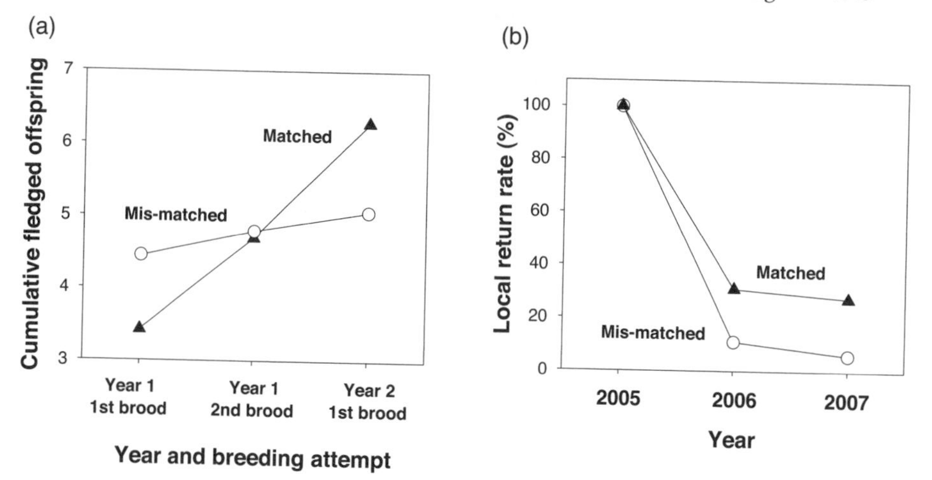

Fig. 6.19. Lower-quality (feather-clipped) female European starlings have higher fitness as measured by (a) cumulative fledgling production and (b) local survival (return rate) when maternal quality and offspring demand are *matched* via corticosterone-mediated mechanisms compared with *mis-matched* females. (Based on data in Love and Williams 2008a.)

with chick-rearing demand matched to their own parental quality via corticosterone-mediated mechanisms had higher overall fitness (the combination of reproductive success and survival) than mismatched females (Love and Williams 2008a). These results support the idea of a complex, context-dependent relationship between corticosterone, maternal condition, and parental care during chick-rearing.

6.6. Conclusions

Parental care has, without doubt, been the focus of more research than any other breeding activity or breeding stage in birds since Lack (1947) highlighted the importance of parental chick-rearing or provisioning ability. Nevertheless, there are still major knowledge gaps in our understanding of the causes and functional consequences of individual variation in parental effort, and there are surprisingly few data to support some of the key conceptual foundations that have underpinned studies of parental care for decades. For example, parental effort is widely assumed to be positively correlated with fitness primarily because higher provisioning rates should lead to higher fledging mass, size, or quality, and higher fledgling qulaity in turn often predicts post-fledging survival and recruitment. Thus, it is intuitively attractive to think that better parents, or higher-quality individuals, *work harder* to rear more or larger offspring. There is strong evidence for the consequence of this assumption: a positive relationship between late-nestling quality and recruitment, a finding that has been well documented (Schwagmeyer and Mock 2008). However, few studies support the assumption that parental work load (at least as measured by provisioning rate) is positively related to fledgling quality or brood size, even though there is marked individual variation in provisioning rates (see figs. 6.4 and 6.5). One reason for this lack of empirical support might be that the provisioning rate itself, although the most commonly used metric in studies of chick-rearing, does not reliably measure parental effort or work load. This might be in part because individuals have considerable flexibility to adjust chick-provisioning in so many other ways (e.g., load size, prey type, foraging distance from the nest; Wright et al. 1998). Alternatively, better, or higher-quality, parents might work *more efficiently*, rather than simply working harder (Daunt et al. 2006; Lescroël et al. 2010). More recent work on parental care during incubation has also provided little evidence to support the assumption that individuals incubating more eggs are working harder or that doing so affects hatching success or other aspects of incubation behavior. Clearly, assumptions about simple, linear or positive relationships between parental effort, workload, and breeding success or offspring quality need to be revaluated in the context of variability in individual quality (T. Williams and Vézina 2001).

The assumption that rearing more or larger offspring or incubating more eggs requires parents to *work harder* has also driven ideas about the potential physiological mechanisms underlying variation in parental care. Where long-term costs of reproduction have been demonstrated, it has been suggested that rearing more offspring results in parents becoming physiologically exhausted or fatigued. However, as we have seen, few empirical data support the idea that female body mass or physiological condition at the end of chick-rearing is negatively affected by parental effort, as would be predicted if there were large energetic (or nutritional) costs. More important, there is "less than compelling" evidence for a positive relationship between parental effort and *physiological work*, as measured by metabolic rate or DEE: most studies have failed to find any relationship between DEE and brood size or provisioning rate (T. Williams and Vézina 2001). The disconnect between theory and data could, in part, be due to technical issues with the doubly labeled water method (especially when used to obtain *individual* values), but it could also be because some of the basic assumptions underpinning this work to date are flawed. For example, perhaps the most compelling explanation is that different individuals adopt flexible, "individually variable" strategies to deal with the costs and benefits of investment in self-maintenance or offspring when faced with increased workload or increased parental effort (de Heij et al. 2008; Vézina et al. 2006; T. Williams et al. 2009). Thus, future experiments should be designed and data collected specifically to incorporate this concept of *individually variable* energetic, physiological, and behavioral strategies (T. Williams 2008).

More recent research on the hormonal regulation of parental care, which does not necessarily depend on ideas of energetic or nutritional constraint per se, offers promise in identifying mechanisms that characterize good, high-quality parents and poor, low-quality parents. Nevertheless there is a long way to go. The long-standing idea that prolactin-based mechanisms regulate incubation behavior is surprisingly poorly supported, especially with experimental studies on free-living birds (Sockman et al. 2006) and in relation to *individual* variation in incubation behavior and the fitness consequences of this variation. Adkins-Regan (2005) concluded that despite the picture that has emerged in mammals, "it is premature to assume that prolactin is the hormone responsible for post-incubation care of offspring in birds generally." Given the little empirical data available to date, it is hard to argue with this conclusion. Indeed, there would appear to be more evidence to support the idea that corticosterone plays a key role in linking physiological condition, foraging behavior, and parental care to reproductive success and survival (Angelier, Giraudeau, et al. 2009; Angelier, Shaffer, et al. 2007; Love and Williams 2008a). However, future studies should focus on the day-to-day role of *baseline* corticosterone in regulating metabolic homeostasis and foraging behavior (cf. acute "stress" effects) in the context of variation in parental care. There is also a need for more experimental studies on a wider range of avian taxa (not just seabirds) to see to what extent current ideas can provide a general explanation for physiological and hormonal mechanisms underlying the variation in parental effort.

6.7. Future Research Questions

- Does individual variation in incubation behavior (e.g., incubation duration, frequency and duration of incubation bouts and off-nest periods) contribute to fitness? Is this variation repeatable and what aspects of a female's phenotype determine this individual variation?
- Do physiological or morphological characteristics of the female's brood patch (size, stage of development, blood flow, etc.) explain individual variation in the *effectiveness* and duration of incubation?
- Given the marked individual variation in provisioning rate (feedings visits per chick), why are there only weak, inconsistent relationships between feeding rate and chick growth, fledging mass, or fledging

success? Do load size, foraging distance from the nest, and the type and size of prey delivered to the nest explain individual variation in parental effort or the number and size of chicks at fledging? What components of female phenotype explain parental quality or repeatable, individual variation in the provisioning rate?

- If parents rearing large broods and large chicks "work harder" but pay a cost of reproduction because they become exhausted or fatigued, what are the physiological mechanisms that link workload and survival?
- Why is there no relationship between brood size or provisioning rate and *physiological work* (energy expenditure) during chick-rearing? Are there *individually variable energy-management strategies* or behavioral strategies by which individuals balance self-maintenance and investment in offspring during incubation or chick-rearing?
- What role do prolactin and corticosterone play in breeding failure (nest or chick abandonment)? Is individual variation in plasma prolactin levels positively correlated with individual variation in parental effort (e.g., provisioning rate) or parental behaviors? Do corticosterone-mediated mechanisms play a general role in regulating individual variation in parental care and, in particular, foraging or provisioning effort, among successfully breeding birds?
- Does the maternal/offspring match hypothesis provide a general explanation for the relationship between individual quality (condition-dependent variation in corticosterone), parental care, and offspring and maternal fitness, via early (egg production and hormonally mediated maternal effects) and late (chick provisioning) effects of parental effort?

CHAPTER 7

Trade-Offs and Carry-Over Effects

Reproduction does not occur in isolation; rather, it is axiomatic that an individual's fitness is dependent on successful integration of multiple life-history stages (non-breeding, breeding, molt, migration), that is, the individual's complete life history. Our understanding of how activities during one life-history stage can impact other life-history stages through carry-over effects and trade-offs, and especially our understanding of the *physiological mechanisms* that mediate such linkages, remains rudimentary. In large part this is due to the difficulties of following individuals across multiple life stages. However, recent technical developments, such as use of stable isotopes (Rubenstein and Hobson 2004) and increasingly miniaturized automated transmitters and data loggers (Daunt et al. 2006; Mackley et al. 2010; Wikelski et al. 2007), are making this work increasingly feasible. A predominant conclusion from early studies was that a bird's annual cycle is organized so as to *segregate* the major energy-demanding functions of molt, migration, and reproduction (Helms 1968; King 1972); for example, separation of breeding and molt was regarded as an adaptation for spreading the energy demands of these activities (Murton and Westwood 1977). This idea of reproduction, molt, and migration as "mutually exclusive energetic activities" that are "sequentially orchestrated," where overlap between successive stages is minimized (fig. 7.1a), remains widespread (Bradshaw and Holzapfel 2007; A. Dawson 2008). However, even if different life stages are temporally separated, it is becoming clear that seasonal interactions among different phases of the annual cycle are common. As one example, environmental conditions and/or behavior in wintering areas can influence breeding events weeks or even months later; there can be long-term "carry-over effects" (X. Harrison et al. 2010; Marra et al. 1998; Sorensen et al. 2009; S. Wilson et al. 2007). Furthermore, there *can* be direct temporal overlap between life-history stages (fig. 7.1b), such as between migration and reproduction (Bauchinger et al. 2007; Crossin et al. 2010) and between breeding and molting (Hemborg and Lundberg 1998; Neto and Gosler 2006).

In this chapter I will first review the evidence for interactions between life-history stages, centering this around reproduction itself: my intention is to put reproduction in the context of the complete life-cycle. I will

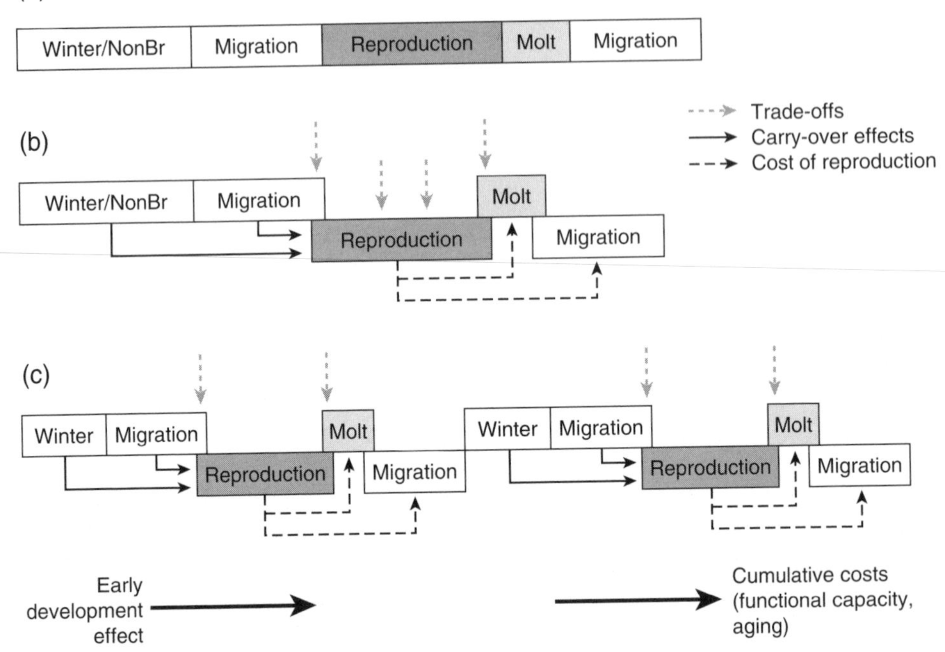

Fig. 7.1. Alternate models for organization and physiological analysis of life histories: (a) *sequentially orchestrated* stages, in which overlap between successive life-history stages is minimized; (b) an annual cycle with overlap between successive stages integrating trade-offs, costs of reproduction, and carry-over effects among life-history stages; and (c) integration over a complete life cycle, including early and late development effects (e.g., senescence).

consider how the wintering and pre-breeding period, including spring migration, can influence reproductive decisions, and in turn how reproductive decisions can influence subsequent post-breeding life stages such as molt, fall migration, and over-winter survival. I consider the costs of reproduction, which we have encountered throughout this book, simply as a more specific example of general carry-over effects, especially from a mechanistic point of view, with potentially common underlying mechanisms. As we have seen, there are costs of reproduction associated with each breeding stage (egg production, incubation, and chick-rearing; chapters 4, 5, and 6), and these costs can be manifest (a) in the short-term, through negative effects on subsequent breeding stages in the same breeding attempt, (b) in subsequent breeding attempts in the same year (e.g., second broods), or (c) through long-term negative effects on future fecundity and survival many months after reproduction. As elsewhere in this book, I will focus on studies linking life stages in *individuals* and in *females* and highlight the role of individual quality in carry-over effects. Second, I will consider the potential mechanisms underlying trade-offs and carry-over effects, focusing on the fact that these mechanisms need to explain interactions (both positive and negative) between life-history stages that extend over many months, if not years. I will argue that short-term energy or nutrient "debts" and resource-allocation trade-offs provide unsatisfactory models for such long-term carry-over effects or costs of reproduction and will instead emphasize potential "non-resource based" mechanisms (sensu T. Williams 2005; Harshman and Zera 2007; Zera and Harshman 2001).

7.1. Carry-Over Effects between Winter, the Pre-breeding Period, and Reproduction

An accumulating number of studies have now shown that the non-breeding habitat, or an individual's activity during the non-breeding or pre-breeding period, can have important carry-over effects that influence the subsequent reproductive period. In neo-tropical migrant passerines, isotope signatures of winter habitats (wet forest, mangrove, citrus or scrub) in their feathers indicate that the quality of the habitat individuals occupy correlates with their body condition, spring departure dates, timing of arrival on the temperate breeding grounds, and subsequent breeding success (Marra et al. 1998; Norris, Marra, Kyser, et al. 2004). Norris, Marra, Kyser, et al. (2004) predicted that, theoretically, female American redstarts occupying high-quality winter habitat would produce more than two additional young and fledge offspring up to a month earlier compared with females wintering in poor-quality habitat. In migratory passerines (R. Smith and Moore 2003) and geese (Bety et al. 2003; Bety et al. 2004; Prop et al. 2003), body condition (fat stores) during pre-migration, on spring staging areas, or on arrival at the breeding grounds can be correlated with the timing of egg-laying, size, and number of eggs laid, or overall reproductive success (fig. 7.2). In one of the few experimental studies to date, Robb et al. (2008) provided supplemental food ad libitum at 10 sites in deciduous woodland during winter, from November to March, then stopped feeding 6 weeks before the first recorded laying date (thus ruling out any *direct* effect of food on reproduction). Blue tits at supplemented sites laid an average of 2.5 days earlier than birds at control sites and had higher fledging success (controlling for laying date), even though there was no difference in clutch size or brood size at hatching.

A simple explanation for many of these results is that use of higher-quality wintering or migratory habitat allows some individuals to

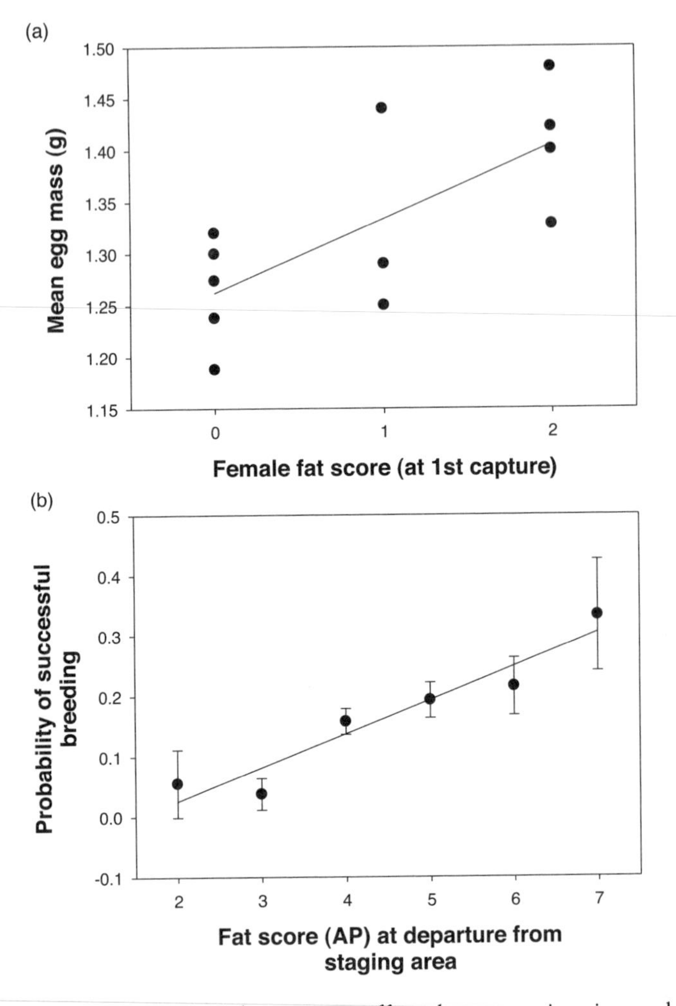

Fig. 7.2. Potential examples of carry-over effects between migration and breeding: (a) positive relationship between fat score on arrival at the breeding grounds and mean egg mass in female American redstarts (based on data in R. Smith and Moore 2003); and (b) relationship between fat score at migratory departure (based on abdominal profile) and probability of successful breeding in barnacle geese (based on data in Prop et al. 2003).

maintain better "body condition" or higher nutrient reserve levels, which then contribute directly to reproduction. Robb et al. (2008) suggested that food supplementation during winter enabled females to enter the breeding season in better body condition, with "improved parental condition enabling the birds to care better for chicks" (although the possibility that higher-quality, more competitive birds had settled and bred in food-supplemented sites could not be completely ruled out in this study). Carry-over effects might therefore reflect simple cycles of resource acquisition (during the non-breeding period) and utilization (during breeding). However, for a number of reasons I think this idea is probably oversimplified, at least in species that do not rely heavily on endogenous (stored) reserves for reproduction (see section 7.5). First, studies linking the quality of the non-breeding habitat and reproduction rarely report strong or consistent relationships with direct measures of body condition (Norris, Marra, Kyser, et al. 2004; Sorensen et al. 2009). Second, as I have discussed at length (chapters 4 and 5), it is not clear if, or how, simple nutrient availability or body condition determines reproductive effort. Third, some carry-over effects appear counterintuitive, and it seems likely that individual quality is a main driver of many of these effects (as we have seen so often throughout this book). For example, Daunt et al. (2006) measured variation in over-winter foraging behavior and subsequent breeding phenology in the European shag (*Phalacrocorax aristotelis*) and showed that individuals that spent less time foraging in February (6 weeks prior to egg-laying) had earlier laying dates. This relationship was much stronger in females than males, and individual females had almost twofold variation in daily foraging duration (fig. 7.3). Daunt et al. (2006) suggested that there were "intrinsic differences" in foraging efficiency among individuals, enabling high-quality females to achieve their daily food requirements in less time, and that high-quality females then also had earlier laying dates. Similarly, Gunnarsson et al. (2005) tracked black-tailed godwits (*Limosa l. islandica*) throughout their migratory range and made a stable-isotope analysis of habitat quality to show that individuals that occupied higher-quality breeding sites, and that had higher breeding success, also used higher-quality wintering sites.

These studies suggest that the quality or use of the winter/migration habitat and improved reproduction do not have a simple, direct cause-and-effect relationship, but rather that they both might be correlates of some other component of individual phenotype that determines individual quality. If so, as Gunnarsson et al. (2005) point out, such "seasonal matching" might inflate the differences in individual fitness over the complete annual cycle, and the fitness measured in one season might underestimate true differences in fitness among individuals.

7.2. Costs of Reproduction

The concept of cost of reproduction is an integral and long-standing component of life-history theory (Harshman and Zera 2007; Stearns 1992;

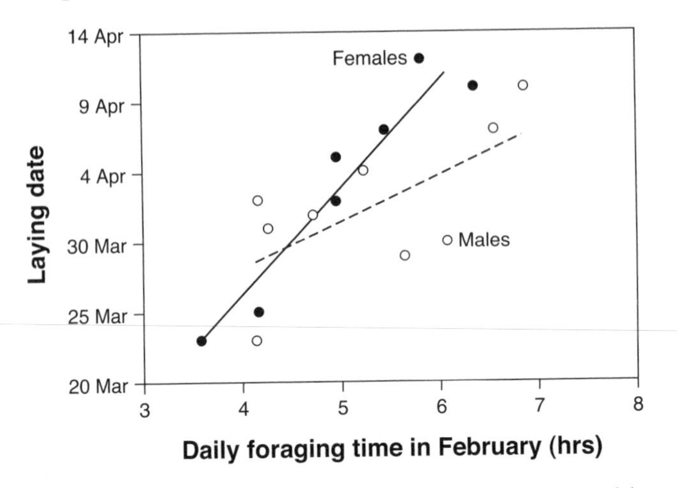

Fig. 7.3. Relationship between time spent foraging in February and laying date in the subsequent breeding season in male (open circles) and female (closed circles) European shags. (Reproduced from Daunt et al. 2006, with permission of Springer.)

G. Williams 1966). Assuming that evolution maximizes individual fitness by maximizing the number of successful offspring produced during an individual's lifetime (an assumption we have encountered throughout this book), then reproduction makes a positive contribution to fitness (Reznick et al. 2000). However, if resources are finite, then reproduction must come at some cost because resources directed toward reproduction cannot be used for other somatic functions that contribute to self-maintenance (the *resource-allocation hypothesis*). Thus, increased reproductive effort in a current breeding attempt should lead to some measureable reduction in survival or future fecundity, that is, a decline in residual reproductive value (G. Williams 1966). Although cost-of-reproduction trade-offs are assumed to be ubiquitous, there continues to be considerable debate about the universality of such costs (M. T. Murphy et al. 2000; Reznick et al. 2000; Shutler et al. 2006). Several reasons have been proposed for the absence of, or failure to detect, the costs of reproduction, most of which revolve around individual quality. If individuals differ in their access to or acquisition of resources, then although *each individual* will face a trade-off decision, it will not necessarily result in a negative relationship between traits *among individuals*, because some individuals will have more resources to allocate to all aspects of their life history (Reznick et al. 2000; van Noordwijk and de Jong 1986). Furthermore, if individuals are able to optimize their reproductive investment to maximize fitness over their lifetime, then any negative correlations between traits such as reproduction and survival could be masked. Experimental manipulations, in which individuals are forced to raise (or reduce) their current reproductive investment, are more likely to reveal trade-offs than will phenotypic correlations (Lessells 1991; Reznick 1985), in part because such experiments can uncouple individual quality and current reproductive effort. Finally, under favorable conditions, all individuals might have the ability to increase their intake of resources and compensate for costs of reproduction, so costs should only be evident when resources are limited or population density is high (Erikstad et al. 1998). Although these explanations focus on the importance of individual variation in resource availability, it has also been suggested that costs of reproduction could be masked by other components of phenotypic variation among individuals, for example, differences in previous experience, social rank, or age, such that "high quality females always do better" over their whole lifetime (Hamel et al. 2009; Lescroël et al. 2009; Sanz-Aguilar et al. 2008).

There is good evidence for costs of reproduction in birds from studies that have experimentally manipulated the reproductive effort (chapters 4, 5, and 6). A common assumption has been that these costs arise from the most energetically demanding breeding activity, which in birds has long been held to be the chick-rearing period (though Clutton-Brock [1991] called this "a dangerous assumption"). However, as we have seen, there is experimental evidence that costs of reproduction are associated with all breeding stages, from egg production through chick-rearing, an association consistent with the idea that all components of the life cycle contribute to overall fitness. Greater reproductive effort during egg production can be associated with reduced chick growth and survival in the same breeding attempt, which is mediated by a reduction in parental provisioning abilities (Monaghan et al. 1995; Monaghan et al. 1998), as well as by diminished physiological condition and reduced survival up to 1 year after breeding (Descamps et al. 2009; Nager et al. 2001; Visser and Lessells 2001). While there is little evidence that the manipulation of incubation effort results in short-term costs of incubation, such as changes in incubation behavior or reduced hatching success (chapter 6), there is accumulating evidence for long-term costs of incubation in terms of offspring recruitment, future fecundity, and survival (de Heij et al. 2006; Visser and Lessells 2001). Similarly, while there is surprisingly little consistent evidence that variation in parental effort during chick-rearing is associated with short-term costs in terms of chick growth or fledging success (chapter 6), some studies have shown that there are longer-term, negative effects of increased chick-rearing efforts on offspring recruitment (Gustafsson and Sutherland 1988) or future fecundity and/or survival of the parents (Daan, Deerenberg, and Dijkstra 1996; Nur 1984a; W. Reid 1987; Roskaft 1985).

A common pattern that emerges from experimental studies is that the costs of reproduction are often not manifest in the first, manipulated, breeding attempt but are deferred and become apparent only during subsequent breeding attempts in the same season or post-breeding, through reduced future fecundity and survival, weeks to months after the reproductive effort was incurred. This long-term nature of the costs of reproduction, which carry over into subsequent life stages, is of critical importance when considering the potential physiological mechanisms underlying these costs, and I will return to this issue in sections 7.4 and 7.5.

7.3. Carry-Over Effects between Reproduction and Post-breeding Life Stages

7.3.1. Post-fledging Parental Care

Many studies use fledging mass or number of chicks at fledging as a surrogate of individual "fitness," with fledging considered to mark the end of the period of parental care. However, a lengthy period of post-fledging care of offspring, which can often be longer than the nestling period, is common in birds (Wheelwright et al. 2003). Continuing post-fledging care can result in trade-offs *among broods* in the same individual in those species making multiple breeding attempts or trade-offs between parental care and subsequent life-history stages (e.g., molt or migration; see below). In great tits, parental care continued for 20 days (range 10–32 days) after fledging (Verhulst and Hut 1996), the duration of post-fledging care was positively correlated with the interval between the first and second clutch, and the initiation of second clutches significantly reduced the female's contribution to post-fledging care of the first brood. In wood thrushes (*Hylocichla mustelina*), the young fledge by 11 days of age but they achieve independence only between 17 and 25 days post-fledging (Rivera et al. 2000). Similarly, Byle (1990) showed that although dunnock chicks fledge at 12–13 days of age, they remained completely dependent on their parents for food in the first week post-fledging and only became independent (defined as the time that the chicks were self-reliant for food) between 22 and 30 days post-hatching. In brown thornbills (*Acanthiza pusilla*), the relative duration of these phases of parental care is even more extreme: the nestling period lasts 14–18 days, but post-fledging care continues for up to 6–8 weeks (D. Green and Cockburn 2001).

In some species, females contribute less to post-fledging care than males; for example, females and males attended young for a further 13 and 19 days respectively in wood thrushes (Rivera et al. 2000). In contrast, in Savannah sparrows (*Passerculus sandwichensis*), the duration of post-fledging care and the number of fledglings cared for were the same for male and female parents (Wheelwright et al. 2003). Similarly in the dunnock, both parents contribute to feeding the fledglings, but there was evidence for brood division, that is, the splitting of a brood into smaller family units fed exclusively by one parent. The occurrence of brood division between parents following fledging can be highly variable within species and can depend on whether the parents are renesting or not (Rivera et al. 2000). In thornbills, individual fledglings were usually fed almost entirely by only one parent, either because the females re-nested and the males consequently cared for all fledglings (8 of 26 broods), a single parent cared for all fledglings (5 of 26 broods), or the broods were divided, with each parent providing more than 90% of all food delivered to specific fledglings (12 of 26 broods; D. Green and Cockburn 2001).

These studies show that it is common for the duration of post-fledging parental care to exceed that of the nestling phase, and this care represents a substantial, though rarely studied, component of reproductive investment. Furthermore, Wheelwright et al. (2003) showed in Savannah sparrows that caring for fledglings can be costly: local survivorship (return rate) of parents declined as a function of the duration of post-fledging care and the number of fledglings cared for. Direct overlap in reproductive investment between successive broods might be especially significant in females if there are physiological costs of egg production (e.g., reproductive anemia; see section 7.7.3) that females incur when initiating their second clutch while they are still contributing to provisioning of their first brood (Willie et al. 2010).

7.3.2. Breeding-Molt Overlap

Although it is still commonly assumed that breeding and molt are separated temporally, in fact overlap of breeding and molt is widespread in birds (G. Morton and Morton 1990; Svensson and Nilsson 1997). In pied flycatchers breeding in Sweden (59°N), an average of 67% of males and 41% of females started molting before their young fledged (Hemborg 1999a), and at a more northerly site (68°N), an average of 41% of males and 25% of females showed breeding-molt overlap (Hemborg 1999b). During a 5-year period, the annual variation in the proportion of individuals with molt-breeding overlap varied considerably (fig.7.4), with greater overlap in males in years with a later mean laying date. In the same species at a more southerly breeding location (central Spain, 40°N), where egg-laying starts about 2 weeks earlier than in Sweden, only 24% of the males and 15% of the females showed breeding-molt overlap during chick-feeding (Sanz 1997). This suggests that breeding-molt overlap varies with latitude, perhaps as a mechanism for adjusting

Fig. 7.4. Annual variation in the proportion of male and female pied flycatchers showing breeding-molt overlap over a 5-year period. (Based on data in Hemborg 1999a.)

the length of the breeding season in the face of different time constraints. In most species the males initiate molt earlier than the females, while there are still dependent young, but in mute swans, the males initiate molt after their female partners. McCleery et al. (2007) suggested that male swans delay molt because it is important for a breeding pair with young to have at least one parent fully winged for the protection of the brood. Breeding-molt overlap appears to be especially common in avian species from the New World and the African tropics (Foster 1975). Foster (1974) proposed that molt-breeding overlap in the tropics was an adaptation to prolong the duration of the breeding season, increasing the potential for renesting in situations when nesting success from individual breeding attempts was low.

Breeding-molt overlap is most common during chick-rearing, but some species initiate molt at even earlier breeding stages. In white-crowned sparrows, 2% of males (n = 110) began molt before their mates had finished laying, 14% of males and 8% of females (n = 96) began molt before incubation was completed, and overall, 33% of males and 19% of females had initiated post-nuptial molt before their chicks had fledged (G. Morton and Morton 1990). Some longer-lived species, such as southern fulmars (*Fulmarus glacialoides*) and giant petrels (*Macronectes* spp.), also show breeding-molt overlap during incubation (Barbraud and Chastel 1998; Hunter 1984), although this is not characteristic of all petrels, some of which delay molting their primary feathers until the young are near fledging or after departure from the breeding colony (Beck 1970; R. Brown 1988). In the earlier-breeding giant petrel species (*M. halli*), males start molting from mid-incubation to hatch, and females from 3 weeks post-incubation. In contrast, in *M. giganteus*, which breeds 6 weeks later, males initiate primary molt at the time of egg-laying, and females at the time of hatch (60 days later), even though females molt at the same rate as males. Hunter (1984) suggested that the more extensive overlap between molt and breeding in giant petrels is the consequence of very abundant and easily available summer food supplies (mainly seal carrion), and that delaying molting until after fledging, when food availability declines in Antarctic waters, might be too costly and could have a negative influence on winter survival or future reproduction. However, to my knowledge only one study, in a very different species, the pied flycatcher, has examined the effects of food availability on breeding-molt overlap experimentally (Siikamaki 1998; see below).

The widely held assumption that both breeding and molt are energetically expensive (an assumption not well supported by empirical data; see section 7.5) suggests that breeding-molt overlap might involve a trade-off between adult somatic maintenance (feather renewal) and parental care. Some studies have shown that individuals that molt while breeding have reduced chick-feeding rates, lower chick fledging mass and/or smaller brood size at fledging (G. Morton and Morton 1990; Svensson and Nilsson 1997), but other studies have found little or no effect of the molt status of females, or their male partners, on the outcome of the current breeding attempt (Hemborg and Merilä 1998; Morales et al. 2007; Siikamaki 1998). Hemborg (1999b) argued that the fact that females whose mates initiated molt before fledging *did not* have lower fledging success suggested that females had to increase their reproductive effort to compensate for the reduced parental care by their molting male partner. This predicts that there might be longer-terms costs of breeding-molt overlap in females. Indeed, in collared flycatchers, female survival was independent of the female's own molt status just before the fledging of the young, but females mated to molting males had a significantly lower return rate as compared with females mated to a non-molting male (fig. 7.5a). Furthermore, females mated in the previous year with males showing molt-breeding overlap started breeding significantly later than females that were aided by a non-molting male; that is, breeding-molt overlap in the male partner negatively affected future fecundity (Hemborg and Merilä 1998). In contrast, in migratory pied flycatchers, females with molt-breeding overlap were actually in better condition at the end of the season, and a higher proportion of molting females returned to the breeding grounds in the following season compared with non-molting ones (Morales et al. 2007; fig. 7.5b). Morales et al. (2007) suggested that

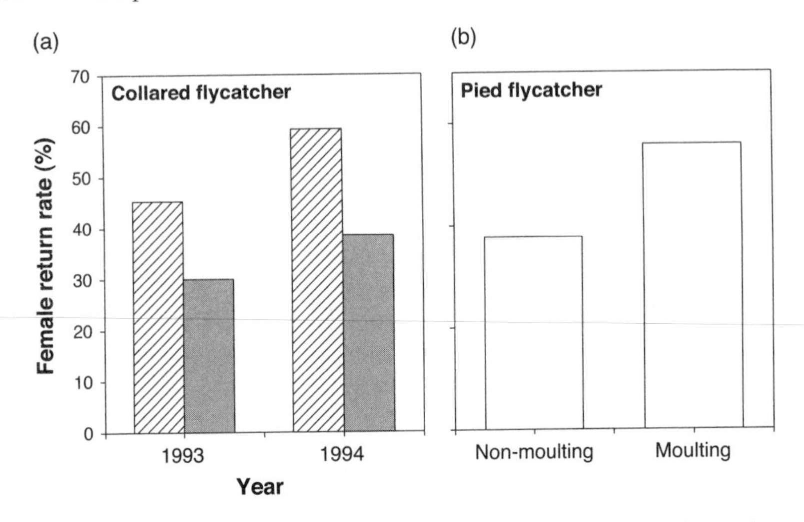

Fig. 7.5. Local return rate of females (a) in relation to breeding-molt overlap of their mates in collared flycatchers (data from Hemborg and Merilä 1998; females mated with non-molting males, hatched bars; females with molting males, open bars) and (b) in relation to female molt status in pied flycatchers (data from Morales et al. 2007).

overlapping molt and breeding might reflect a shift in resource allocation from current reproduction toward increased self-maintenance that would, in turn, lead to increased survival and/or future fecundity. Flinks et al. (2008) found that in migratory European stonechats (*Saxicola rubicola*) there was no effect of late breeding, nor of an increased probability of breeding-molt overlap, on survival, and they concluded that birds could display considerable plasticity in their molting and breeding schedules.

Although breeding-molt overlap might have costs, delaying molt due to late breeding can also have negative consequences if there are subsequent time constraints, for example, autumn territoriality or preparation for, or onset of, dispersal or migration. Later-breeding albatrosses, which have less time for molt after reproduction, replace fewer primaries each year, and this might have long-term consequences on future reproduction, since a year of breeding may need to be skipped to replace the accumulated worn-out primaries (Langston and Rohwer 1996; Rohwer et al. 2011). Nilsson and Svensson (1996) and Siikamaki (1998) experimentally delayed hatching in blue tits and pied flycatchers, respectively, and showed that delayed birds were more likely to initiate molt while they were still feeding chicks, supporting the idea that time constraints are important in determining molt schedules of late breeders. The delayed onset of molt in late breeders can be associated with a faster molt, especially in late-breeding females (G. Morton and Morton 1990) and

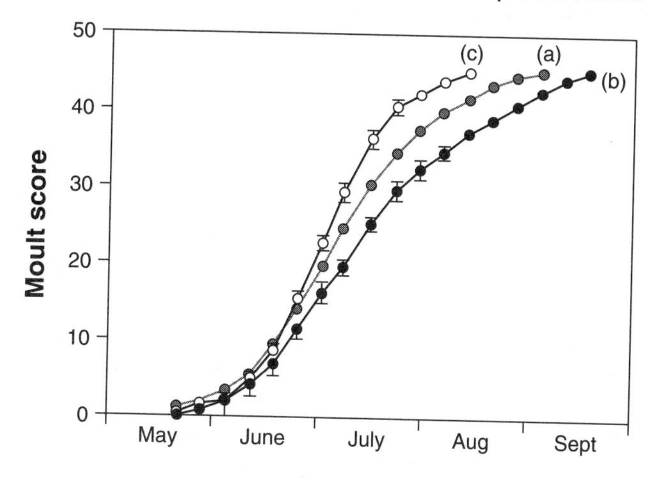

Fig. 7.6. Effect of day length on molt. Birds were held on (a) natural changes in day length (gray circles) or (b) constant day length (18L:6D; solid circles) or (c) were exposed to day lengths that decreased by 1 h each week from the beginning of molt, in mid-May, until day length was 12 h, 6 weeks later (open circles). (Reproduced from Dawson et al. 2000, with permission of the Royal Society.)

late-fledging offspring (Bojarinova et al. 1999). Dawson et al. (2000) suggested that later-molting birds molt more rapidly because the rate of molt is influenced by the seasonal change in day length. The start of molt is triggered by long days, but once it has started, the decreasing day length increases the rate of molt (Dawson 1994, 1998). In European starlings, molt duration under natural changes in day length is 85 ± 2 days (fig. 7.6). However, birds held under constant 18L day lengths molted more slowly (103 ± 4 days), and birds exposed to day lengths that decreased by 1 hour each week from the beginning of molt molted more rapidly (73 ± 3 days). Since most species of birds at temperate latitudes molt after the summer solstice, individuals molting later in the year will be subject to shorter days and a rapidly decreasing day length during their molt and should respond by molting more rapidly (Dawson et al. 2000).

A faster moult can be detrimental to feather quality. Newly grown primary feathers of birds that molt more rapidly are shorter and more asymmetrical, weigh less, have a thinner rachis, and are softer and less rigid than the feathers of birds that molt more slowly (Dawson et al. 2000; K. Hall and Fransson 2000). Furthermore, the feathers produced in a rapid molt can be less resistant to wear so that the differences in mass and asymmetry increase with time (Dawson et al. 2000). These effects of molt speed on feather quality might be especially important for species in which migration imposes time constraints; for example, birds in migrant populations of blackcaps (*Sylvia atricapilla*) grew their feathers faster but produced lighter feathers than birds in sedentary populations (De La Hera et al. 2009). However, the long-term consequences of the effects of molt speed on feather quality remain unknown. It has been suggested that poorer feather quality might increase subsequent thermoregulatory costs (Nilsson and Svensson 1996), decrease flight performance (Swaddle et al. 1996), or constrain the process of feather pigmentation and subsequent expression of carotenoid-based feather ornaments that are important in sexual signaling (Norris, Marra, Kyser, et al. 2004; Serra et al. 2007). Nilsson and Svensson (1996) experimentally delayed breeding in blue tits, removing the first clutch after its completion and causing the birds to lay replacement clutches, thus delaying fledging and onset of molt. When these delayed individuals were recaptured the following winter (January–March), they had 15% higher thermoregulatory costs than control birds, a finding that Nilsson and Svensson suggested was due to the production of low-quality feathers during the late, delayed molt. In addition, in 1 of 2 years, delayed birds had lower survival (20%) than control birds (37%; although it is not clear this was due to feather quality per se, since delayed birds had also had to produce an additional clutch and potentially reared the chicks under sub-optimal feeding conditions).

Nilsson and Svensson's study (1996), coupled with that of Dawson et al. (2000), strongly suggests that a loss of feather quality due to more rapid molt in late-breeding birds might be one mechanism by which the costs of breeding are deferred beyond the end of the current reproductive attempt, via effects on insulation, thermoregulation, and energy expenditure. This is one of the few explicit mechanisms linking breeding, the costs of reproduction, and survival in birds, and it is very surprising, therefore, that there has not been far more work conducted on this potentially very important mechanism.

7.4. Physiological Mechanisms Underlying Trade-Offs and Carry-Over Effects

The physiological basis of trade-offs has been of long-standing interest in life-history studies (Calow 1979; Fisher 1930; Zera and Harshman 2001), but we still know very little about the specific physiological mechanisms underlying any trade-off, including the costs of reproduction (Harshman and Zera 2007; Ketterson and Nolan 1999; T. Williams 2005), or the mechanisms underlying other carry-over effects (X. Harrison et al. 2010; Norris and Marra 2007). Trade-offs are, by definition, negative relationships between traits or life-history stages, whereas carry-over effects can be both positive and negative. Individuals occupying high-quality winter

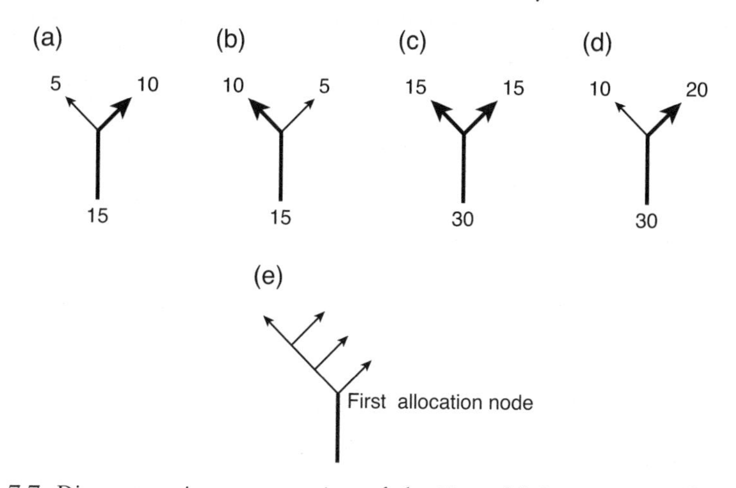

Fig. 7.7. Diagrammatic representation of the Y model for resource-allocation trade-offs. Trees in (a) and (b) illustrate the standard trade-off; tree in (c) indicates the potential confounding effect of increased resource input on a trade-off; tree in (d) shows that an individual with increased resources can still face a trade-off if allocation to both traits is sub-optimal and there are benefits to increased investment in either trait; tree in (e) shows a more complex allocation scheme. (Modified from de Jong 1993 and Zera and Harshman 2001.)

habitats might obtain benefits that enhance subsequent breeding productivity, whereas other individuals might have to expend more resources in over-winter survival in low-quality habitats such that they have fewer resources (a resource "debt") for the subsequent breeding season. Nevertheless, it seems parsimonious to argue that common physiological mechanisms, or at least common currencies, might mediate different trade-offs and carry-over effects. Traditionally, physiological explanations for trade-offs have focused almost exclusively on the competition for, and differential allocation of, limited energy and nutrients among different functions such as reproduction, self-maintenance, metabolism, or growth (Harshman and Zera 2007; Stearns 1992). This emphasis on resource-based mechanisms, or "macronutrient supply," has also been highlighted in relation to carry-over effects (X. Harrison et al. 2010). While it might be true that simple resource-allocation mechanisms (the Y model; van Noordwijk and de Jong 1986; Zera and Harshman 2001; fig. 7.7) offer a plausible mechanistic explanation for trade-offs at a single point in time (e.g., for a direct cost for current reproduction, see below), it is far more difficult to see how immediate or short-term competition for resources and transient energy "debts" or deficits could generate trade-offs over longer time periods, such as between breeding and survival, or carry-over effects operating between different life-history stages.

Given this disconnect between resource-allocation mechanisms and the long-term linkages between life-history stages or seasons, I would argue that it is important to consider a much broader set of potential *physiological* mechanisms (Harshman and Zera 2007; T. Williams 2005). There is a growing acceptance that life-history trade-offs might not necessarily be based entirely on a currency involving energy or resource allocation (Dowling and Simmons 2009; Råberg et al. 1998; Zera and Harshman 2001). For example, in the context of costs of reproduction it has been suggested that the reproductive process itself, or the physiological, hormonal, and metabolic regulatory networks controlling reproduction, might generate trade-offs and long-term costs (Barnes and Partridge 2003; Harshman and Zera 2007; Partridge et al. 2005; T. Williams 2005). This idea of trade-offs arising from direct physiological or hormonal "conflict" between different regulatory systems, in which multiple effects of molecular or physiological regulators directly mediate costs (Harshman and Zera 2007), should be generally applicable to longer-term trade-offs between non-reproductive life-history traits, as well as to carry-over effects. Hormones have also been recognized as strong candidates for *generating* (not just mediating) such trade-offs, because a fundamental characteristic of hormones is that they have both positive and negative pleiotropic effects on multiple physiological systems (Finch and Rose 1995; Ketterson and Nolan 1999; Zera et al. 2007). Can hormonal pleiotropy—a single hormone having both positive and negative effects on multiple physiological systems—or hormonal or physiological "conflict" between different regulatory systems required for multiple overlapping functions provide common, non-resource-based mechanisms to explain phenotypic variation in life-history traits and for trade-offs and carry-over effects between life-history stages or physiological states (see fig. 7.1b)?

Here I will first consider some mechanistic aspects of, and problems with, the general concept of resource-allocation-based trade-offs that have been the focus of most avian studies to date (Gustafsson et al. 1994; Houston et al. 2007; Houston et al. 1995a; Kullberg, Houston, and Metcalfe 2002; L. Martin et al. 2003). I will then consider the idea of direct hormonal or physiological conflict between the regulatory systems underlying migration, breeding, or molt as potential resource-independent mechanisms for trade-offs. In both cases I will extend these ideas to consider potential mechanisms for other carry-over effects (cf. X. Harrison et al. 2010). Finally, I will consider a complementary idea: rather than there being specific mechanisms mediating linkages between specific life-history stages, trade-offs and carry-over effects are driven by more "integrated" physiological systems that determine individual quality and provide common mechanisms linking multiple life-history stages. From a physiological perspective, this alternate viewpoint might further diminish the likelihood of some mechanisms (e.g., resource allocation) providing general explanations for trade-offs and carry-over effects. I will conclude by briefly describing three examples of the more integrative physiological mechanisms that might provide a more satisfactory explanation for long-term trade-offs or carry-over effects: (a) stress, allostasis, and reactive scope (McEwen and Wingfield 2003; Romero et al. 2009); (b) oxidative stress (Costantini 2008; Monaghan et al. 2009); and (c) oxygen-transport, hematology, and anemia. Harshman and Zera (2007) review the evidence for an even broader range of potential physiological and molecular mechanisms for trade-offs including (a) hormonal regulation, (b) intermediary metabolism and allocation, (c) immune function, (d) reproductive proteins, and (e) defenses against stress and toxicity. However, the evidence for some of these mechanisms comes principally from studies of insects, and they have not been explored in birds. Therefore, the aim here is to focus on common mechanisms that might potentially underpin the transitions and linkages between different life-history stages throughout an organism's life cycle and that form the basis of trade-offs, costs of reproduction, carry-over effects, and age-dependent transitions, including early developmental effects and senescence (fig. 7.1c).

7.5. Resources and Resource-Allocation Mechanisms

Several potential trade-offs that are assumed to involve resource-allocation based mechanism have been identified in birds. These include a trade-off between reproduction and flight ability in egg-producing females (Kullberg, Houston, and Metcalfe 2002), immunosuppression associated with parental effort during both incubation and chick-rearing (Ardia et al. 2003; Hanssen et al. 2005), and energetic constraints on molt-breeding overlap (Siikamaki 1998). Unfortunately, these examples highlight the problems of identifying trade-offs without a knowledge of the underlying mechanisms (Zera and Harshman 2001): (a) it is often *assumed* that specific traits or physiological functions are costly and that these energy or nutrient costs are sufficient for investment in one trait or function to cause an *obligate reduction* in investment in a second trait or function; typically, empirical data on the actual energy or nutrient cost of specific functions or activities is weak or lacking; (b) little or no information is available on the specific nature of the *functional physiological link* between the negatively associated traits or functions (see below); and (c) it is unclear, and rarely demonstrated, how transient, short-term energy or resource "debts" that arise due to allocation decisions can have prolonged, long-term effects. Some examples will illustrate these points.

It has been suggested that the impairment of flight ability in egg-laying females is associated with the high protein demands of egg production (see chapter 4), which lead to depletion of protein reserves from the pectoral (flight) muscle and results in decreased pectoral muscle mass (Houston et al. 1995b; Houston, Donnan, and Jones 1995). Interestingly, these original studies did not refer to trade-offs and, in fact, Houston et al. (1995b) suggested that the pectoral muscle may function as a reserve that enables females to provide amino acids for egg production "without impairing normal muscle function." However, subsequent studies explicitly linked decreases in pectoral muscle mass during egg production with a cost-of-reproduction trade-off through effects on compromised flight performance, decreased chick-rearing ability, and potentially higher predation risk (Kullberg, Houston, and Metcalfe 2002; Lee et al. 1996; Veasey et al. 2008). So what do we know about the *currency* and the physiological *mechanism* of this putative trade-off? Initial studies suggested that this trade-off involved competition for specific sulphur-rich amino acids (e.g., lysine, threonine, and methionine; Houston et al. 1995b) that might be a limiting resource for egg production (Ramsay and Houston 1998). However, a more detailed analysis of pectoral muscle showed that the protein band most reduced over the course of egg-laying did not contain elevated levels of these "limiting" amino acids (Cottam et al. 2002). Rather, it was suggested that egg production might involve the breakdown of specific myofibrillar (contractile) proteins or that the "removal" of protein is of a general nature not confined to any specific proteins (Cottam et al. 2002; Veasey et al. 2008). However as we saw in chapters 4 and 5, changes in pectoral muscle mass are not exclusively associated with egg production; there can be marked phenotypic plasticity in body composition (organ size and composition) associated with different stages of reproduction (Vézina and Williams 2003), and in some species egg production does not involve changes in body or pectoral muscle mass (Christians 2000a; Woodburn and Perrins 1997). Furthermore, there are equally plausible alternative explanations, including putative adaptive explanations, for changes in pectoral muscle mass during egg-laying (Lee et al. 1996; T. Williams 2005). Thus, in the absence of specific physiological or mechanistic data it is difficult to conclude unequivocally that such losses represent a true resource-allocation trade-off with protein as the currency and a mechanism involving direct transfer of protein(s) from muscle to eggs.

Another, similar, example is provided by the idea that immunosuppression represents a resource-allocation trade-off. This also starts with the assumption that both reproduction and immune function are sufficiently costly that there is competition between these functions for some limited resource, such that individuals making a large reproductive effort cannot invest as much in immune function (and vice versa). It is widely assumed that the limiting resource or currency for this trade-off is energy, but there is only weak and contradictory evidence that variation in immune function is driven by energy or resource costs per se (Harshman and Zera 2007; Viney et al. 2005). Indeed, this would seem to be a very naive and overly simplistic view of the immune system, which is, without doubt, very complex and has plenty of scope for non-resource-based interactions with other physiological systems, including hormonal regulatory mechanisms (Adelman and Martin 2009; Klasing 1998). L. Martin et al. (2003) estimated the cost of an elevated cell-mediated immune response as 29% of resting metabolic rate in captive, non-breeding house sparrows and equivalent to the cost of production of half an egg (although the latter was not empirically determined); this paper is widely cited in support of the high costs of immune function. In contrast, Klasing (1998) estimated that the resource requirements to support the immune system, including the proliferation of leukocytes and the production of antibodies during an infectious challenge, was very small (1%–4%) relative to the demands of growth or reproduction (although the protein demands of the acute phase response might be more significant). Once again, there is currently little unequivocal evidence that immunosuppression represents a true resource-allocation trade-off in which energy is the currency, and we know virtually nothing about the functional, physiological linkages between reproduction and immune function.

A third example involves the assumption that energy limitation mediates a trade-off between breeding and molt (Hemborg 1999b; Siikamaki 1998; Svensson and Nilsson 1997) or molt and other physiological functions (Moreno-Rueda 2010). It is commonly stated that molt is energetically expensive; however, M. E. Murphy (1996) concluded that the maximum amount of energy deposited daily in the integument of rapidly molting species for feather synthesis and replacement was equivalent to less than 6% of the BMR and less than 2% of the DEE. The main cost of molt appears to come from other physiological processes required to support feather replacement, for example, increased body protein turnover, skeleton restoration, and increased blood volume (M. E. Murphy 1996; Vézina et al. 2009; and references therein). These other costs can potentially be captured in estimates of total energy expenditure or oxygen consumption of molting birds, but these estimates, relative to BMR, vary widely: from no significant change (C. Brown and Bryant 1996), to an approximately 10% increase (M. E. Murphy and King 1992; Vézina et al. 2009), up to an 80%–111% increase over BMR (Lindström et al. 1993; Portugal et al. 2007). Nevertheless, M. E. Murphy (1996) suggested that birds could offset these energy costs by reductions in locomotor activity (e.g., see Portugal et al. 2007). The protein demands of molt, and particularly the requirements for sulphur-rich amino acids, are relatively greater compared with maintenance levels than are energy demands, but there is still little evidence for nutritional constraints on molt, and most birds can meet these increased resource demands over a wide range of diets without any increase in selective foraging or increased intake of metabolizable energy (M. E. Murphy 1996; M. E. Murphy and King 1984). Even severe food deprivation, which is sufficient to cause a 20% reduction in body mass, has been shown to have no effect on the overall duration of molt (Meijer 1991; Swaddle and Witter 1997). Interestingly, more than 15 years ago M. E. Murphy (1996) also concluded that "non-nutritional" processes might play a fundamental role in shaping the molt dynamics in birds, but there has been relatively little interest in this idea since.

The best way to confirm resource-allocation mechanisms and to uncover the currency of the resource that is differentially allocated is to identify the specific physiological, biochemical, and molecular pathways or negative *functional* interactions that are implicit in the use of the term "trade-off" (Zera and Harshman 2001). The Y model of resource-allocation requires that we identify the causal mechanisms involved in changes in the flow of metabolites through pathways of intermediary metabolism at the "allocation node" (see fig. 7.7; de Jong 1993) as well as the regulatory signals of nutrient sensing, nutrient input, and feedback that mediate the "outputs" of somatic and reproductive function (Harshman and Zera 2007). To my knowledge, such work has not been done for any of the putative trade-offs identified in birds (as described above). However, endocrine, biochemical, and metabolic studies on allocation tradeoffs using *Drosophila* mutations and artificial selection in wing-polymorphic crickets provide excellent examples of this approach and have provided detailed information on the functional causes of resource-based allocation mechanisms that underlie the key life-history trade-offs between reproduction and life span, and flight and reproduction, respectively, in insects (Harshman and Zera 2007; Tatar et al. 2001; Zera 2005; Zera and Zhao 2006). The conceptual Y model of resource allocation might be a simplification, and more complex "allocation trees" are probably the case in real organisms (fig. 7.7). However, the first major allocation in the pathway after acquisition of resources is the one most likely to cause negative covariance between traits (de Jong 1993), so Y-models can provide a useful way of thinking about and identifying critical allocation mechanisms.

Even if resource-allocation mechanisms do form the basis for some trade-offs, they are only likely to provide an explanation for long-term costs under certain, specific circumstances, for example, at critical points in development where allocation decisions lead to permanent changes in an individual's physiology and phenotype. The most obvious example in birds would be during the nestling period where "developmental stress" associated with inadequate resources might force chicks to make allocation decisions between different aspects of growth and development (e.g., structural size versus immune function), each of which might have different short- and long-term costs and benefits (Dubiec et al. 2006; Rowland et al. 2007). In fully grown adults, which typically reach somatic maturity at fledging, critical "developmental" phases in which resource allocation could generate long-term effects are inherently less common. However, the annual molt does provide one example as described above (section 7.3.2). Another transient, more temporarily variable example might be if individuals are able to mount only a "sub-maximal" immune response to infection or disease exposure, due to inadequate resources, which is sufficient for short-term survival but then has long-term effects on subsequent survival (L. Martin 2009). Nevertheless, in all cases it is important to *unequivocally demonstrate* that the negative long-term costs (e.g., lower survival) are a *direct* consequence of the immediate, short-term allocation "decision," rather than simply assuming this to be the case.

The long-term nature of carry-over effects that operate over weeks to months between life-history stages or seasons presents a similar problem with regard to if, and how, resources per se could mediate these effects. A simple mechanism for carry-over effects suggests that individuals make the transition between seasons, or stages, in different "condition," that is, with different levels of resources (macro- or micronutrients), which then directly affect performance (reproduction, survival) in the subsequent period (X. Harrison et al. 2010). However, direct evidence to support this simple model is currently lacking: only of 3 of 16 studies of carry-over effects linking wintering habitats, migration, and breeding cited in X. Harrison et al. (2010) included any measure of condition (as body mass, fat mass, or mass change (A. Baker et al. 2004; Ebbinge and Spaans 1995; Marra et al. 1998). Furthermore, none of these studies showed that improved breeding performance was the direct consequence of a better resource state (a more general issue that was dealt with at length in chapters 4 and 5). Recent studies have confirmed that even long-distance migrants primarily use locally obtained, exogenous resources for egg production rather than endogenous reserves that could have been carried from wintering or staging areas (Gauthier et al. 2003; Klaassen et al. 2001; Yohannes et al. 2010). In fact there is no a priori reason why, for example, improved breeding performance has to be causally related to an improved condition or resource state on arrival at the breeding grounds. If the use of high-quality winter habitats allowed birds to maintain a better condition, which in turn allowed for an earlier departure for migration or more rapid migration, then the main benefit might be earlier arrival at the breeding grounds, which itself should lead to greater reproductive success (chapter 3). In other words, *time* could be the critical factor, not resources per se. As with trade-offs in general, it has also been proposed that energy or macronutrients are not the only currency driving carry-over effects and that these effects might also be mediated by micronutrients such as antioxidants, essential amino acids, and minerals (X. Harrison et al. 2010). However, as I have discussed at length elsewhere in this book, it is far from clear if any of these factors are actually limiting in relation to reproduction, and our understanding of the mechanisms involved are sufficiently rudimentary that it is premature, and perhaps unnecessary, to extend a key role for micronutrients to carry-over effects. One simple example serves to illustrate the point. For the majority of birds studied to date, there is no evidence of long-term storage of calcium in the medullary bone in advance of egg production. Rather, the calcium required for shell formation is obtained from the daily dietary intake during the period of egg formation, that is, on the breeding grounds (Graveland and Berends 1997; Wilkin et al. 2009). So, in my view, transport of micronutrients from wintering to breeding areas is unlikely to provide a general mechanism to explain carry-over effects. Once again I would argue that an exclusive focus on resource-based mechanisms for carry-over effects might, at this stage, be counterproductive.

7.6. Mechanisms Arising from Direct Physiological or Hormonal "Conflict" between Overlapping Functions

When different life-history stages, activities, or functions directly overlap (see fig. 7.1b), their coincidence creates the potential for direct "physiological conflict" between different regulatory systems. This conflict could involve pleiotropic effects of single molecular or physiological regulators (Harshman and Zera 2007), or it might reflect some other incompatibility or constraint between different regulatory systems when they are required to function simultaneously. Hormones and endocrine systems are probably important in this regard (Ketterson and Nolan 1999; Ketterson et al. 1996; Sinervo 1999; Zera et al. 2007), but there is no a priori reason why pleiotropic effects could not be caused by a wide range of other regulatory molecules and cellular or biochemical mechanisms. As with resource allocation, physiological conflict mechanisms would most obviously generate short-term trade-offs or direct costs at a single point in time. However, they could also generate long-term effects if the immediate physiological conflict or constraint directly affects, for example, the pattern or level of reproductive investment or the quality of feathers produced during molt, the "cost" of which is not paid until later. Examples of this type of mechanism are still relatively rare simply because we have not focused on phases of *over-lapping* life-history stages or functions. Nevertheless, as described above, overlaps of migration and breeding and of breeding and molt can be common and provide an opportunity for this type of mechanism to operate. Here I will present some examples to illustrate this idea of non-resource-based, physiological conflict mechanisms.

Do females of migratory species initiate reproductive development, vitellogenesis, and yolk formation before arrival at the breeding grounds, while they are still actively migrating? Overlap between breeding and migration has been clearly documented in a pelagic seabird, the macaroni penguin (Crossin et al. 2010). Egg formation takes about 22–23 days in macaroni penguins, but females initiate egg-laying between 7 and 14 days after arrival at the colony. Therefore, a substantial period of egg formation occurs during the inward migration, while birds are returning to the breeding colony. Evidence that egg formation during migration involves a physiological constraint comes from the fact that females with shorter intervals between arrival and laying (i.e., with a greater migration-reproduction overlap) appear less "reproductively ready" at arrival, having significantly lower plasma VTG levels (fig. 7.8a). Moreover, greater overlap between migration and egg production also has an impact upon subsequent patterns of reproductive investment in that these females produce clutches with significantly greater intra-clutch egg-size dimorphism (fig. 7.8b). How general might overlap between migration and egg production be? Reliable data on arrival dates and intervals between arrival and laying can be hard to obtain, but in several other penguin species the minimum interval between arrival and egg-laying is also less than the time required for egg formation, suggesting that the birds are forming eggs at sea, and this might be widespread among migratory pelagic seabirds in which RYD can take up to 20–30 days (Astheimer and Grau 1990). In a shorebird, the Arctic-breeding red phalarope (*Phalaropus fulicaria*), the interval between arrival and laying can be as little as two days, far less than the duration of yolk formation in this species (Schamel and Tracy 1987), and in female surf scoters (*Melanitta perspicillata*), 25% of females ($n = 12$) had plasma VTG levels indicative of the onset of yolk-precursor production on their northward spring migration while still up to 1,200 km from their breeding grounds (J. Takekawa, M. Wilson, and T. Williams, unpublished data). These examples suggest that overlap between migration and female reproductive development might be common, although there are equally clear examples in which females are very unlikely to initiate egg production until after arrival at the breeding ground. In barnacle geese (*Branta leucopsis*) the minimum time between arrival and laying in Spitsbergen, Norway, is 12 days, and for most individuals this interval is much longer (Prop et al. 2003), whereas the time required for egg development is about 12–13 days. Similarly, in

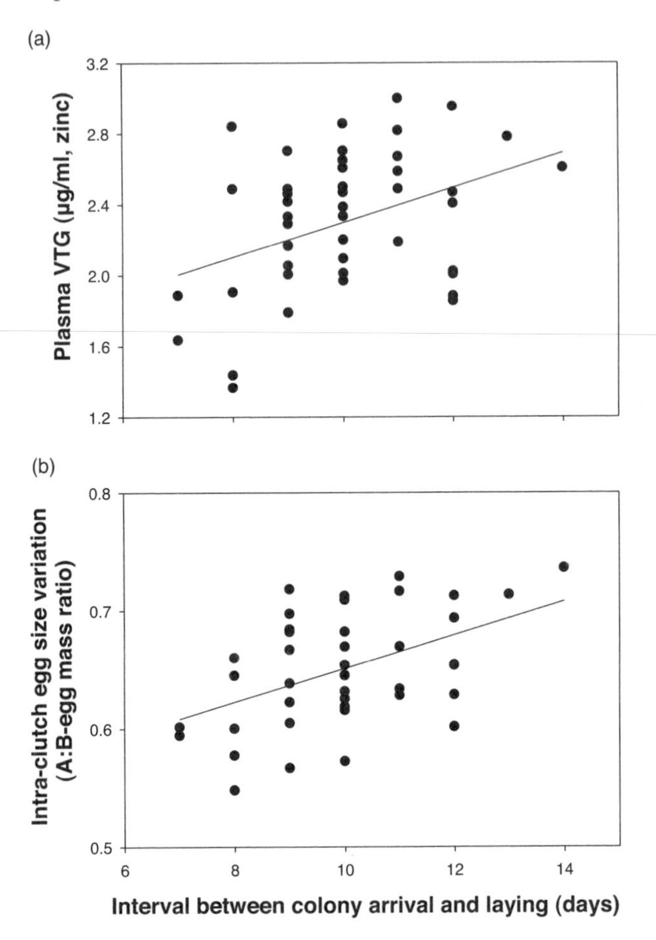

Fig. 7.8. An example of physiological conflict between migration and reproduction in macaroni penguins. As the interval between arrival and laying increases (reflecting decreased migration-reproduction overlap), arriving females have (a) high reproductive readiness (plasma VTG levels) and (b) produce less dimorphic eggs within a clutch. (Based on data from Crossin et al. 2010.)

smaller-bodied birds, RYD can be initiated as little as 4–6 days before laying (chapter 2), which might suggest that overlap between follicle development and migration is less likely. However, in some species there has been a long-term decline in the interval between arrival and laying that is associated with climate change (Both and Visser 2001), such that either migratory overlap or an increasingly rapid transition from a migratory to reproductive state might be becoming more common (Weidinger and Kral 2007).

A number of putative hormonal mechanisms could drive *functional, physiological* conflicts during migration-breeding overlap. Elevated estradiol levels are required for yolk-precursor production (chapter 2), but estradiol is anti-erythropoietic and causes reproductive anemia or a lower hematocrit (see section 7.7.3). It has been suggested that a higher hematocrit is an adaptation for migration (Piersma et al. 1996; Prats et al. 1996), thus the early onset of estrogen-dependent vitellogenesis might cause a premature decline in hematocrit with negative consequences for aerobic capacity and migratory flight performance. The onset of vitellogenesis and yolk formation also involves a major estrogen-dependent shift in lipoprotein metabolism from the synthesis of generic VLDL, which is used to meet the female's own metabolic requirements, to VLDLy (chapter 2). If this process is the functional basis of an "allocation node" (see fig. 7.7), then females might have a limited ability to maintain an adequate supply of generic VLDL to support migratory activity once this major metabolic shift occurs and directs resources to egg production as VLDLy. Thus, a trade-off here could arise solely from the characteristics of the lipoprotein metabolism pathway and from biochemical limitations on the way lipoproteins can be allocated, rather than by the *amount* of lipoprotein resource per se. As a third example, if corticosterone is involved in the physiological adaptations for migration (Holberton et al. 2008), is this function compatible with corticosterone's role in metabolic changes associated with onset of reproduction? In common eiders, females that have higher plasma corticosterone levels at arrival at the breeding grounds lay later and lay fewer, smaller eggs than females arriving with lower plasma corticosterone levels (Oliver Love, pers. comm.).

Another situation in which incompatibility, or "conflict," between regulatory, physiological systems might generate a trade-off is in the overlap between breeding and molt. We still know very little about the hormonal control of molt, although there is evidence for involvement of prolactin, thyroid hormones, and potentially, gonadal steroids (Dawson 2006; Kuenzel 2003). The onset of forced molt in hens is associated with a decrease in T4 and an increase in T3 (Kuenzel 2003), but experimental work in tree sparrows (*Spizella arborea*) suggests that T3 has little effect on feather replacement, and T4 has been proposed as the tissue-active hormone programming post-nuptial molt (Reinert and Wilson 1997), perhaps through a direct effect on the development of feather papillae (Kuenzel 2003). However, since thyroid hormones can interact with and influence endogenous prolactin secretion, it is not clear if the effect of T4 is direct or prolactin-mediated (Dawson 2006). As we have seen, prolactin is thought to play a role in parental care and so would predict a requirement for elevated plasma prolactin levels to support incubation and chick-rearing (see chapter 6). It has also been suggested that prolactin is required for the induction of post-nuptial molt (Kuenzel 2003), and in many species peak plasma prolactin levels occur at, or just before, the start of molt (Dawson 2006 and references therein). However, high prolactin levels per se cannot cause molt because incubating birds do not initiate molt, and these events are markedly temporally dissociated in multiple-brooded species (Dawson 2006; see fig. 6.18). Indeed, as we have seen, reproductive activity can actually delay the onset of molt, especially in females. Dawson (2006) suggested, instead, that molt is initiated not once a particular threshold prolactin level has been attained but, rather, when prolactin starts to decrease. Thus, elevated plasma prolactin levels of breeding birds might actually inhibit the onset of molt and could explain why, in general, molt starts later in breeding birds, in which prolactin cannot decline until the cessation of parental care, than in non-breeding birds (Dawson 2008; Dawson et al. 2009).

The pleiotropic effects of prolactin might then be manifested in two ways: (a) birds might be selected to minimize breeding-molt overlap or to delay molting until after parental care (Dawson et al. 2009), as seen in females, although this could still generate long-term costs associated with more rapid molt, and thus the production of lower-quality feathers; or (b) if breeding-molt overlap does occur, there could be direct, negative pleiotropic effects of prolactin on the molt process. Either way, the timing of molt and the consequences of breeding-molt overlap would arise from the potential pleiotropic effects of prolactin on two different physiological systems, one regulating parental care and one regulating molt itself. The pleiotropic effects of corticosterone provide yet another potential mechanism for breeding-molt overlap. Baseline corticosterone is maintained at very low levels during post-breeding molt (see fig. 2.14), perhaps as a requirement to avoid the hormone's potential degradative effects on proteins and corticosterone-dependent inhibition of protein synthesis during feather growth (DesRochers et al. 2009). But if low corticosterone levels are a requirement for molt, what effect does this have on corticosterone's role in mediating the homeostatic regulation of metabolism, especially when individuals are faced with variable energy demands due to environmental challenges during molt? In individuals with molt-breeding overlap, how do birds reconcile the need to up-regulate baseline corticosterone levels during chick-rearing to support foraging, while down-regulating corticosterone to avoid negative effects on molt? For both prolactin and corticosterone there are, therefore, potential "conflicts" in the hormonal regulation of overlapping functions of molt, parental care, and self-maintenance.

7.7. "Integrated" Physiological Mechanisms and Individual Quality

As we have seen throughout this book, certain high-quality individuals on average express higher fitness-related trait values (e.g., earlier breeding dates, larger clutch sizes) and have greater reproductive success (offspring recruitment). Yet these same individuals often have higher future fecundity and survival and appear not to pay a greater cost for their increased reproductive effort. Numerous studies support the idea that individuals optimize their reproductive investment not just in terms of fitness outcomes of the current breeding stage (e.g., egg-laying or incubation or chick-rearing), but over the entire breeding cycle. In long-lived birds this argument can be extended to the optimization of reproductive effort over multiple breeding attempts, and it is likely that birds must somehow integrate investment decisions, costs, and benefits across all life-history stages. As a consequence, a relatively small number of high-quality individuals produce the majority of the offspring in the next generation, and the majority of individuals that fledge or reach independence die without producing young themselves (see fig. 3.8; Newton 1989a; Clutton-Brock 1988).

How can we conceptualize this long-term, life-history perspective in terms of potential underlying physiological and hormonal regulatory mechanisms? Rather than search for specific mechanisms that might explain the linkages between specific, adjacent life-history stages, it might be more profitable to identify "integrated" physiological mechanisms that could operate over an individual's entire life span and that could provide a *common* mechanism for linkages between multiple different stages and transitions. Such mechanisms might then provide an explanation for, and mechanistic links between, maternal effects and other development effects in offspring as well as for reproductive investment decisions, trade-offs, and carry-over effects in adults. These integrated physiological mechanisms should also provide insight into the components of phenotype that determine variations in individual quality. It is, admittedly, much harder to do physiology at this life-history scale, but it would be analogous to the approach taken in long-term ecological and evolutionary studies (Clutton-Brock and Sheldon 2010). Indeed, the best approach would be to incorporate *comprehensive* physiological studies into existing long-term studies (doing so could provide the continued "novel" contributions from these studies that seems to be required by funding agencies, Clutton-Brock and Sheldon 2010). What is important here, however, is that with this integrated view, the types of physiological systems investigated might be different from those that are the focus of current work. Here I will present a brief and simplified description of three such integrated physiological systems that potentially fulfill the criteria outlined above: they involve (a) marked individual variation in physiological traits, (b) hormonal regulation and the potential for pleiotropic effects generated by these regulatory systems, and (c) functional significance of the physiological system at multiple life-history stages throughout the complete life-cycle in relation to phenotypic variation and individual quality.

7.7.1. Stress, Allostasis, and Reactive Scope

There is an intuitive appeal to the idea that "stress," or the mechanisms underlying responses to stress, might provide an integrated framework for understanding the physiology of life histories. A common assumption is that an increased workload is associated with higher stress levels and that higher stress levels can lead to greater morbidity and mortality. However, this assumption reveals an immediate problem in that the word "stress" has many meanings and is often used rather ambiguously with little explicit reference to the underlying physiological mechanisms (Dallman 2003; Romero 2004). In contrast to the previous statement, an appropriate physiological stress response should be adaptive and increase the chance that an individual will cope with, and survive, a transient, environmental stressor. There has been no shortage of studies of stress physiology in birds (Romero 2004; Wingfield et al. 1998; Wingfield and Romero 2000), but much of this work has focused solely on corticosterone-mediated mechanisms and on patterns of variation in, and the functional consequences of, *maximum* or *stress-induced* corticosterone levels that occur in response to acute stressors or unpredictable events (e.g., a predator attack or a storm). This acute stress response mediates fairly rapid, short-term physiological and behavioral changes (the "emergency life-history stage," Wingfield et al. 1998) to allow individuals to cope with environmental stressors. More recent attempts to place this work in the context of life histories has highlighted the small number of studies that have linked variation in glucocorticoids to variation in fitness, fecundity, reproductive success or survival, either for baseline corticosterone (Bonier et al. 2009) or stress-induced corticosterone levels (Breuner et al. 2008; see section 6.4.2). Variation in corticosterone at any single point in time can be positively, negatively, or non-significantly related to surrogates of fitness, and the relationships can (not surprisingly) vary within individuals at different times in their life history. These inconsistent results strongly suggest that a more comprehensive, more integrated view of stress physiology is required in a life-history context.

In what could provide the required paradigm shift, two related concepts—"allostasis" (McEwen and Wingfield 2003; McEwen and

Wingfield 2010) and "reactive scope" (Romero et al. 2009)—have been applied in an ecological context to provide a more integrated "framework for the organization and management of life cycles" (McEwen and Wingfield 2003). A major strength of these new models is that they try to address how entire life cycles are integrated and controlled by physiological mediators of stress for the "predictable" life-history stages (breeding, migration, molt), not just for unpredictable events (even though, as currently formulated, both models continue to place considerable emphasis on "acute responses to stressors" [Romero et al. 2009] or on the "emergency life history stage" [Wingfield et al. 1998]). The reactive scope model, as presented by Romero et al. (2009), also allows the incorporation of early developmental effects and maternal effects which, as we have seen, might have long-term effects on adult phenotype. In other words, these models potentially provide an integrated framework for considering an individual's complete life history in the context of physiological mechanisms and coping strategies regulating the responses to stress.

Allostasis is defined as "achieving stability through change" and is regulated by a wide range of physiological mediators (e.g., glucocorticoids, catecholamines, cytokines; McEwen and Wingfield 2003). Energy is central to the allostasis model and is considered in a very general sense as "all of the energy and nutrients an organism needs for ... daily and seasonal routines as the life cycle progresses and to deal with unpredictable events" (McEwen and Wingfield 2010). An individual's allostatic state refers to the variable, elevated, and sustained activity of the physiological mediators that integrate physiological and behavioral responses to changing environments. The cumulative result of the allostatic state is the allostatic load, which, within limits, is considered to be the "adaptive" response to variation in seasonal energy demands. Thus, this model should be relevant to the acquisition, utilization, and storage of energy reserves that are widely assumed to be critical for lifetime fitness. Allostatic overload occurs when the energy demands of a particular life-history stage exceed the energy available (including stored reserves). In the short term, elevated levels of physiological mediators of allostasis are thought to have positive effects; that is, they are "adaptive" and should "enhance overall fitness," but if prolonged, they can generate negative effects (McEwen and Wingfield 2003). This allostasis model would therefore appear to represent a good example of pleiotropic effects of hormonal regulators, in this case of glucocorticoids and catecholamines.

Romero et al. (2009) define "reactive scope" as the range of values of the regulatory, physiological mediators of stress required to maintain *normal physiological function in a healthy animal* (thereby getting away from the more restrictive concept of acutely stressed animals). Reactive scope encompasses a range of values for physiological mediators that vary with time and are required for individuals to make adjustments to cope with predictable challenges (e.g., reproduction, migration), a process termed predictive homeostasis (fig. 7.9). Reactive homeostasis then reflects the elevated level of mediators required to establish homeostasis after an unpredictable challenge. Homeostatic overload, which is equivalent to allostatic overload, occurs when the mediator itself begins to disrupt normal function, which would, again, appear to be an example of a pleiotropic effect. The reactive scope model (Romero et al. 2009) extends and develops the concept of allostasis in a number of important ways. First, the model downplays a critical role for energy as the currency for, or a simple measure of, allostatic load (Walsberg 2003). Second, whereas McEwen and Wingfield (2003) highlight the connection between glucocorticoids and energy, Romero et al.'s model (2009) incorporates a much wider range of physiological mediators in addition to glucocorticoids, for example, immune factors (T-cells, cytokines), cardiovascular responses (heart rate, catecholamines), behavior, and central nervous system regulation, dealing with a much more comprehensive, integrated regulatory system (see also Schulkin 2003). Third, Romero et al. (2009) link allostatic load directly to the concept of cumulative "wear and tear," which is defined as the cost of maintaining and using the physiological systems that support an individual's ability to cope with the variable demands of life history and environment (which might parallel ideas of cellular damage discussed in section 7.7.2). Reactive scope therefore explicitly incorporates the short-term and long-term effects of multiple, repeated stressors, including those arising from predictable events and different life-history stages. Romero et al.'s (2009) model still places considerable emphasis on the occurrence of homeostatic overload leading to pathology. However, the physiological modulation of homeostasis within the "normal reactive scope" (including predictive and reactive homeostatic changes; fig. 7.9) would appear to be more relevant for most individuals during the different life-history stages composing their complete life cycle.

From an ecological and evolutionary perspective, both allostasis and reactive scope are largely theoretical concepts at present, and they have been the subject of a largely semantic debate (Dallman 2003; McEwen and Wingfield 2010). What is needed now is experimental work in free-living birds to test the predictions of these models. Nevertheless, they do provide a promising framework for integrating studies of specific physiological mechanisms and mediators (glucocorticoids, catecholamines, cytokines) underlying stress and stress responses in a long-term, life-history context. As Romero et al. (2009) point out, they presented their reactive scope model with a top-down approach "more typical of an ecological approach to addressing physiological mechanism" that did not require

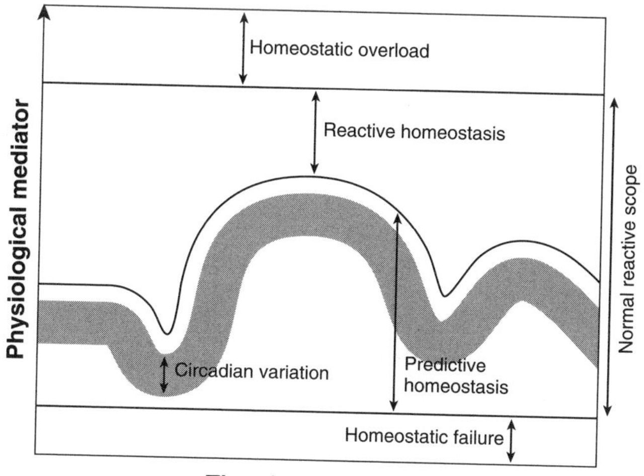

Fig. 7.9. Graphical model of variation in concentrations of different physiological mediators over time showing the normal reactive scope (including predictive and reactive homeostasis) and homeostatic overload. (Reproduced from Romero et al. 2009, with permission of Elsevier.)

detailed knowledge of physiological mechanisms. However, the challenge will be to use the conceptual and predictive value of these models to formulate specific, testable hypotheses that can be addressed with empirical, physiological, and experimental data in an ecological and evolutionary context.

7.7.2. Oxidative Stress

Oxidative stress, driven by the damaging effects of free radicals, or reactive oxygen species (ROS), has been proposed as "a primary and universal constraint in life-history evolution" (Dowling and Simmons 2009; but see Isaksson et al. 2011). The free radical theory of ageing, which was developed largely from studies of model organisms such as yeast, *Drosophila*, and *Caenorhabditis elegans*, is currently thought to represent the most likely candidate mechanism to explain the ageing process across a wide range of species (Finkel and Holbrook 2000; Nemoto and Finkel 2004). However, there is accumulating evidence that the regulatory systems underlying oxidative stress might also provide a framework for the integration of early developmental effects, costs of reproduction and other trade-offs, effects of increased physical activity or workload, and age-dependent functional capacity, in addition to longevity and senescence (Costantini 2008; Metcalfe and Alonso-Alvarez 2010; Monaghan et al. 2009). Oxidative stress reflects an imbalance between the production of ROS and antioxidant compounds. ROS arise as by-products of normal metabolic activity and can have deleterious effects on membranes, proteins, lipids, and DNA (Finkel and Holbrook 2000; Nemoto and Finkel 2004). Since cells continuously produce ROS through routine metabolism, oxidative-stress homoeostasis can only be maintained if antioxidants are present to delay, prevent, or remove oxidative damage (Barja 2004; Costantini 2008). Many antioxidants have been selected and conserved during animal evolution including enzymes (e.g., superoxide dismutase, glutathione peroxidase) and non-enzyme antioxidants of both endogenous and dietary origin (e.g., glutathione, vitamins A, C, E, carotenoids, uric acid; Barja 2004; Costantini 2008). The effects of oxidative stress will therefore depend on the individual's metabolic rate and ROS production, efficiency of antioxidant production and function, and susceptibility to, and repair and rate of accumulation of, cellular oxidative damage.

It has long been assumed that resource-allocation decisions play a fundamental role in mediating the effects of oxidative stress through the allocation of resources to antioxidant production and to maintenance and repair of oxidative damage (Kirkwood 2005). Kirkwood (1977) and Kirkwood and Austed (2000) argued, in the context of age-dependent decreases in function (senescence), that the rate of deterioration of an individual reflects an "optimised" balance between resource allocation to self-maintenance (tissue repair, antioxidant production) and to competing activities such as reproduction. However, there is accumulating evidence that ROS also play important roles in cell signaling and homeostasis, including the regulation of smooth muscle function and blood flow, and in immune function (D'Autreaux and Toledano 2007). Although these ROS signaling molecules are thought to be produced in a more controlled and compartmentalized manner, there is nevertheless the potential for *functional conflict* whereby increased antioxidant production, which would lower oxidative stress, might also have negative effects on ROS signaling and other required functions of ROS. Thus, the link between oxidative stress, life-history traits, and trade-offs might involve resource-allocation decisions (Kirkwood 1977; Kirkwood and Austad 2000). However, it might also occur simply because of functional interactions (sensu Zera and Harshman 2001) whereby one activity or process generates negative consequences for other traits or activities independent of resources per se (Costantini 2008; Monaghan et al. 2009).

Production of ROS occurs continuously via routine metabolism, but periods of increased workload and higher metabolic demand might involve increased production of ROS or a decrease in available antioxidants, thereby leading to an increase in oxidative stress (Costantini 2008;

Leeuwenburgh and Heinecke 2001; Monaghan et al. 2009). Although ROS levels and oxygen consumption are related, this relationship can be complex, and several mechanisms have been identified that may alter the levels of ROS produced or released per molecule of oxygen consumed (Finkel and Holbrook 2000; Leeuwenburgh and Heinecke 2001). Nevertheless, some studies in birds have provided preliminary evidence for a relationship between oxidative stress, or antioxidant status, and workload during several adult life-history stages, for example, reproduction (Alonso-Alvarez et al. 2004; Bertrand, Alonso-Alvarez, et al. 2006; Bize et al. 2008; Wiersma et al. 2004) and migration (Costantini et al. 2008). Alonso-Alverez et al. (2004) suggested that increased susceptibility to oxidative stress might be a general form of a cost of reproduction (see also Dowling and Simmons 2009), but this idea needs to be confirmed with additional data. In addition to variation among adult life-history stages, early developmental conditions and growth rate can influence various components of oxidative stress, including antioxidant status, rates of ROS production, and telomere dynamics (Alonso-Alvarez et al. 2007; Blount et al. 2003; M. E. Hall et al. 2010; M. E. Hall et al. 2004). These early developmental effects might also be associated directly with ROS-mediated functional trade-offs and not necessarily with resource-allocation trade-offs (De Block and Stoks 2008). Finally, the long-term, cumulative effects of the "hidden" costs of ROS production, those associated, for example, with increased reproductive effort, and/or inadequate resource allocation to antioxidants or cellular repair, could lead to faster accumulation of oxidative damage (Monaghan et al. 2009). Lower functional capacity (senescence) and survival could therefore result from the accumulation of unrepaired cellular and molecular damage, which occurs throughout life from the earliest time that somatic cells and tissue first begin to form (Kirkwood 2005; Metcalfe and Alonso-Alvarez 2010; Monaghan et al. 2008). The role of one particular component of cellular damage, telomere shortening, in the evolution of life histories and as a constraint underpinning life-history trade-offs has become an important focus of avian studies (Monaghan 2010). Telomeres, the caps at the chromosome ends that function in maintaining genome stability, are vulnerable to attacks from ROS, and the accelerated reduction in telomere length that results from oxidative stress can hasten cell senescence (Richter and Zglinicki 2007). In several avian species, individuals with the shortest telomere lengths, or the highest rate of loss of telomere length, have the lowest survival (Bize et al. 2009; M. E. Hall et al. 2004; Salomons et al. 2009; fig. 7.10).

Thus, oxidative stress can potentially operate at multiple life-history stages throughout the complete life-cycle of birds. However, although extensive studies in model organisms provide a sound basis for linking oxidative stress, reproduction, life span, and fitness (Finkel and Holbrook

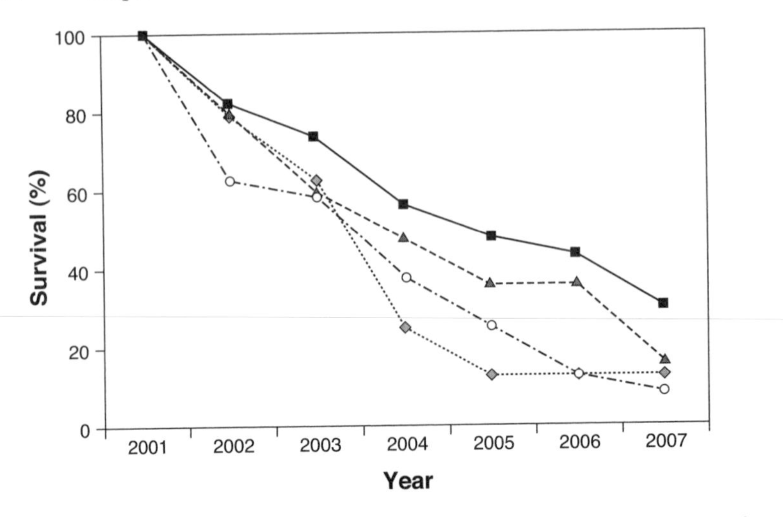

Fig. 7.10. Survival curves for adult Alpine swifts (n = 96) over 6 years in relation to their telomere length. Data are divided into quartiles on the basis of average telomere length in 2001: squares, 4th quartile, long telomeres; triangles, 3rd quartile; diamonds, 2nd quartile; circles, 1st quartile, short telomeres. (Reproduced from Bize et al. 2009, with permission of the Royal Society.)

2000; Nemoto and Finkel 2004), many issues have not yet been critically examined in free-living birds, especially in an ecological and evolutionary context (Monaghan et al. 2009). Most avian studies to date have applied a relatively simple physiological approach, often measuring a single biochemical marker or a single antioxidant, and no studies have directly measured oxidative damage or shown that it then causes (long-term) negative effects on function in free-living birds (Costantini 2008; Metcalfe and Alonso-Alvarez 2010). Physiological mechanisms underpinning oxidative stress are, undoubtedly, complex, and this complexity requires the use of a combination of different measurements; ideally all four components of oxidative stress (free-radical production, antioxidant defenses, oxidative damage, and repair mechanisms) need to be measured simultaneously (Monaghan et al. 2009). Finally, little attention has so far been focused on the role that hormonal regulatory mechanisms and hormonal pleiotropy might play in oxidative stress homeostasis. The insulin/IGF-1 pathway is important in the regulation of nutritional state, metabolism, development, and ageing (Tatar et al. 2003) and has also been implicated in the regulation of ovarian function in birds (Onagbesan et al. 2009; Onagbesan et al. 1999), with evidence for interaction between these two systems in non-avian taxa (Hsin and Kenyon 1999). Some work has been conducted on metabolic hormones (IGF-1, growth hormone, thyroid hormone) in poultry in relation to growth (Baeza et al. 2001; Carew et al. 2003) and reproduction (Bruggeman et al. 1997; Scanes et al. 1999), but not in the context of oxidative stress. Mechanisms underlying oxidative-stress homeostasis therefore provide a good example of an integrated physiological system that operates at multiple developmental and life-history stages. In addition, there is evidence for marked individual variation in the rate of declining functional capacity and the rate of ageing (Stadtman 2002; A. Wilson et al. 2008). Homeostatic and regulatory mechanisms underlying oxidative stress might therefore also represent components of phenotype that determine variation in individual quality.

7.7.3. Oxygen-Transport Systems, Hematology, and Anemia

Whole-organism aerobic capacity is one of the main predictors of endurance, or the ability to sustain a high workload, and it in turn depends on the oxygen-carrying capacity of the blood (Calbet et al. 2006; P. Wagner 1996). Clearly oxygen transport involves a complex, highly integrated, multi-functional physiological system, and all the components of this system, from oxygen uptake in the lungs to oxygen utilization in muscle tissue, contribute to aerobic capacity and maximum oxygen consumption (VO_2max; e.g., J. Jones 1998). One key step in oxygen transport is convection through the vascular system, which is determined by the rate of blood flow and the concentration of the major respiratory pigment, hemoglobin (Hb; P. Wagner 1996). Flucturations in the blood Hb concentration can cause fluctuations in the aerobic capacity and VO_2max, even when the circulating blood volume is maintained at the same level, via changes in the oxygen-carrying capacity of the blood (Calbet et al. 2006). Hemoglobin also has other functions; for example, it plays a role in oxygen off-loading and transfer from the capillaries to the mitochondria in muscle tissue (P. Wagner 1996). The blood-flow rate is inversely proportional to blood viscosity, and hematocrit (Hct; packed cell volume, the relative volume of RBCs compared with total blood volume) is one of the main factors affecting viscosity (Birchard 1997). RBCs are thought to have other functions beyond the simple transport of oxygen; they might sense tissue oxygen conditions as they pass through capillaries and release vasodilatory compounds (ATP, nitric oxide) that enhance the blood flow to hypoxic tissues and match tissue oxygen delivery with local oxygen demand (Jensen 2009). Importantly, Hct has dual, but opposing, effects on oxygen transport: an increase in Hct results in a linear increase in the oxygen-carrying capacity of blood and an exponential increase in blood viscosity. Consequently, the relationship between oxygen transport and Hct is parabolic (fig. 7.11), and oxygen transport should be maximized at some particular hematocrit, with a lower Hct resulting

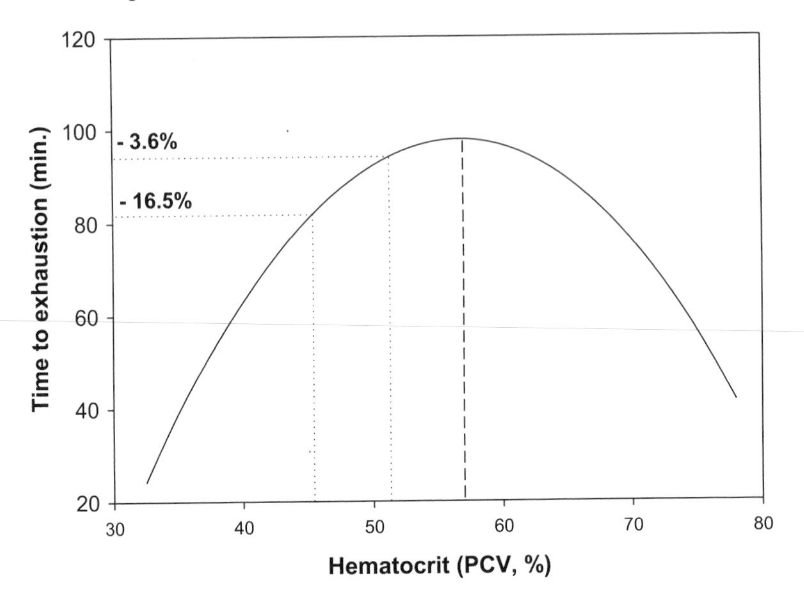

Fig. 7.11. The concept of optimal hematocrit in mice with estimates of decrease in performance (time to exhaustion) based on reductions in hematocrit of 10% and 20% from the optimal value. (Based on data in Schuler et al. 2010.)

in decreased oxygen transport due to reduced oxygen-carrying capacity, and higher Hct resulting in increased viscosity (Birchard 1997). This idea of an optimal Hct that maximizes endurance or performance has recently been confirmed in mice (Schuler et al. 2010). Thus, two easily measurable traits, Hct and Hb concentration, are potentially key determinants of oxygen transport, aerobic capacity, and the ability to sustain high workloads. Can this relatively simple idea be extended to provide another example of an integrated, physiological system that might link phenotype, individual quality, and fitness in a life-history context (fig. 7.1c)? How variable are Hct and Hb, and is there evidence that this individual variation might be functionally significant at multiple life-history stages, for example, during embryo and nestling development as well as in adults?

Hematocrit varies markedly throughout the annual cycle in birds (Davey et al. 2000; Hõrak et al. 1998; M. Morton 1994): in migratory white-crowned sparrows, the highest mean Hct was approximately 60% on arrival at the breeding grounds, and the lowest values (~50%) occurred during post-nuptial molt. Furthermore, despite the predictions of an optimal Hct, there is marked individual variation in both Hct and Hb in apparently healthy, free-living (fig.7.12) and captive birds (E. Wagner, Prevolsek, et al. 2008). Hematocrit varied between 33% and 54%

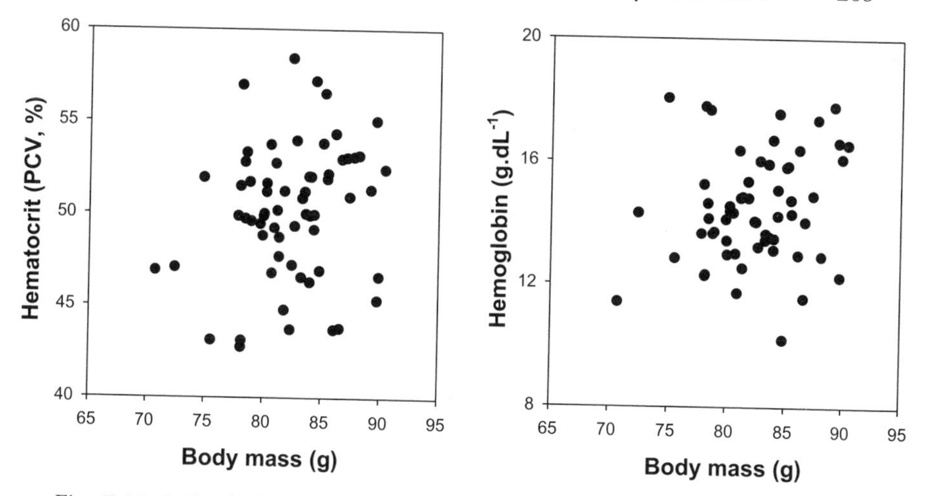

Fig. 7.12. Individual and mass-independent variation in (left) hematocrit and (right) plasma hemoglobin concentrations in female European starlings.

in chick-rearing tree swallows (Burness et al. 2001) and between 40% and 60% in chick-rearing great tits (Ots et al. 1998). Some studies have shown that this variation can be repeatable in free-living adult birds across breeding stages (r = 0.3–0.4; R. Fronstin and T. Williams, unpublished data) and across years (Potti 2007; Potti et al. 1999) and in captive-breeding birds (Hct, r = 0.63; Hb, r = 0.36; E. Wagner, Prevolsek, et al. 2008). E. Wagner, Prevolsek, et al. (2008) also showed that individual changes in hematocrit in relation to reproductive anemia associated with egg production were repeatable (r = 0.32; see below). However, there is also a significant environmental component to variation in adult Hct and Hb (Norte et al. 2009; Potti 2007).

Changes in hematocrit can occur as a result of changes in total blood or plasma volume (hemodilution or dehydration), that is, osmoregulatory adjustments, as might be the case during molt (Chilgren and deGraw 1977) or migration (Jenni et al. 2006; but see Prats et al. 1996). However, variation in Hct can also occur due to changes in the number or size of RBCs per unit volume of blood, as is the case during reproduction (see below), fasting (Le Maho et al. 1981), and winter acclimatization (Swanson 1990). Birds probably have a more limited ability to make short-term adjustments to RBC number than other vertebrates since the avian spleen does not store RBCs (John 1994). Other components of oxygen transport might also be less plastic in birds compared with mammals, for example, Hb-oxygen binding affinity (Swanson 1990). Rather, in birds longer-term (seasonal) changes in hematocrit are probably met by increased production of RBCs (erythrocytes) in the bone marrow or thymus, a process referred to as erythropoiesis, resulting in an increased number of immature RBCs (reticulocytes). The life-span of avian erythrocytes (30–40 days; Beuchat and Chong 1998) also suggests that endogenous changes in Hct and Hb through cell turnover and erythropoiesis will be a relatively slow process, consistent with repeatability within individuals.

One of the major perturbations to hematological homeostasis in adult females is reproductive anemia, which occurs during egg production and which results in a reduction in Hct of 5%–10% on average (Davey et al. 2000; P. Jones 1983; M. Morton 1994). T. Williams, Challenger, et al. (2004) suggested reproductive anemia might provide a mechanism for the long-term costs of reproduction, such as the observed declines in chick-provisioning ability, re-nesting probability, or survival of breeding females (see chapter 4). Reproductive anemia is a complex process, the effects of which can extend well beyond the actual period of vitellogenesis and egg formation (fig. 7.13). With the onset of vitellogenesis, there are rapid decreases in Hct and Hb (fig. 7.13a,b). Hct and Hb remain low at clutch completion but generally return to pre-breeding levels around the time of hatching. However, in some years Hct can remain low throughout incubation and into chick-rearing (T. Williams, Challenger et al. 2004). More importantly, physiological recovery from reduced Hct is prolonged: the mean RBC volume is significantly higher at clutch completion (fig. 7.13c), and the proportion of new immature RBCs (reticulocytes) increases only around hatching, 10–12 days after clutch completion (fig. 7.13d). The latter changes are consistent with more prolonged regenerative erythropoiesis or reticulocytosis: the enhanced production and release of larger, immature RBCs into the circulation following a transient suppression of erythropoiesis (E. Wagner, Prevolsek, et al. 2008; E. Wagner, Stables, and Williams 2008). These changes represent the classic physiological response to recovery from anemia (Fernandez and Grindem 2006).

The initial decline in Hct with the onset of egg production is probably the result of hemodilution: the osmotic movement of water from extracellular spaces into the blood, which increases the plasma volume in order to maintain plasma osmolarity in the face of marked increases in plasma lipids and proteins associated with yolk-precursor production (de Graw et al. 1979; Reynolds and Waldron 1999). However, reproductive anemia also involves transient estrogen-dependent suppression of erythropoiesis and, subsequently, increased reticulocytosis (fig. 7.13d; E. Wagner, Prevolsek, et al. 2008; E. Wagner, Stables, and Williams 2008). In quail and mice, E2 suppresses bone hematopoiesis (Clermont and Schraer 1979) and causes atrophy of the thymus (Olsen and Kovacs 1996), both of which are important sites for erythropoiesis in adult birds. Reproductive anemia is therefore associated with negative pleiotropic effects of

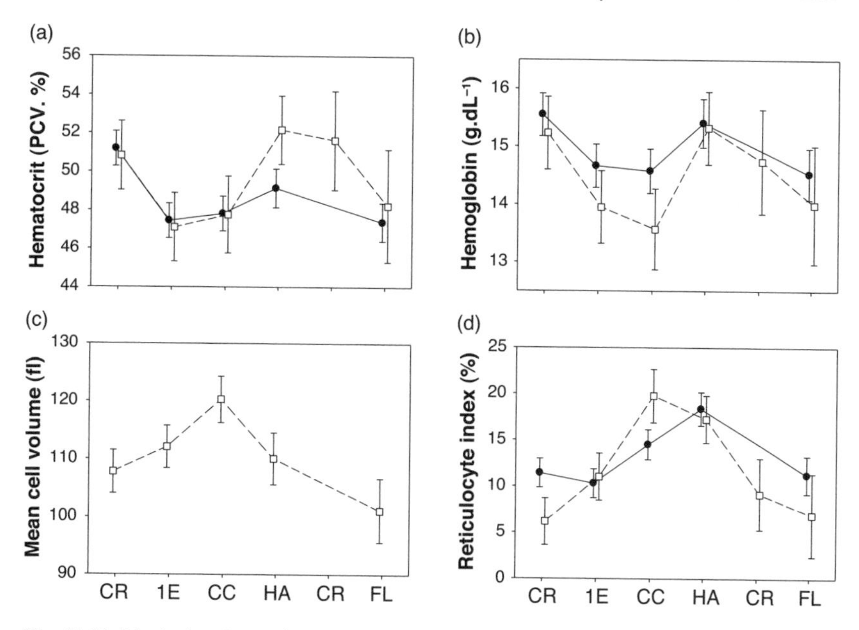

Fig. 7.13. Variation in (a) hematocrit, (b) plasma hemoglobin, (c) mean red blood cell volume, and (d) reticulocyte index during the reproductive cycle in female zebra finches. The two lines indicate data from two different, replicate experiments. (Based on data in E. Wagner, Stables, et al. 2008 and on unpublished data from T. Willliams.)

the essential reproductive hormone estrogen, which causes the long-term effects of, and the slower physiological recovery from, reduced Hct that persist beyond the period of egg production itself (E. Wagner, Prevolsek, et al. 2008).

Kalmbach et al. (2004) suggested that the extent of reproductive anemia might be proportional to reproductive effort, such that females that lay more or larger eggs or that lay eggs more frequently (e.g., in replacement clutches) would become more anemic and, therefore, would potentially pay higher costs for egg production. Kalmbach et al. (2004) experimentally increased egg production in female great skuas (from the normal two-egg clutch to six eggs) by egg removal, and this increase was associated with a greater reduction in Hct and RBC number compared with control females. Furthermore, some of these effects appeared to persist for up to 1 year. However, experimental work on captive-breeding zebra finches suggests that females might also regulate their hematological status to *maintain* Hct, Hb, and RBC number at some minimum functional level, and that this maintenance may occur at a cost of reduced reproductive investment (see also Garcia et al. 1986). Female zebra finches breeding on a low-quality diet do not have elevated levels of anemia, a result that suggests anemia is resource independent, but they do lay smaller eggs and smaller clutches, potentially trading off reproduction for hematological status (E. Wagner, Stables, and Williams 2008). Similarly, in female zebra finches forced to lay either two or three successive clutches, with or without a recovery period during incubation, there was no difference in the extent of reproductive anemia (fig. 7.14). Furthermore, there was no relationship between Hct and egg production among females laying a total of between 7 and 21 eggs over these multiple clutches (Willie et al. 2010). In this scenario, which might be common, for example, in females making multiple re-nesting attempts due to nest predation, females maintained reduced Hct and Hb levels for a prolonged period (20–35 days). Females that initiate egg-laying for a natural second clutch around the time of fledging of their first brood also undergo a second phase of reproductive anemia (E. Wagner, Stables, and Williams 2008; Willie et al. 2010), which might have important consequences for post-fledging parental care. Putative adaptive changes in Hct (between 5% and 10%) that are thought to be associated with varying demands for aerobic capacity and oxygen transport occur during many other adult life-history stages or states, for example, during migration (Bairlein and Totzke 1992; Landys-Ciannelli et al. 2002; Piersma et al. 1996), during winter acclimatization (Swanson 1990), and with increasing altitude (Clemens 1990; Prats et al. 1996; Ruiz et al. 1995). Some studies provide support for a link between Hct and aerobic capacity or flight ability (Hammond et al. 2000) or between Hct and measures of individual female quality, such as clutch size, egg size, and laying date (Dufva 1996; Norte et al. 2010). However, much more experimental work is required to firmly establish the relationships between hematological status, aerobic capacity, workload, individual quality, and trade-offs, including the costs of reproduction.

Finally, a key requirement for an integrated physiological system, as defined above, is that it can provide a common mechanism operating at multiple life-history stages. In this regard, both Hct and Hb change markedly, and systematically, during embryo and nestling development, and individual variability appears to be even greater at these life-history stages than in adults. In the domestic hen embryo, the functional capacity of the chorioallantoic membrane is thought to be at a maximum by day 14 of development, even though embryonic demand for oxygen continues to increase up to hatching, around day 21 (Starck 1998). Oxygen-diffusing capacity increases sixfold between days 10 and 18 of incubation due to increases in embryonic Hct and Hb levels, as well as increases in total blood volume, capillary volume, and total blood flow (Vleck and Bucher 1998). However, superimposed on this general ontogenetic change, there is marked individual variation in embryo Hct levels

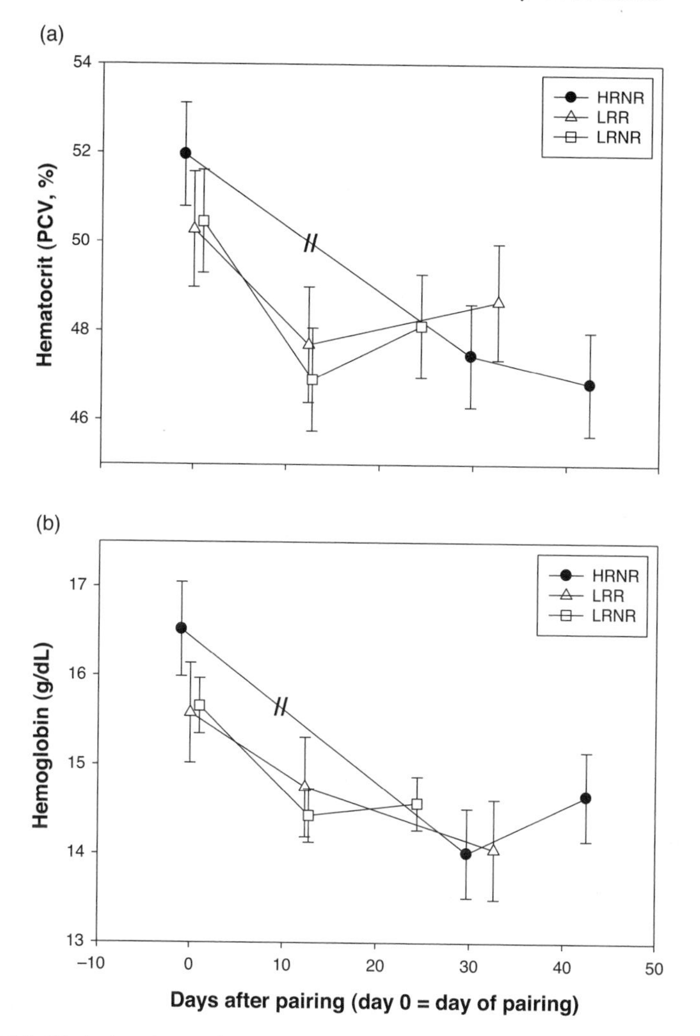

Fig. 7.14. Variation in (a) hematocrit and (b) plasma hemoglobin concentration over time in female zebra finches in relation to reproductive effort: high renesting/no recovery (HRNR) females laid three successive clutches with no recovery; low renesting/recovery (LRR) females laid two successive clutches with 10 days for physiological recovery during incubation; low renesting/no recovery (LRNR) females laid two successive clutches with no recovery. (Reproduced from Willie et al. 2010.

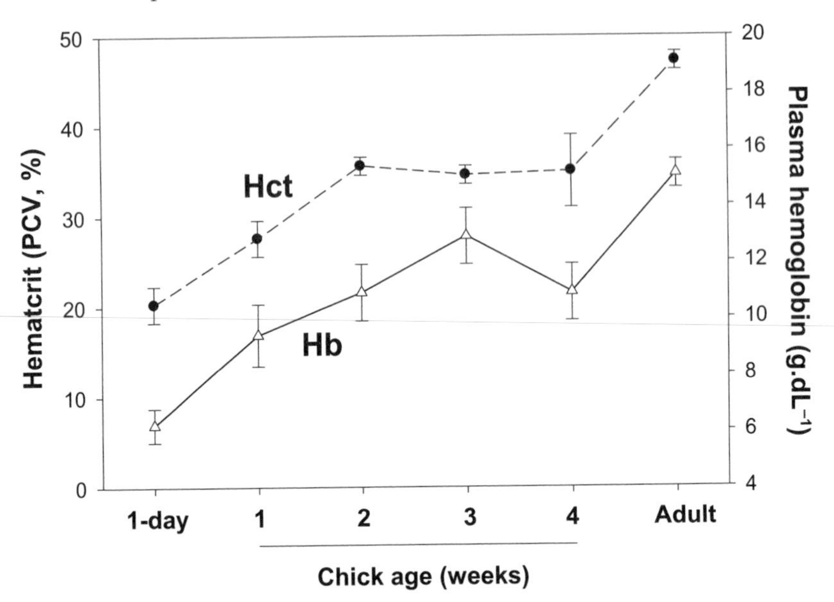

Fig. 7.15. Age-dependent changes in hematocrit and hemoglobin in pigeons during chick growth and in relation to adult values (Based on data in Gayathri et al. 2004.)

when controlling for developmental stage, even just prior to hatching; for example, in individual chick embryos, Hct varied between 24% and 39% (Khorrami et al. 2008). The importance of this individual variation in Hct and Hb for oxygen transport, growth, and hatching success of avian embryos is still largely unknown, especially in non-poultry species (Deeming 2002b; Khorrami et al. 2008) There is some limited evidence for adaptive variation in Hct in relation to hypoxia and embryo development at high altitude (Carey et al. 1994). Furthermore, at least some components of phenotypic variation in adult respiratory function and oxygen transport appear to be determined during ontogeny (B. Williams and Kilgore 1992).

After hatching, during nestling development, Hct and Hb continue to increase (fig. 7.15), driven by an increase in the mass of erythropoietic tissue, and the increase in Hct appears to be quite plastic among populations of the same species (Kostelecka-Myrcha et al. 1997). Once again, there is marked individual variation in nestling Hct and Hb close to fledging, for example 21%–52% in 13-day-old tree swallow chicks (Morrison et al. 2009) and 27%–52% in 17-day-old European starling chicks. Mean Hct and, in particular, blood Hb levels at fledging are also significantly lower than adult values, for example, Hct, 42.7 ± 0.7% vs. 50.8 ± 0.6%, and Hb 11.1 ± 0.2 vs. 15.9 ± 0.4 g/dL in European starling chicks and adults respectively (R. Fronstin and T. Williams, unpublished data). Hct has been reported to have high heritability (0.46–0.80) in the domestic hen (Shlosberg et al. 1998). However, numerous studies in free-living birds have reported only low (h^2 = 0.08–0.14; Potti et al 1999; Christe et al. 2000) or no heritability of Hct (Morrison et al. 2009; Potti 2007). Rather, Hct appears to be a phenotypically plastic trait with most of the variation explained by the environment (nest) where the bird is reared during the developmentally sensitive growth phase (Potti 2007; Simon et al. 2005). Interestingly, the effect of rearing environment on chick Hct is not reflected in body mass, since Hct is independent of fledging mass or condition (Morrison et al. 2009). What is the functional significance of this individual variation in Hct, in the context of developmental or rearing conditions, for determining offspring phenotype, quality, and post-fledging survival?

In summary, Hct and blood Hb, key components of an individual's oxygen-transport system, are highly variable at multiple life-history stages (in embryos, nestlings, and adults). This physiological variation, as well as changes in hematological status, can be repeatable but also has a significant environmental component, and birds might have a more limited ability to make short-term adjustments in oxygen transport because this adjustment occurs through cell turnover and erythropoiesis in birds, not release of stored RBCs. Given the suggestion that, on average, changes in Hct of 5%–10% can have adaptive functional significance (e.g., during migration, during acclimatization, or for flight ability and aerobic capacity in general), then the equivalent natural variation in Hct during reproduction or between years (T. Williams, Challenger et al. 2004) and the much greater individual variation in Hct at all life stages (20%–30%) are likely to be functionally significant. Finally, as with the other physiological systems considered here, hormones play a fundamental role in regulating hematological homeostasis. The glycoprotein erythropoietin is the erythropoiesis-stimulating hormone, and molecular regulation of erythropoietin is quite well understood, at least during hypoxia, with its gene expression controlled by hypoxia-inducible transcription factors in response to various molecular oxygen sensors (Jelkmann 2007). Other aspects of humoral regulation of erythropoiesis are less well understood but appear to involve a wide range of reproductive (estrogens, prolactin) and metabolic hormones (growth hormone, IGF-1, thyroid hormones). Erythropoietin itself also has non-erythropoietic functions, including the stimulation of angiogenesis in the reproductive tract, which also appear to be regulated by estrogens (Fandrey 2004). Thus, there appears to be potential for hormonal pleiotropy, or hormonal "conflict," to be important in this system.

7.8. Conclusion

Reproduction, the events between onset of seasonal reproductive development and fledging of chicks from the nest, or chick independence (chapters 3, 4, 5, 6), clearly needs to be viewed as only one part of an individual's integrated annual cycle (fig. 7.1b) and its complete life-history (fig. 7.1c). There can be complex interactions between over-lapping life-history stages (e.g., migration-reproduction, reproduction-molt) and complex effects of transitions or linkages between sequential life-history stages (carry-over effects). Earlier stages of the annual cycle (wintering, migration, pre-breeding) can significantly affect reproductive decisions, and reproductive decisions can, in turn, have long-term effects on post-breeding life-history stages, in particular due to costs of reproduction, but also, for example, due to the long-term effects on timing or rate of molt.

If we are to better understand the physiology of life-histories, including the physiological mechanisms underlying variation in phenotypic traits and the mechanisms underlying trade-offs between these traits or between life-history stages, then we need to approach physiological studies with this same long-term, integrated, life-history perspective (fig. 7.1c). Simple resource-allocation mechanisms would appear to provide plausible explanations, at best, for relatively short-term trade-offs, and even then we need far better empirical data on the resource "currency" and on the functional, physiological nature of the interaction or allocation decision (the Y model) before accepting these ideas. Just because trade-offs and carry-over effects could be mediated by resource-based processes does not mean that they *have to be* resource-based processes. As Zera et al. (2007) point out, detailed information on proximate mechanisms will help evaluate competing evolutionary hypotheses for trade-offs. There is emerging evidence for, and examples of, other plausible, non-resource-based mechanisms, in which the currency of the trade-off does not involve energy or resource allocation and in which the nature of the regulatory (physiological) networks themselves generates trade-offs and carry-over effects due to cellular, metabolic, or hormonal conflict, incompatibility, or pleiotropy. Stress, allostasis, and reactive scope; oxidative stress; and oxygen transport, hematology, and anemia provide three examples of integrated physiological mechanisms that could operate over an individual's entire life span and that might provide common mechanisms for linkages between multiple different stages and transitions. Such mechanisms might then provide an explanation for, and mechanistic links between, maternal effects and other development effects in offspring as well as reproductive investment decisions, trade-offs, and carry-over effects in adults. These integrated physiological mechanisms should also provide insight into the components of phenotype that determine variation in individual quality. While it will be much harder to work on physiology at the life-history scale, which will involve measurement of *multiple* components of complex physiological systems (Travers et al. 2010; T. Williams 2008; Wingfield et al. 2008), such research would be analogous to the multivariate, long-term approach more typical of ecological and evolutionary studies—and should be our goal for future studies in evolutionary physiology.

7.9. Future Research Questions

- Does "individual quality" provide a common explanation for positive relationships between winter/non-breeding habitats, or activities, and reproductive success? Is foraging efficiency a better measure of individual quality than absolute foraging effort in the context of winter-breeding carry-over effects (sensu Daunt et al. 2006)?
- If "high quality females always do better" over their entire lifetime (Hamel et al. 2009), then what morphological, behavioral, and physiological components of phenotype determine individual quality?
- If it is common for the duration of post-fledging parental care to exceed that of the nestling phase, how does this substantial, but rarely studied, component of reproductive investment affect estimates of "fitness" or breeding success that are determined at fledging?
- What are the long-term consequences of delayed molt, and more rapid molt, on feather quality, flight performance, thermoregulation, winter energetics, and over-winter survival?
- For putative resource-allocation-based trade-offs (and carry-over effects), it is necessary to explicitly identify the "currency" of the limited resource that is being allocated and the nature of the functional (i.e., physiological) link between the negatively associated traits or functions (e.g., flow of metabolites through pathways of intermediary metabolism at the "allocation node," the regulatory signals of nutrient sensing, nutrient input, and feedback [Harshman and Zera 2007]). Otherwise such mechanisms should not be assumed to be correct.
- Does hormonal pleiotropy—a single hormone having both positive and negative effects on multiple physiological systems—or hormonal or physiological "conflict" between different regulatory systems required for multiple overlapping functions, provide common non-resource-based mechanisms to explain phenotypic variation, trade-offs, and carry-over effects?

- Do the allostasis and reactive scope models provide a strong conceptual framework for integrating studies of specific physiological mechanisms and mediators (glucocorticoids, catecholamines, cytokines) underlying stress and stress responses in a long-term, life-history context?
- Does oxidative stress provide "a primary and universal constraint in life-history evolution" (Dowling and Simmons 2009), and can it provide an integrated, conceptual framework for understanding phenotypic variation, individual quality, trade-offs, and carry-over effects?
- Do oxygen transport systems, including variation in hematological status, provide a conceptual framework for integration of aerobic capacity, workload, individual quality, and trade-offs, including costs of reproduction, across multiple life-history stages (in embryos, nestlings, and adults)?
- Can we work on physiology at this life-history scale by integrating mechanistic studies with on-going long-term ecological and evolutionary studies (thus providing the continued "novel" contributions from these studies that seem to be required by funding agencies)?

CHAPTER 8

Conclusions

FIFTY YEARS ON from David Lack's seminal contributions to breeding in birds (Lack 1966, 1968), the main questions I have tried to address in this book are, Are there any grand challenges left in avian reproductive biology? and, if so, What are the big, unresolved questions that we *should* be focusing our research on? The answer to the first question, based on my interpretation of the material presented here with its physiological or mechanistic focus, is an emphatic Yes. We have only a very rudimentary, and in some cases non-existent, understanding of the physiological, metabolic, energetic, and hormonal mechanisms that underpin individual, phenotypic variation in the major reproductive life-history traits; trade-offs between these traits (and other physiological systems); and linkages or carry-over effects between different life-history stages. Only with this information will we be able to fully resolve those very same "big questions" that interested Lack more than 50 years ago: the causes and consequences of individual or phenotypic variation in timing of breeding, clutch size, and parental care that determine individual variation in fitness. Oddly, despite claims of an increased interest in the integration of behavioral ecology, evolutionary biology, and physiology in recent years, true integration, and progress, has been slow (Harshman and Zera 2007; Isaksson et al. 2011). The key life-history traits of timing of breeding, egg size and number, and parental ability have been the focus of very little of this recent "integrated" research; a large amount of research has instead considered more specific topics such as yolk hormones, eco-immunology, antioxidants, sex-ratio, etc. In my opinion we have, rather uncritically, continued to accept widely held but poorly supported and very general explanations for the variation in reproductive traits. It would be a brave (or foolish) person indeed, to argue that energy has *nothing* to do with reproduction, and it is premature to suggest that the general importance of resources or nutritional constraint will turn out to be wrong. However, at present these ideas constitute little more than "conventional wisdom" that has a surprising lack of support from empirical, experimental studies, or with much critical thought about the assumptions of these conceptual models, especially when applied to individual variation and physiological mechanism. To continue with my exemplars, if nutritional constraint *is* a major driver of variation in avian reproduction, we need to understand the (digestive) physiology to confirm the underlying mechanisms. Similarly, in relation to the conventional wisdom about prolactin and parental care, if we are going to argue that prolactin is a major regulator of individual variation in incubation then we must undertake the experimental hormonal studies needed to confirm this. An uncritical, myopic focus on energy or resources is too limiting, and we should be open to exploring alternative hypotheses and mechanisms. Refocusing even a quarter of the current research effort in avian reproduction back on the physiological basis of the main life-history traits, such as breeding phenology, clutch size, and parental care, would be a very good thing! In this regard, the different chapters in this book have attempted to provide a comparison of the development and state of the field for these different life-history traits, integrating evolutionary biology and physiology:

- In relation to timing of breeding (chapter 3) there has been an abundance of work on the photoperiodic control of seasonal breeding, but the overwhelming male-bias in these studies has left us remarkably short on information on how environmental cues are integrated in the HPG axis to regulate *female* reproductive development (ovarian and oviduct development, vitellogenesis, and yolk uptake) and the timing of egg-laying. Time and again in this book I have been drawn to the conclusion that the *female* is the key to our understanding of timing of breeding, as are the physiological changes that occur in the last few days prior to egg-laying (fig. 3.10). Previous work on male birds has solved many of the issues underlying the general endocrine control of seasonal breeding , the circadian nature of the timing mechanisms, and the central nervous system pathways involved. However, the current "working model" with day length as an initial predictive cue and supplemental factors such as temperature and food availability fine-tuning the actual timing of egg-laying (fig. 3.11) might be too simplistic. Certainly at present this model is surprisingly poorly supported by empirical, experimental data for *female* birds, and it does not currently provide a proper, well-founded mechanistic explanation for *individual variation* in the timing of breeding in females.
- In the case of clutch size (chapter 5), there has been a dearth of experimental work in non-poultry species, even though an explicit, predictive model for hormonally regulated clutch-size determination has been around for 15–20 years (fig. 5.11). This model would appear to have the potential to explain individual, date-dependent (seasonal) and date-independent variation in clutch size (at least in

single-brooded species). Clearly, the mechanisms of clutch-size determination should be a research priority, and these models need to be extended to explain seasonal variation in clutch size in multiple-brooded species.

- Ironically, we are perhaps closest to identifying the key regulatory mechanisms underlying individual variation for egg size, a life-history trait that is highly individually variable but appears to contribute little to variation in lifetime fitness (chapter 4). Here, a putative cell-level, receptor-mediated process (yolk uptake) appears to explain "large-egg" and "small-egg" phenotypes. However, this raises a number of interesting conceptual issues: How does variation in receptor expression or activity fit with traditional, long-standing ideas such as resource- or condition-dependence, nutritional constraint, or resource-allocation trade-offs as drivers of reproductive investment?
- It remains true that relatively little work has been conducted on the physiological mechanisms underlying individual variation in incubation behavior (chapter 6), but it would be timely to reconsider the widely held assumptions about energetic constraints driving behavioral variation and the idea that prolactin necessarily plays a central role in regulating individual variation in these behaviors.
- Finally, I would argue that we do not even have a viable working model for the physiological basis of individual variation in parental care during chick-rearing (chapter 6). Some key conceptual foundations have underpinned the studies of parental care for decades: that parental effort should be positively correlated with fitness, since higher provisioning rates should lead to higher-quality offspring with higher survival and recruitment, and that better parents or higher-quality individuals work harder to rear more or larger offspring but then risk becoming physiologically "*exhausted*" or "*fatigued*," generating long-term costs of reproduction. These ideas are currently very poorly supported by empirical data, and the physiological regulation of provisioning behavior, parental care, and foraging in general represents another fertile area for future experimental, mechanistic, and physiological studies.

Most of the key reproductive life-history traits considered in this book, and those traits that contribute most to variation in fitness, are *female-specific* traits (e.g., the timing of egg-laying, egg size or number) or traits that rely directly on components of female physiology. Even for the later stages of breeding, sex-specific differences in parental care, breeding-molt overlap, or sex-specific survival are routinely attributed to the differential costs of early reproductive decisions in females, highlighting the need for a better understanding of these costs and their underlying mechanisms. As a consequence, a much greater, more explicit, female-specific perspective is required in future mechanistic studies to address the disconnect in avian biology between physiological research (at least on non-poultry species) and the vast bulk of ecological or evolutionary studies that *have* focused on these female-specific traits. We have the non-genomic and genetics/genomic tools to tackle female reproduction, from the cell-molecular to the whole-organism level. However, we need to design and structure future experimental studies based on the realization that female reproductive development probably has a different temporal pattern than testis development does in males (fig. 3.10) and equally might respond differently to environmental cues or cue values. Future experiments need to utilize ecologically relevant photoperiodic and temperature conditions that fully reflect this sex difference. In addition, downstream of the pituitary, the physiological and hormonal regulation of female reproduction is very different from that of males and has unique female-specific attributes (one example being reproductive anemia). Experimental protocols, and the physiological and reproductive endpoints measured in experiments, need to keep this in mind. Well-designed studies of captive birds, such as the zebra finch, where females *will* lay eggs and rear chicks, and where short-generation times make it possible to do experiments in a longer-term, life-history context (Love, Salvante, et al. 2008; Roberts et al. 2007; von Engelhardt et al. 2006) will continue to be valuable here. Alternatively, experimental field studies are the method of choice *provided* we select study systems where some long-term endpoints (at least cumulative fecundity, or local survival, or even reproductive success) can be, *and are*, measured, and the "ecological context" of the experimental treatment can be assessed. We need far fewer short-term experimental studies, which are generally carried out in single years without known ecological or life-history contexts and in which "significant" experimental perturbations of physiology, condition, or reproductive investment are simply *assumed* to have long-term (negative) consequences. In both cases we need to integrate molecular, genetics, and genomics tools into such studies (as well as taking advantage of other technological advances, such as ever smaller, more sophisticated data loggers), but we should not ignore, or lose sight of, the *physiology* that provides the *functional* link between genotype and phenotype; and we have to apply these new tools to the right questions.

Finally, I have argued elsewhere that individual variation remains an underutilized resource in terms of our understanding of endocrine systems (T. Williams 2008), and this book makes clear that this argument can be extended to the physiological, metabolic, and energetic mechanisms underpinning variation in reproductive life-history traits in general.

Importantly, a greater appreciation of the magnitude of individual variation in reproductive traits forces us to *explicitly* address the mechanistic basis of phenotypic variation and phenotypic plasticity. In this regard, four important, overriding concepts have emerged from this book:

- First, a common conclusion of many ecological and evolutionary studies is that there are *intrinsic* differences between individuals such that certain high-quality females lay many, large eggs, with a relatively early laying date, rear large broods, and recruit more offspring, while also having higher future fecundity and survival, thus apparently "avoiding" underlying the trade-offs that would predict negative correlations between these traits. However, all studies to date have followed this conclusion with a statement to the effect that the components of the female phenotype that determine variation in reproductive investment, or the parameters that provide reliable estimates of "individual quality," remain unknown (e.g., Garant, Hadfield et al. 2007; Erikstad et al. 2009). Identifying the *physiological* mechanisms that determine individual quality in relation to variation in fitness should therefore be a research priority.
- Second, most individuals *do not* reproduce successfully; rather, the majority of individuals do not reproduce at all, while a few individuals contribute a large proportion of the offspring making up the next generation (Clutton-Brock 1988; Newton 1989a). This is a very important point because it means that unsuccessful birds, with zero lifetime fitness, represent most of the individuals that many of us would classify as "successful" in our experimental field studies in which fledging is taken as the end point. This highly skewed variation in fitness highlights the need for physiological studies to be integrated into long-term population studies in which individual variation in fitness can be reliably assessed (or at least the need for work to be conducted in systems where future fecundity or local return rates can be measured).
- Third, at the physiological level, just as much as at the ecological and behavioral level, different individuals are likely to adopt very flexible, "*individually variable*" strategies to deal with the costs and benefits of investment in self-maintenance or offspring when faced with increased workload or increased parental effort (de Heij et al. 2008; Vézina et al. 2006; T. Williams et al. 2009). Thus, experiments have to be designed and data collected specifically to embrace this concept of individually variable energetic and physiological strategies, for example, using the repeated-measures, "reaction norm" approach common in ecological studies (T. Williams 2008). Unless the negative effects of experimental perturbations of physi-

ological or reproductive end points are directly demonstrated, these might simply reflect strategic, individually variable adjustments to maximize or optimize that individual's overall fitness in the face of altered conditions.

- Fourth, reproduction clearly needs to be viewed as only one part of an individual's integrated annual cycle and its complete life-history. Individuals make decisions about reproductive investment not just in terms of fitness outcomes of the current breeding stage (e.g., egg-laying or incubation or chick-rearing) but over the entire breeding cycle, and in longer-lived birds investment decisions and costs and benefits will be influenced by other life-history stages (e.g., winter and the pre-breeding period, molt, and migration). Therefore, if we are to better understand the *physiology of life histories*, including the physiological mechanisms underlying variation in phenotypic traits and the trade-offs or carry-over effects between these traits or between life-history stages, then we need to approach physiological studies with this same long-term, integrated, life-history perspective (fig. 7.1c). Identifying and studying such "integrated" physiological mechanisms that can operate over an individual's entire life span might provide common mechanisms for links between early developmental effects, maternal effects, and reproductive investment decisions, trade-offs, and carry-over effects in adults, and should provide insight into the components of phenotype that determine variation in individual quality. While it will admittedly be much harder to work on physiology at this life-history scale, which entails the measurement of multiple components of complex physiological systems, it is analogous to the multivariate, long-term approach more typical of ecological and evolutionary studies. The best approach would be to incorporate detailed, integrated physiological studies into these existing long-term studies, generating a comprehensive new synthesis of physiology, ecology, and evolutionary biology.

Bibliography

Abzhanov, A., W. P. Kuo, C. Hartmann, B. Grant, P. R. Grant, and C. J. Tabin. 2006. The calmodulin pathway and evolution of elongated beak morphology in Darwin's finches. *Nature* 442:563–567.

Abzhanov, A., M. Protas, B. R. Grant, P. R. Grant, and C. J. Tabin. 2004. *Bmp4* and morphological variation of beaks in Darwin's finches. *Science* 305:1462–1465.

Addison, B., Z. M. Benowitz-Fredericks, J. M. Hipfner, and A. S. Kitaysky. 2008. Are yolk androgens adjusted to environmental conditions? A test in two seabirds that lay single-egg clutches. *General and Comparative Endocrinology* 158:5–9.

Addison, B., R. E. Ricklefs, and K. C. Klasing. 2010. Do maternally derived antibodies and early immune experience shape the adult immune response? *Functional Ecology* 24:824–829.

Adelman, J. S., and L. B. Martin. 2009. Vertebrate sickness behaviors: Adaptive and integrated neuroendocrine immune responses. *Integrative and Comparative Biology* 49:202–214.

Adkins-Regan, E. 2005. *Hormones and Animal Social Behavior*. Princeton, NJ: Princeton University Press.

Adkins-Regan, E., M. A. Ottinger, and J. Park. 1995. Maternal transfer of estradiol to egg yolks alters sexual differentiation of avian offspring. *Journal of Experimental Zoology* 271:466–470.

Aldrich, T. W., and D. G. Raveling. 1983. Effects of experience and body weight on incubation behavior of Canada geese. *Auk* 100:670–679.

Alonso-Alvarez, C., S. Bertrand, G. Devevey, J. Prost, B. Faivre, and G. Sorci. 2004. Increased susceptibility to oxidative stress as a proximate cost of reproduction. *Ecology Letters* 7:363–368.

Alonso-Alvarez, C., S. Bertrand, B. Faivre, and G. Sorci. 2007. Increased susceptibility to oxidative damage as a cost of accelerated somatic growth in zebra finches. *Functional Ecology* 21:873–879.

Alonso-Alvarez, C., L. Perez-Rodríguez, J. T. Garcia, and J. Vinuela. 2009. Testosterone-mediated trade-offs in the old age: A new approach to the immunocompetence handicap and carotenoid-based sexual signalling. *Proceedings of the Royal Society B: Biological Sciences* 276:2093–2101.

Alonso-Alvarez, C., L. Perez-Rodríguez, R. Mateo, O. Chastel, and J. Vinuela. 2008. The oxidation handicap hypothesis and the carotenoid allocation trade-off. *Journal of Evolutionary Biology* 21:1789–1797.

Amundsen, T. 1995. Egg size and early nestling growth in the Snow Petrel. *Condor* 97:345–351.

Amundsen, T., S.-H. Lorentsen, and T. Tveraa. 1996. Effects of egg size and parental quality on early nestling growth: An experiment with the Antarctic petrel. *Journal of Animal Ecology* 65:545–555.

Amundsen, T., and J. N. Stokland. 1990. Egg size and parental quality influence nestling growth in the Shag. *Auk* 107:410–413.

Ancel, A., L. Petter, and R. Groscolas. 1998. Changes in egg and body temperature indicate triggering of egg desertion at a body mass threshold in fasting incubating blue petrels (*Halobaena caerulea*). *Journal of Comparative Physiology* 168A:533–539.

Angelier, F., C.-A. Bost, M. Giraudeau, G. Bouteloup, S. Dano, and O. Chastel. 2008. Corticosterone and foraging behavior in a diving seabird: The Adelie penguin, *Pygoscelis adeliae*. *General and Comparative Endocrinology* 156:134–144.

Angelier, F., and O. Chastel. 2009. Stress, prolactin and parental investment in birds: A review. *General and Comparative Endocrinology* 163:142–148.

Angelier, F., C. Clément-Chastel, G. W. Gabrielsen, and O. Chastel. 2007. Corticosterone and time-activity budget: An experiment with Black-legged kittiwakes. *Hormones and Behavior* 52:482–491.

Angelier, F., C. Clément-Chastel, J. Welcker, G. W. Gabrielsen, and C. O. 2009. How does corticosterone affect parental behaviour and reproductive success? A study of prolactin in black-legged kittiwakes. *Functional Ecology* 23:784–793.

Angelier, F., M. Giraudeau, C.-A. Bost, F. Le Bouard, and O. Chastel. 2009. Are stress hormone levels a good proxy of foraging success? An experiment with King Penguins, *Aptenodytes patagonicus*. *Journal of Experimental Biology* 212:2824–2829.

Angelier, F., S. A. Shaffer, H. Weimerskirch, C. Trouve, and O. Chastel. 2007. Corticosterone and foraging behavior in a pelagic seabird. *Physiological and Biochemical Zoology* 80:283–292.

Angelier, F., H. Weimerskirch, S. Dano, and O. Chastel. 2007. Age, experience and reproductive performance in a long-lived bird: A hormonal perspective. *Behavioral Ecology and Sociobiology* 61:611–621.

Angelier, F., J. C. Wingfield, H. Weimerskirch, and O. Chastel. 2010. Hormonal correlates of individual quality in a long-lived bird: A test of the 'corticosterone-fitness hypothesis'. *Biology Letters* 6:846–849.

Ankney, C. D. 1977. Feeding and digestive organ size in breeding Lesser Snow Geese. *Auk* 94:275–282.

Ankney, C. D., A. D. Afton, and R. T. Alisauskas. 1991. The role of nutrient reserves in limiting waterfowl reproduction. *Condor* 93:1029–1032.

Ankney, C. D. and D. M. Scott. 1988. Size of digestive organs in breeding Brown-headed Cowbirds, *Molothrus ater*, relative to diet. *Canadian Journal of Zoology* 66:1254–1257.

Appleby, B. M., S. J. Petty, J. K. Blakey, P. Rainey, and D. W. MacDonald. 1997. Does variation of sex ratio enhance reproductive success of offspring in tawny owls (*Strix aluco*). *Proceedings of the Royal Society of London B: Biological Sciences* 264:1111–1116.

Arcese, P., and J.N.M. Smith. 1988. Effects of population density and supplemental food on reproduction in song sparrows. *Journal of Animal Ecology* 57:119–136.

Ardia, D. R., and E. D. Clotfelter. 2007. Individual quality and age affect responses to an energetic constraint in a cavity-nesting bird. *Behavioral Ecology* 18:259–266.

Ardia, D. R., C. B. Cooper, and A. A. Dhondt. 2006. Warm temperatures lead to early onset of incubation, shorter incubation periods and greater hatching asynchrony in tree swallows *Tachycineta bicolor* at the extremes of their range. *Journal of Avian Biology* 37:137–142.

Ardia, D. R., J. H. Pérez, E. K. Chad, M. A. Voss, and E. D. Clotfelter. 2009. Temperature and life history: Experimental heating leads female tree swallows to modulate egg temperature and incubation behaviour. *Journal of Animal Ecology* 78:4–13.

Ardia, D. R., K. A. Schat, and D. W. Winkler. 2003. Reproductive effort reduces long-term immune function in breeding tree swallows (*Tachycineta bicolor*). *Proceedings of the Royal Society of London B: Biological Sciences* 270:1679–1683.

Arnold, K. E., K. J. Orr, and R. Griffiths. 2003. Primary sex ratios in birds: Problems with molecular sex identification of undeveloped eggs. *Molecular Ecology* 12:3451–3458.

Arnold, T. W. 1992. Variation in laying date, clutch size, egg size, and egg composition of yellow-headed blackbirds (*Xanthocephalus xanthocephalus*): A supplemental feeding experiment. *Canadian Journal of Zoology* 70:1904–1911.

———. 1993. Factors affecting egg viability and incubation time in prairie dabbling ducks. *Canadian Journal of Zoology* 71:1146–1152.

———. 1994. Effects of supplemental food on egg production in American Coots. *Auk* 111:337–350.

———. 1999. What limits clutch size in waders? *Journal of Avian Biology* 30:216–220.

Arnold, T. W., and C. D. Ankney. 1997. The adaptive significance of nutrient reserves to breeding American Coots: A reassessment. *Condor* 99:91–103.

Arnold, T. W., and F. C. Rohwer. 1991. Do egg formation costs limit clutch size in waterfowl? A skeptical view. *Condor* 93:1032–1038.

Astheimer, L. B. 1985. Long laying intervals: A possible mechanism and its implications. *Auk* 102:401–409.

———. 1986. Egg formation in Cassin's Auklet. *Auk* 103:682–693.

Astheimer, L. B., W. A. Buttemer, and J. C. Wingfield. 1992. Interactions of corticosterone with feeding, activity and metabolism in passerine nirds. *Ornis Scandinavica* 23:355–365.

Astheimer, L. B., and C. R. Grau. 1985. The timing and energetic consequences of egg formation in the Adelie Penguin. *Condor* 87:256–268.

———. 1990. A comparison of yolk growth rates in seabird eggs. *Ibis* 132:380–394.

Bacon, W. L., B. Leclercq, and J. C. Blum. 1978. Difference in metabolism of very low density lipoprotein from laying chicken hens in comparison to immature chicken hens. *Poultry Science* 57:1675–1686.

Bacon, W. L., and H.-K. Liu. 2004. Progesterone injection and egg production in turkey hens. *Biology of Reproduction* 71:878–886.

Bacon, W. L., M. A. Musser, and K. I. Brown. 1974. Plasma free fatty acid and neutral lipid concentrations in immature, laying and broody turkey hens. *Poultry Science* 53:1154–1160.

Badman, M. K., and J. S. Flier. 2005. The gut and energy balance: Visceral allies in the obesity wars. *Science* 307:1909–1914.

Badyaev, A. V. 2009. Evolutionary significance of phenotypic accommodation in novel environments: An empirical test of the Baldwin effect. *Philosophical Transactions of the Royal Society B: Biological Sciences* 364:1125–1141.

———. 2011. How do precise adaptive features arise in development? Examples with evolution of context-specific sex ratios and perfect beaks. *Auk* 128:467–474.

Badyaev, A. V., D. A. Acevado Seaman, K. J. Navarra, G. E. Hill, and M. T. Mendonça. 2006. Evolution of sex-biased maternal effects in birds. III. Adjustment of ovulation order can enable sex-specific allocation of hormones, carotenoids, and vitamins. *Journal of Evolutionary Biology* 19:1044–1057.

Badyaev, A. V., M. L. Beck, G. E. Hill, and L. A. Whittingham. 2003. The evolution of sexual size dimorphism in the house finch. V. Maternal effects. *Evolution* 57:384–398.

Badyaev, A. V., and R. A. Duckworth. 2005. Evolution of plasticity in hormonally-integrated parental tactics: An example with the house finch. In *Functional Avian Endocrinology*, ed. A. Dawson and P. J. Sharp, 375–386. New Delhi, India: Narosa Publishing House.

Badyaev, A. V., T. L. Hamstra, K. P. Oh, and D. A. Acevado Seaman. 2006. Sex-biased maternal effects reduce ectoparasite-induced mortality in a passerine bird. *Proceedings of the National Academy of Sciences USA* 103:14406–14411.

Badyaev, A. V., G. E. Hill, and L. A. Whittingham. 2002. Population consequences of maternal effects: Sex-bias in egg-laying order facilitates divergence in sexual dimorphism between bird populations. *Journal of Evolutionary Biology* 15:997–1003.

Badyaev, A. V., and K. P. Oh. 2008. Environmental induction and phenotypic retention of adaptive maternal effects. *BMC Evolutionary Biology* 8:3.

Badyaev, A. V., K. P. Oh, and R. Mui. 2006. Evolution of sex-biased maternal effects in birds. II. Contrasting sex-specific oocyte clustering in native and recently established populations. *Journal of Evolutionary Biology* 19:909–921.

Badyaev, A. V., H. Schwabl, R. L. Young, R. A. Duckworth, K. J. Navara, and A. F. Parlow. 2005. Adaptive sex differences in growth of pre-ovulation oocytes in a passerine bird. *Proceedings of the Royal Society B: Biological Sciences* 272:2165-2172.

Badyaev, A. V., and T. Uller. 2009. Parental effects in ecology and evolution: Mechanisms, processes and implications. *Philosophical Transactions of the Royal Society B: Biological Sciences* 364:1169–1177.

Badyaev, A. V., R. L. Young, G. E. Hill, and R. A. Duckworth. 2008. Evolution of sex-biased maternal effects in birds. IV. Intra-ovarian growth dynamics can link sex determination and sex-specific acquisition of resources. *Journal of Evolutionary Biology* 21:449–460.

Baeza, E., J. Williams, D. Guemene, and M. J. Duclos. 2001. Sexual dimorphism for growth in Muscovey ducks and changes in insulin-like growth factor I (IGF-1), growth hormone (GH) and triiodothyronine (T3) plasma levels. *Reproduction, Nutrition, and Development* 41:173–179.

Bahr, J. M., S. C. Wang, M. Y. Huang, and F. O. Calvo. 1983. Steroid concentrations in isolated theca and granulosa layers of preovulatory follicles during the ovulatory cycle of the domestic hen. *Biology of Reproduction* 29:326–334.

Bairlein, F., and U. Totzke. 1992. New aspects on migratory physiology of trans-Saharan passerine migrants. *Ornis Scandinavica* 23:244–250.

Baker, A. J., P. M. González, T. Piersma, L. J. Niles, I. L. S. do Nascimento, P. W. Atkinson, N. A. Clark, et al. 2004. Rapid population decline in red knots: Fitness consequences of decreased refuelling rates and late arrival in Delaware Bay. *Proceedings of the Royal Society B: Biological Sciences* 271:875–882.

Baker, J. R. 1938. The evolution of breeding seasons. In *Evolution: Essays on Aspects of Evolutionary Biology*, ed. G. R. de Beer, 161–177. London: Oxford University Press.

Balakrishnan, C. N., S. V. Edwards, and D. F. Clayton. 2010. The Zebra Finch genome and avian genomics in the wild. *Emu* 110:233–241.

Ball, G. F. 2007. The ovary knows more than you think! New views on clock genes and the positive feedback control of luteinizing hormone. *Endocrinology* 148:3029–3030.

Ball, G. F., and E. D. Ketterson. 2008. Sex differences in the response to environmental cues regulating seasonal reproduction in birds. *Philosophical Transactions of the Royal Society B* 363:231–246.

Ball, I. J., M. J. Artmann, and S. T. Hoekman. 2002. Does Mallard clutch size vary with landscape composition? *Wilson Bulletin* 114:404–406.

Balthazart, J., and E. Adkins-Regan. 2002. Sexual differentiation of brain and behavior in birds. In *Hormones, Brain and Behavior*, ed. D. W. Pfaff, A. P. Arnold, A. M. Etgen, S. E. Fahrbach, and R. T. Rubin, 223–301. Amsterdam: Academic Press.

Baptista, L. F., and L. Petrinovich. 1986. Egg production in hand-raised White-crowned Sparrows. *Condor* 88:379–380.

Barber, D. L., E. J. Sanders, R. Aebersold, and W. J. Schneider. 1991. The receptor for yolk lipoprotein deposition in the chicken oocyte. *Journal of Biological Chemistry* 266:18761–18770.

Barbraud, C., and O. Chastel. 1998. Southern Fulmars moult their primary feathers while incubating. *Condor* 100:563–566.

Barja, G. 2004. Aging in vertebrates, and the effect of caloric restriction: A mitochondrial free radical production–DNA damage mechanism? *Biological Reviews* 79:235–251.

Barnes, A. I., and L. Partridge. 2003. Costing reproduction. *Animal Behaviour* 66:199–204.

Barnes, D. M. 2001. Expression of P-glycoprotein in the chicken. *Comparative Biochemistry and Physiology A: Molecular and Integrative Physiology* 130:301–310.

Battley, P. F., and T. Piersma. 2005. Adaptive interplay between feeding ecology and features of the digestive system in birds. In *Physiological and Ecological Adaptations to Feeding in Vertebrates*, ed. J. M. Starck and T. Wang, 201–228. Enfield, N.H.: Science Publishers.

Bauchinger, U., T. Van't Hof, and H. Biebach. 2007. Testicular development during long-distance spring migration. *Hormones and Behavior* 51:295–305.

Beck, J. R. 1970. Breeding seasons and moult in some smaller Antarctic petrels. In *Antarctic Ecology*, ed. M. W. Holdgate, 542–550. London: Academic Press.

Beebe, K., G. E. Bentley, and M. Hau. 2005. A seasonally breeding tropical bird lacks absolute photorefractoriness in the wild, despite high photoperiodic sensitivity. *Functional Ecology* 19:505–512.

Beissinger, S. R., M. I. Cook, and W. J. Arendt. 2005. The shelf life of bird eggs: Testing egg viability using a tropical climate gradient. *Ecology* 86:2164–2175.

Bencina, D., M. Narat, A. Bidovec, and O. Zorman-Rojas. 2005. Transfer of maternal immunoglobulins and antibodies to *Mycoplasma gallisepticum* and *Mycoplasma synoviae* to the allantoic and amniotic fluid of chicken embryos. *Avian Pathology* 34:463–472.

Bennett, A. F. 1987. Interindividual variability: An underutilized resource. In *New Directions in Ecological Physiology*, ed. M. E. Feder, A. F. Bennett, W. W. Burggren, and R. B. Huey, 147–169. Cambridge: Cambridge University Press.

Bennett, P. M., and I.P.F. Owens. 2002. *Evolutionary Ecology of Birds*. Oxford Series in Ecology and Evolution. Oxford: Oxford University Press.

Bensadoun, A., and A. Rothfeld. 1972. The form of absorption of lipids in the chicken, *Gallus domesticus*. *Proceedings of the Society of Experimental Biology and Medicine* 141:814–817.

Bentley, G. E., J. P. Jensen, G. J. Kaur, D. W. Wacker, K. Tsutsui, and J. C. Wingfield. 2006. Rapid inhibition of female sexual behavior by gonadotropin-inhibitory hormone (GnIH). *Hormones and Behavior* 49:550–555.

Bentley, G. E., I. T. Moore, S. A. Sower, and J. C. Wingfield. 2004. Evidence for a novel gonadotropin-releasing hormone in hypothalamic and forebrain areas in songbirds. *Brain, Behavior and Evolution* 63:34–46.

Bentley, G. E., K. Tsutsui, and W. J.C. 2007. Endocrinology of reproduction. In *Reproductive Biology and Phylogeny of Birds*, ed. B.G.M. Jamieson, 181–242. Enfield, N. H.: Science Publishers.

Bentley, G. E., T. Ubuka, N. L. McGuire, V. S. Chowdhury, Y. Morita, T. Yano, I. Hasunuma, et al. 2008. Gonadotropin-inhibitory hormone and its receptor in the avian reproductive system. *General and Comparative Endocrinology* 156:34–43.

Bergink, E. W., R. A. Wallace, J. A. Van de Berg, E. S. Bos, M. Gruber, and A. B. Geert. 1974. Estrogen-induced synthesis of yolk proteins in roosters. *American Zoologist* 14:1177–1193.

Bernardo, J. 1996. The particular maternal effect of propagule size, especially egg size: Patterns, models, quality of evidence and interpretations. *American Zoologist* 36:216–236.

Berthouly, A., A. Cassier, and H. Richner. 2008. Carotenoid-induced maternal effects interact with ectoparasite burden and brood size to shape the trade-off between growth and immunity in nestling great tits. *Functional Ecology* 22:854–863.

Bertrand, S., C. Alonso-Alvarez, G. Devevey, B. Faivre, J. Prost, and G. Sorci. 2006. Carotenoids modulate the trade-off between egg production and resistance to oxidative stress in zebra finches. *Oecologia* 147:576–584.

Bertrand, S., B. Faivre, and G. Sorci. 2006. Do carotenoid-based sexual traits signal the availability on non-pigmentary antioxidants? *Journal of Experimental Biology* 209:4414–4419.

Bety, J., G. Gauthier, and J. F. Giroux. 2003. Body condition, migration, and timing of reproduction in snow geese: A test of the condition-dependent model of optimal clutch size. *American Naturalist* 162:110–121.

Bety, J., J. F. Giroux, and G. Gauthier. 2004. Individual variation in timing of migration: Causes and reproductive consequences in greater snow geese (*Anser caerulescens atlanticus*). *Behavioral Ecology and Sociobiology* 57:1–8.

Beuchat, C. A., and C. R. Chong. 1998. Hyperglycemia in hummingbirds and its consequences for hemoglobin glycation. *Comparative Biochemistry and Physiology A: Molecular and Integrative Physiology* 120:409–416.

Beukeboom, L. W., C. Dijkstra, S. Daan, and T. Meijer. 1988. Seasonality of clutch size determination in the Kestrel, *Falco tinnunculus*: An experimental study. *Ornis Scandinavica* 19:41–48.

Biard, C., P. F. Surai, and A. P. Møller. 2007. An analysis of pre- and post-hatching maternal effects mediated by carotenoids in the blue tit. *Journal of Evolutionary Biology* 20:326–339.

Biebach, H. 1981. Energetic costs of incubation on different clutch sizes in Starlings (*Sturnus vulgaris*). *Ardea* 69:141–142.

Birchard, G. F. 1997. Optimal hematocrit: Theory, regulation and implications. *American Zoologist* 37:65–72.

Birkhead, T. R., and A. P. Møller. 1992. *Sperm Competition in Birds*. London: Academic Press.

Biro, P. A., and J. A. Stamps. 2008. Are animal personality traits linked to life-history productivity? *Trends in Ecology and Evolution* 23:361–368.

———. 2010. Do consistent individual differences in metabolic rate promote consistent individual differences in behavior? *Trends in Ecology and Evolution* 25:653–659.

Bize, P., F. O. Criscuolo, N. B. Metcalfe, L. Nasir, and P. Monaghan. 2009. Telomere dynamics rather than age predict life expectancy in the wild. *Proceedings of the Royal Society B: Biological Sciences* 276:1679–1683.

Bize, P., G. Devevey, P. Monaghan, B. Doligez, and P. Christe. 2008. Fecundity and survival in relation to resistance to oxidative stress in a free-living bird. *Ecology* 89:2584–2593.

Bize, P., A. Roulin, and H. Richner. 2002. Covariation between egg size and rearing condition determines offspring quality: An experiment with the alpine swift. *Oecologia* 132:231–234.

Blackburn, T. M. 1991. The interspecific relationship between egg size and clutch size in waterfowl. *Auk* 108:209–211.

Bleeker, M., S. A. Kingma, I. Szentirmai, T. Székely, and J. Komdeur. 2005. Body condition and clutch desertion in penduline tit *Remiz pendulinus*. *Behavior* 142:1465–1478.

Blomqvist, D., O. C. Johansson, and F. Gotmark. 1997. Parental quality and egg size affect chick survival in a precocial bird, the Lapwing *Vanellus vanellus*. *Oecologia* 110:18–24.

Blondel, J., P. C. Dias, P. Perret, M. Maistre, and M. M. Lambrechts. 1999. Selection-based biodiversity at a small spatial scale in a low-dispersing insular bird. *Science* 285:1399–1402.

Blount, J. D., D. C. Houston, and A. P. Møller. 2000. Why egg yolk is yellow. *Trends in Ecology and Evolution* 15:47–49.

Blount, J. D., D. C. Houston, P. F. Surai, and A. P. Møller. 2004. Egg-laying capacity is limited by carotenoid pigment availability in wild gulls *Larus fuscus*. *Proceedings of the Royal Society B: Biological Sciences* 271:S79–S81.

Blount, J. D., N. B. Metcalfe, K. E. Arnold, P. F. Surai, G. L. Devevey, and P. Monaghan. 2003. Neonatal nutrition, adult antioxidant defences and sexual attractiveness in the zebra finch. *Proceedings of the Royal Society B: Biological Sciences* 270:1691–1696.

Blount, J. D., P. F. Surai, R. G. Nager, D. C. Houston, A. P. Møller, M. L. Trewby, and M. W. Kennedy. 2002. Carotenoids and egg quality in the lesser black-backed gull *Larus fuscus*: A supplemental feeding study of maternal effects. *Proceedings of the Royal Society B: Biological Sciences* 269:29–36.

Boersma, P. D., N. T. Wheelwright, M. K. Nerini, and E. S. Wheelwright. 1980. The breeding biology of the Fork-tailed Storm-Petrel (*Oceanodroma furcata*) *Auk* 97:268–282.

Bojarinova, J. G., E. Lehikoinen, and T. Eeva. 1999. Dependence of postjuvenile moult on hatching date, condition and sex in the Great Tit. *Journal of Avian Biology* 30:437–446.

Bolton, M. 1991. Determinants of chick survival in the lesser black-backed gull: Relative contributions of egg size and parental quality. *Journal of Animal Ecology* 60:949–960.

Bolton, M., D. Houston, and P. Monaghan. 1992. Nutritional constraints on egg formation in the lesser black-backed gull: An experimental study. *Journal of Animal Ecology* 61:521–532.

Bonier, F., P. R. Martin, I. T. Moore, and J. C. Wingfield. 2009. Do baseline glucocorticoids predict fitness? *Trends in Ecology and Evolution* 24:634–642.

Bonisoli-Alquati, A., D. Rubolini, M. Romano, G. Boncoraglio, M. Fasola, and N. Saino. 2007. Effects of egg albumen removal on yellow-legged gull chick phenotype. *Functional Ecology* 21:310–316.

Bonneaud, C., J. Burnside, and S. V. Edwards. 2008. High-speed developments in avian genomics. *BioScience* 58:587–595.

Bortolotti, G. R., J. J. Negro, P. F. Surai, and P. Prieto. 2003. Carotenoids in eggs and plasma of red-legged partridges: Effects of diet and reproductive output. *Physiological and Biochemical Zoology* 76:367–374.

Boswell, T., and S. Takeuchi. 2005. Recent developments in our understanding of the avian melanocortin system: Its involvement in the regulation of pigmentation and energy homeostasis. *Peptides* 26:1733–1743.

Both, C., S. Bouwhius, C. M. Lessells, and M. E. Visser. 2006. Climate change and population declines in a long-distance migratory bird. *Nature* 441:81–83.

Both, C., J. M. Tinbergen, and M. E. Visser. 2000. Adaptive density dependence of avian clutch size. *Ecology* 81:3391–3403.

Both, C., and M. E. Visser. 2001. Adjustment to climate change is constrained by arrival date in a long-distance migrant bird. *Nature* 411:296–298.

———. 2005. The effect of climate change on the correlation between avian life-history traits. *Global Change Biology* 11:1606–1613.

Boulinier, T., and V. Staszewski. 2008. Maternal transfer of antibodies: Raising immuno-ecology issues. *Trends in Ecology and Evolution* 23:282–288.

Bourgault, P., D. Thomas, P. Perret, and J. Blondel. 2010. Spring vegetation phenology is a robust predictor of breeding date across broad landscapes: A multi-site approach using the Corsican blue tit (*Cyanistes caeruleus*). *Oecologia* 162:885–892.

Bourgeon, S., F. Criscuolo, Y. Le Maho, and T. Raclot. 2006. Phytohemagglutinin response and immunoglobulin level decrease during incubation fasting in female Common Eiders. *Physiological and Biochemical Zoology* 79:793–800.

Bourgeon, S., and T. Raclot. 2006. Corticosterone selectively decreases humoral immunity in female eiders during incubation. *Journal of Experimental Biology* 209:4957–4965.

Boutin, S. 1990. Food supplementation experiments with terrestrial vertebrates: Patterns, problems and the future. *Canadian Journal of Zoology* 68:203–220.

Boyce, M. S., and C. M. Perrins. 1987. Optimizing great tit clutch size in a fluctuating environment. *Ecology* 68:142–153.

Boyle-Roden, E., and R. L. Walzem. 2005. Integral apolipoproteins increase surface-located triacylglycerol in intact native apo-protein B100 containing lipoproteins. *Journal of Lipid Research* 46:1624–1632.

Bradshaw, W. E., and C. M. Holzapfel. 2007. Evolution of animal photoperiodism. *Annual Review of Ecology Evolution and Systematics* 38:1–25.

Breuner, C. W., S. H. Patterson, and T. P. Hahn. 2008. In search of relationships between the acute adrenocortical response and fitness. *General and Comparative Endocrinology* 157:288–295.

Brinkhof, M.W.G. 1997. Seasonal variation in food supply and breeding success in European Coots *Fulica atra*. *Ardea* 85:51–65.

Brinkhof, M.W.G., A. J. Cavé, S. Daan, and A. C. Perdeck. 2002. Timing of current reproduction directly affects future reproductive output in European coots. *Evolution* 56:400–411.

Brommer, J. E., J. Merilä, B. C. Sheldon, and L. Gustafsson. 2005. Natural selection and genetic variation for reproductive reaction norms in a wild bird population. *Evolution* 59:1362–1371.

Brommer, J. E., H. Pietiainen, and H. Kokko. 2002. Cyclic variation in seasonal recruitment and the evolution of the seasonal decline in Ural owl clutch size. *Proceedings of the Royal Society B: Biological Sciences* 269:647–654.

Brommer, J. E., and K. Rattiste. 2008. "Hidden" reproductive conflict between mates in a wild bird population. *Evolution* 62:2326–2333.

Brommer, J. E., K. Rattiste, and A. J. Wilson. 2008. Exploring plasticity in the wild: Laying date-temperature reaction norms in the common gull *Larus canus*. *Proceedings of the Royal Society B: Biological Sciences* 275:687–693.

Brooke, M., and T. Birkhead. 1991. *The Cambridge Encyclopedia of Ornithology*. Cambridge: Cambridge University Press.

Brown, C. R., and M. B. Brown. 1999. Fitness components associated with laying date in the Cliff Swallow. *Condor* 101:230–245.

Brown, C. R., and D. M. Bryant. 1996. Energy expenditure during molt in dippers (*Cinclus cinclus*): No evidence of elevated costs. *Physiological Zoology* 69:1036–1056.

Brown, R.G.B. 1988. The wing-moult of fulmars and shearwaters (Procellariidae) in Canadian Arctic waters. *Canadian Field Naturalist* 102:203–208.

Bruggeman, V., D. Vanmontfort, R. Renaville, D. Portetelle, and E. Decuypere. 1997. The effect of food intake from two weeks of age to sexual maturity on plasma growth hormone, insulin-like growth factor-I, insulin-like growth factor-binding proteins, and thyroid hormones in female broiler breeder chickens. *General and Comparative Endocrinology* 107:212–220.

Bryan, S. M., and D. M. Bryant. 1999. Heating nest-boxes reveals an energetic contraint on incubation behaviour in great tits, *Parus major*. *Proceedings of the Royal Society B: Biological Sciences* 266:157–162.

Bryant, D. M. 1975. Breeding biology of the House Martin *Delichon urbica* in relation to aerial insect abundance. *Ibis* 117:180–216.

Buechler, K., P. S. Fitze, B. Gottstein, A. Jacot, and H. Richner. 2002. Parasite-induced maternal response in a natural bird population. *Journal of Animal Ecology* 71:247–252.

Bujo, H., T. Yamamoto, K. Hayashi, M. Hermann, J. Nimpf, and W. J. Schneider. 1995. Mutant oocytic low density lipoprotein receptor gene family member causes atherosclerosis and female sterility. *Proceedings of the National Academy of Sciences USA* 92:9905–9909.

Buntin, J. D. 1996. Neural and hormonal control of parental behavior in birds. *Advances in the Study of Behavior* 25:161–213.

Buntin, J. D., R. M. Hnasko, P. H. Zuzick, D. L. Valentine, and J. G. Scammell. 1996. Changes in bioactive prolactin-like activity in plasma and its relationship to incubation behavior in breeding ring doves. *General and Comparative Endocrinology* 102:221–232.

Burley, R. W., and D. V. Vadehra. 1989. *The Avian Egg: Chemistry and Biology*. New York: John Wiley and Sons.

Burness, G. P., R. C. Ydenberg, and P. W. Hochachka. 2001. Physiological and biochemical correlates of brood size and energy expenditure in tree swallows. *Journal of Experimental Biology* 204:1491–1501.

Byle, P. A. F. 1990. Brood division and parental care in the period between fledging and independence in the dunnock (*Prunella modularis*). *Behaviour* 113:1–20.

Calbet, J. A. L., C. Lundby, M. Koskolou, and R. Boushel. 2006. Importance of hemoglobin concentration to exercise: Acute manipulations. *Respiratory Physiology and Neurobiology* 151:132–140.

Calow, P. 1979. The cost of reproduction: A physiological approach. *Biological Reviews* 54:23–40.

Camfield, A. F., S. F. Pearson, and K. Martin. 2010. Life history variation between high and low elevation subspecies of horned larks *Eremophila* spp. *Journal of Avian Biology* 41:273–281.

Camper, P. M., and W. H. Burke. 1977. The effects of prolactin on reproductive function in female Japanese quail (*Coturnix coturnix japonica*). *Poultry Science* 56:1130–1134.

Carere, C., and J. Balthazart. 2007. Sexual versus individual differentiation: The controversial role of avian maternal hormones. *Trends in Endocrinology and Metabolism* 18:73–80.

Carew, L. B., J. P. McMurtry, and F. A. Alster. 2003. Effects of methionine deficiencies on plasma levels of thyroid hormones, insulin-like growth factors-I and -II, liver and body weights, and feed intake in growing chickens. *Poultry Science* 82:1932–1938.

Carey, C. 1996. Female reproductive energetics. In *Avian Energetics and Nutritional Ecology*, ed. C. Carey, 324–374. New York: Chapman and Hall.

———. 2009. The impacts of climate change on the annual cycles of birds. *Philosophical Transactions of the Royal Society B: Biological Sciences* 364:3321–3330.

Carey, C., O. Dunin-Borkowski, F. Leon-Velarde, D. Espinoza, and C. Monge. 1994. Gas exchange and blood gases of Puna teal (*Anas versicolor puna*) embryos in the Peruvian Andes. *Journal of Comparative Physiology* 163B:649–656.

Caro, S. P., A. Charmantier, M. M. Lambrechts, J. Blondel, J. Balthazart, and T. D. Williams. 2009. Local adaptation of timing of reproduction: Females are in the driver's seat. *Functional Ecology* 23:172–179.

Caro, S. P., M. M. Lambrechts, and J. Balthazart. 2005. Early seasonal development of brain song control nuclei in male blue tits. *Neuroscience Letters* 386:139–144.

Caro, S. P., M. M. Lambrechts, J. Balthazart, and P. Perret. 2007. Non-photoperiodic factors and timing of breeding in blue tits: Impact of environmental and social influences in semi-natural conditions. *Behavioural Processes* 75:1–7.

Caro, S. P., M. M. Lambrechts, O. Chastel, P. J. Sharp, D. W. Thomas, and J. Balthazart. 2006. Simultaneous pituitary-gonadal recrudescence in two Corsican populations of male blue tits with asynchronous breeding dates. *Hormones and Behavior* 50:347–360.

Cassey, P., J. G. Ewen, R. L. Boulton, T. M. Blackburn, A. P. Møller, C. Biard, V. Olson, et al. 2005. Egg carotenoids in passerine birds introduced to New Zealand: Relations to ecological factors, integument coloration and phylogeny. *Functional Ecology* 19:719–726.

Catoni, C., A. Peters, and H. Martin Schaefer. 2008. Life history trade-offs are influenced by the diversity, availability and interactions of dietary antioxidants. *Animal Behaviour* 76:1107–1119.

Challenger, W. O., T. D. Williams, J. K. Christians, and F. Vézina. 2001. Follicular development and plasma yolk precursor dynamics through the laying cycle in the European starling (*Sturnus vulgaris*). *Physiological and Biochemical Zoology* 74:356–365.

Chamberlain, D. E., A. R. Cannon, M. P. Toms, D. I. Leech, B. J. Hatchwell, and K. J. Gaston. 2009. Avian productivity in urban landscapes: A review and meta-analysis. *Ibis* 151:1–18.

Chappell, M. A., C. Bech, and W. A. Buttemer. 1999. The relationship of central and peripheral organ masses to aerobic performance variation in house sparrows. *Journal of Experimental Biology* 202:2269–2279.

Charmantier, A., R. H. McCleery, L. R. Cole, C. Perrins, L. E. B. Kruuk, and B. C. Sheldon. 2008. Adaptive phenotypic plasticity in response to climate change in a wild bird population. *Science* 320:800–803.

Charmantier, A., C. Perrins, R. H. McCleery, and B. C. Sheldon. 2006. Evolutionary response to selection in clutch size in a long-term study of the mute swan. *American Naturalist* 167:453–465.

Chastel, O., H. Weimerskirch, and P. Jouventin. 1995. Body condition and seabird reproductive performance: A study of three petrel species. *Ecology* 76:2240–2246.

Chaurand, T., and H. Weimerskirch. 1994. Incubation routine, body mass regulation and egg neglect in the Blue Petrel *Halobaena caerulea*. *Ibis* 136:285–290.

Chen, S.-E., D. W. Long, K. E. Nestor, R. L. Walzem, V. L. Meuniot, H. Zhu, R. J. Hansen, et al. 1999. Effect of divergent selection for total plasma phosphorous on plasma and yolk very low density lipoproteins and plasma concentrations of selected hormones in laying Japanese quail. *Poultry Science* 78:1241–1251.

Cheng, M.-F. 2005. Audio-vocal pathways controlling GnRH release. In *Functional Avian Endocrinology*, ed. A. Dawson and P. J. Sharp, 113–128. New Delhi: Narosa Publishing House.

Cheng, M.-F., C. Desiderio, M. Havens, and A. Johnson. 1988. Behavioral stimulation of ovarian growth. *Hormones and Behavior* 22:388–401.

Cherel, Y., J. P. Robin, O. Walch, H. Karmann, P. Netchitailo, and Y. Le Maho. 1988. Fasting in king penguin I. Hormonal and metabolic changes during breeding. *American Journal of Physiology* 254:R170–177.

Cheviron, Z. A., A. Whitehead, and R. T. Brumfield. 2008. Transcriptomic variation and plasticity in rufous-collared sparrows (*Zonotrichia capensis*) along an altitudinal gradient. *Molecular Ecology* 17:4556–4569.

Chilgren, J. D., and W. A. deGraw. 1977. Some blood characteristics of White-crowned Sparrows during molt. *Auk* 94:169–171.

Chin, E. H., O. P. Love, J. J. Verspoor, T. D. Williams, K. Rowley, and G. Burness. 2009. Juveniles exposed to embryonic corticosterone have enhanced flight performance. *Proceedings of the Royal Society B: Biological Sciences* 276:499–505.

Christe, P., A. P. Møller, N. Saino, and F. De Lope. 2000. Genetic and environmental components of phenotypic variation in immune response and body size of a colonial bird, *Delichon urbica* (the house martin). *Heredity* 85:75–83.

Christians, J. K. 2000a. Producing extra eggs does not deplete macronutrient reserves in European Starlings *Sturnus vulgaris*. *Journal of Avian Biology* 31:312–318.

———. 2000b. Trade-offs between egg size and number in waterfowl: An interspecific test of the van Noordwijk and de Jong model. *Functional Ecology* 14:497–501.

———. 2002. Avian egg size: Variation within species and inflexibility within individuals. *Biological Reviews* 77:1–26.

Christians, J. K., and T. D. Williams. 1999. Organ mass dynamics in relation to yolk precursor production and egg formation in European starlings *Sturnus vulgaris*. *Physiological and Biochemical Zoology* 72:455–461.

———. 2001. Interindividual variation in yolk mass and the rate of growth of ovarian follicles in the zebra finch (*Taeniopygia guttata*). *Journal of Comparative Physiology* 171B:255–261.

———. 2002. Effects of porcine follicle-stimulating hormone on the reproductive performance of female zebra finches (*Taeniopygia guttata*). *General and Comparative Endocrinology* 125:121–131.

Ciccone, N. A., I. C. Dunn, T. Boswell, K. Tsutsui, T. Ubuka, K. Ukena, and P. J. Sharp. 2004. Gonadotrophin inhibitory hormone depresses gonadotrophin alpha and follicle-stimulating hormone beta subunit expression in the pituitary of the domestic chicken. *Journal of Neuroendocrinology* 16:999–1006.

Ciccone, N. A., P. J. Sharp, P. W. Wilson, and I. C. Dunn. 2005. Changes in reproductive neuroendocrine mRNAs with decreasing ovarian function in ageing hens. *General and Comparative Endocrinology* 144:20–27.

Clark, A. B., and D. S. Wilson. 1981. Avian breeding adaptations: Hatching asynchrony, brood reduction, and nest failure. *Quarterly Review of Biology* 56:253–277.

Clarke, W. C., and H. A. Bern. 1980. Comparative endocrinology of prolactin. In *Hormonal Proteins and Peptides*, ed. C. H. Li, 105–197. New York: Academic Press.

Clemens, D. T. 1990. Interspecific variation and effects of altitude on blood properties of rosy finches (*Leucosticte arctoa*) and house finches (*Carpodacus mexicanus*). *Physiological Zoology* 63:288–307.

Clermont, C. P., and H. Schraer. 1979. Effect of estrogen on rate of Fe-59 uptake by hematopoietic-tissue in Japanese quail. *American Journal of Physiology* 236:E342–E346.

Cloues, R., C. Ramos, and R. Silver. 1990. Vasoactive intestinal polypeptide-like immunoreactivity during reproduction in doves: Influence of experience and number of offspring. *Hormones and Behavior* 24:215–231.

Clutton-Brock, T. H. 1988. *Reproductive Success: Studies of Individual Variation in Contrasting Breeding Systems*. Chicago: University of Chicago Press.

———. 1991. *The Evolution of Parental Care*. Princeton, NJ: Princeton University Press.

Clutton-Brock, T., and B. C. Sheldon. 2010. Individuals and populations: The role of long-term, individual-based studies of animals in ecology and evolutionary biology. *Trends in Ecology and Evolution* 25:562–573.

Cohen, A. A., and K. J. McGraw. 2009. No simple measures for antioxidant status in birds: Complexity in inter- and intraspecific correlations among circulating antioxidant types. *Functional Ecology* 23:310–320.

Cole, L. J. 1930. The laying cycle in the House Wren. *Wilson Bulletin* 42:78.

Conway, C. J., and T. E. Martin. 2000. Evolution of passerine incubation behavior: Influence of food, temperature, and nest predation. *Evolution* 54:670–685.

Cook, M. I., S. R. Beissinger, G. A. Toranzos, and W. J. Arendt. 2005. Microbial infection affects egg viability and incubation behavior in a tropical passerine. *Behavioral Ecology* 16:30–36.

Cook, M. I., S. R. Beissinger, G. A. Toranzos, R. A. Rodriguez, and W. J. Arendt. 2003. Trans-shell infection by pathogenic micro-organisms reduces the shelf life of non-incubated bird's eggs: A constraint on the onset of incubation? *Proceedings of the Royal Society B: Biological Sciences* 270:2233–2240.

Cooke, F., R. F. Rockwell, and D. B. Lank. 1995. *The Snow Geese of La Perouse Bay*. Oxford: Oxford University Press.

Cooke, F., P. D. Taylor, C. M. Frances, and R. F. Rockwell. 1990. Directional selection and clutch size in birds. *American Naturalist* 136:261–267.

Cooper, C. B., W. M. Hochachka, G. Butcher, and A. A. Dhondt. 2005. Seasonal and latitudinal trends in clutch size: Thermal constraints during laying and incubation. *Ecology* 86:2018–2031.

Coppack, T., and F. Pulido. 2004. Photoperiodic response and the adaptability of avian life cycles to environmental change. *Advances in Ecological Research* 35:131–150.

Correa, S. M., E. Adkins-Regan, and P. A. Johnson. 2005. High progesterone during avian meiosis biases sex ratios toward females. *Biology Letters* 1:215–218.

Costantini, D. 2008. Oxidative stress in ecology and evolution: Lessons from avian studies. *Ecology Letters* 11:1238–1251.

Costantini, D., G. Dell'Ariccia, and H.-P. Lipp. 2008. Long flights and age affect oxidative status of homing pigeons (*Columba livia*). *Journal of Experimental Biology* 211:377–381.

Costantini, D., and A. P. Møller. 2008. Carotenoids are minor antioxidants for birds. *Functional Ecology* 22:367–370.

Costantini, D., and S. Verhulst. 2009. Does high antioxidant capacity indicate low oxidative stress? *Functional Ecology* 23:506–509.

Cottam, M., D. Houston, G. Lobley, and I. Hamilton. 2002. The use of muscle protein for egg production in the Zebra Finch *Taeniopygia guttata*. *Ibis* 144:210–217.

Cresswell, W., S. Holt, J. M. Reid, D. P. Whitfield, and R. J. Mellanby. 2003. Do energetic demands constrain incubation scheduling in a biparental species? *Behavioral Ecology* 14:97–102.

Cresswell, W., S. Holt, J. M. Reid, D. P. Whitfield, R. J. Mellanby, D. Norton, and S. Waldron. 2004. The energetic costs of egg heating constrain incubation attendance but do not determine daily energy expenditure in the pectoral sandpiper. *Behavioral Ecology* 15:498–507.

Cresswell, W., and R. McCleery. 2003. How great tits maintain synchronization of their hatch date with food supply in response to long-term variability in temperature. *Journal of Animal Ecology* 72:356–366.

Crick, H.Q.P., C. Dudley, D. E. Glue, and D. L. Thomson. 1997. UK birds are laying eggs earlier. *Nature* 388:526–527.

Crick, H.Q.P., D. W. Gibbons, and R. D. Magrath. 1993. Seasonal changes in clutch size in British birds. *Journal of Animal Ecology* 62:263–273.

Criscuolo, F., O. Chastel, G. W. Gabrielsen, A. Lacroix, and Y. Le Maho. 2002. Factors affecting plasma concentrations of prolactin in the common eider *Somateria mollissima*. *General and Comparative Endocrinology* 125:399–409.

Criscuolo, F., T. Raclot, Y. Le Maho, and G. W. Gabrielsen. 2003. Do T3 levels in incubating eiders reflect the cost of incubation among clutch sizes? *Physiological and Biochemical Zoology* 76:196–203.

Crisóstomo, S., D. Guémené, M. Garreau-Mills, C. Morvan, and D. Zadworny. 1998. Prevention of incubation behavior expression in turkey hens by active immunization against prolactin. *Theriogenology* 50:675–690.

Crisóstomo, S., D. Guémené, M. Garreau-Mills, and D. Zadworny. 1997. Prevention of the expression of incubation behaviour using passive immunisation against prolactin in turkey hens (*Meleagris gallopavo*). *Reproduction, Nutrition, Development* 37:253–286.

Crossin, G. T., P. N. Trathan, R. A. Phillips, A. Dawson, F. Le Bouard, and T. D. Williams. 2010. A carryover effect of migration underlies individual variation in reproductive readiness and extreme egg-size dimorphism in macaroni penguins. *American Naturalist* 176:357–366.

Crossin, G. T., P. N. Trathan and T. D. Williams. In press. Potential mode of clutch size determination and follicle development in Eudyptes penguins. *Polar Biology*.

Cuthbert, N. L. 1945. The ovarian cycle of the ring dove (*Streptopelia risoria*). *Journal of Morphology* 77:35.

Daan, S., C. Deerenberg, and C. Dijkstra. 1996. Increased daily work precipitates natural death in the kestrel. *Journal of Animal Ecology* 65:539–544.

Daan, S., C. Dijkstra, and J. M. Tinbergen. 1990. Family planning in the Kestrel (*Falco tinnunculus*): The ultimate control of covariation of laying date and clutch size. *Behaviour* 114:83–116.

Daan, S., C. Dijkstra, and F. J. Weissing. 1996. An evolutionary explanation for seasonal trends in avian sex ratios. *Behavioral Ecology* 7:426–430.

Daan, S., D. Masman, and A. Groenwald. 1990. Avian basal metabolic rates: Their association with body composition and energy-expenditure in nature. *American Journal of Physiology* 259:R333–R340.

D'Alba, L., P. Monaghan, and R. G. Nager. 2010. Advances in laying date and increasing population size suggest positive responses to climate change in Common Eiders *Somateria mollissima* in Iceland. *Ibis* 152:19–28.

D'Alba, L., M. D. Shawkey, P. Korsten, O. Vedder, S. A. Kingma, J. Komdeur, and S. R. Beissinger. 2010. Differential deposition of antimicrobial proteins in blue tit (*Cyanistes caeruleus*) clutches by laying order and male attractiveness. *Behavioral Ecology and Sociobiology* 64:1037–1045.

Dallman, M. F. 2003. Stress by any other name . . . ? *Hormones and Behavior* 43:18–20.

Darwin, C. R. 1872. *The Origin of Species by Means of Natural Selection, or the Preservation of Favoured Races in the Struggle for Life*. 6th ed. London: John Murray.

Daunt, F., V. Afanasayev, J. R. D. Silk, and S. Wanless. 2006. Extrinsic and intrinsic determinants of winter foraging and breeding phenology in a temperate seabird. *Behavioral Ecology and Sociobiology* 59:381–388.

D'Autreaux, B., and M. B. Toledano. 2007. ROS as signalling molecules: Mechanisms that generate specificity in ROS homeostasis. *Nature Reviews Molecular Cell Biology* 8:813–824.

Davey, C., A. Lill, and J. Baldwin. 2000. Variation during breeding in parameters that influence blood oxygen carrying capacity in shearwaters. *Australian Journal of Zoology* 48:347–356.

Davies, N. B., and A. Lundberg. 1985. The influence of food on time budgets and timing of breeding of the Dunnock *Prunella modularis*. *Ibis* 127:100–110.

Davis, S. D., J. B. Williams, W. J. Adams, and S. L. Brown. 1984. The effect of egg temperature on attentiveness in the Belding's Savannah Sparrow. *Auk* 101:556–566.

Dawson, A. 1986. The effect of restricting the daily period of food availability on testicular growth of Starlings *Sturnus vulgaris*. *Ibis* 128:572–575.

———. 1994. The effects of daylength and testosterone on the initiation and progress of moult in Starlings *Sturnus vulgaris*. *Ibis* 136:335–340.

———. 1998. Photoperiodic control of the termination of breeding and the induction of moult in House Sparrows *Passer domesticus*. *Ibis* 140:35–40.

———. 2002. Photoperiodic control of the annual cycle in birds and comparison with mammals. *Ardea* 90:355–367.

———. 2003. A comparison of the annual cycles in testicular size and moult in captive European starlings *Sturnus vulgaris* during their first and second years. *Journal of Avian Biology* 34:119–123.

———. 2005. The effect of temperature on photoperiodically regulated gonadal maturation, regression and moult in starlings: Potential consequences of climate change. *Functional Ecology* 19:995–1000.

———. 2006. Control of molt in birds: Association with prolactin and gonadal regression. *General and Comparative Endocrinology* 147:314–322.

———. 2007. Seasonality in a temperate zone bird can be entrained by near equatorial photoperiods. *Proceedings of the Royal Society B: Biological Sciences* 274:721–725.

———. 2008. Control of the annual cycle in birds: Endocrine constraints and plasticity in response to ecological variables. *Philosophical Transactions of the Royal Society B* 363:1621–1633.

Dawson, A., B. K. Follett, A. R. Goldsmith, and T. J. Nicholls. 1985. Hypothalamic gonadotrophin-releasing hormone and pituitary and plasma FSH and prolactin during photostimulation and photorefractoriness in intact and thyroidectomised starlings (*Sturnus vulgaris*). *Journal of Endocrinology* 105:71–77.

Dawson, A., and A. R. Goldsmith. 1982. Prolactin and gonadotrophin secretion in wild starlings (*Sturnus vulgaris*) during the annual cycle and in relation to nesting, incubation, and rearing young. *General and Comparative Endocrinology* 48:213–221.

———. 1983. Plasma prolactin and gonadotrophins during gonadal development and the onset of photorefractoriness in male and female starlings (*Sturnus vulgaris*) on artificial photoperiods. *Journal of Endocrinology* 97:253–260.

———. 1984. Effects of gonadectomy on seasonal changes in plasma LH and prolactin concentrations in male and female starlings *Sturnus vulgaris*. *Journal of Endocrinology* 100:213–218.

Dawson, A., S. A. Hinsley, P. N. Ferns, R. H. C. Bonser, and L. Eccleston. 2000. Rate of moult affects feather quality: A mechanism linking current reproductive effort to future survival. *Proceedings of the Royal Society B: Biological Sciences* 267:2093–2098.

Dawson, A., V. M. King, G. E. Bentley, and G. F. Ball. 2001. Photoperiodic control of seasonality in birds. *Journal of Biological Rhythms* 16:365–380.

Dawson, A., C. M. Perrins, P. J. Sharp, D. Wheeler, and S. Groves. 2009. The involvement of prolactin in avian molt: The effects of gender and breeding success on the timing of molt in Mute swans (*Cygnus olor*). *General and Comparative Endocrinology* 161:267–270.

Dawson, A., and P. J. Sharp. 2007. Photorefractoriness in birds: Photoperiodic and non-photoperiodic control. *General and Comparative Endocrinology* 153:378–384.

Dawson, R. D., and R. G. Clark. 2000. Effects of hatching date and egg size on growth, recruitment, and adult size of Lesser Scaup. *Condor* 102:930–935.

Dean, M., and T. Annilo. 2005. Evolution of the ATP-binding cassette (ABC) transporter superfamily in vertebrates. *Annual Review of Genomics and Human Genetics* 6:123–142.

de Ayala, R. M., R. Martinelli, and N. Saino. 2006. Vitamin E supplementation enhances growth and condition of nestling barn swallows (*Hirundo rustica*). *Behavioral Ecology and Sociobiology* 60:619–630.

De Block, M., and R. Stoks. 2008. Compensatory growth and oxidative stress in a damselfly. *Proceedings of the Royal Society B: Biological Sciences* 275:781–785.

Decherf, S., I. Seugnet, S. Kouidhi, A. Lopez-Juarez, M.-S. Clerget-Froidevaux, and B. A. Demeneix. 2010. Thyroid hormone exerts negative feedback on hypothalamic type 4 melanocortin receptor expression. *Proceedings of the National Academy of Sciences USA* 107:4471–4476.

Deeley, R. G., R. A. Burtchwright, C. E. Grant, P. A. Hoodless, A. K. Ryan, and T. J. Schrader. 1993. Synthesis and deposition of egg proteins. In *Manipulation of the Avian Genome*, ed. R. J. Etches and A.M.V. Gibbens, 205–222. Boca Raton: CRC Press.

Deeming, D. C. 2002a. *Avian Incubation*. Oxford: Oxford University Press.

———. 2002b. Embryonic development and utilisation of egg components. In *Avian Incubation*, ed. D. C. Deeming, 43–53. Oxford: Oxford University Press.

Deeming, D. C., and M.W.J. Ferguson. 1992. Physiological effects of incubation temperature on embryonic development in reptiles and birds. In *Egg Incubation: Its Effects on Embryonic Development in Birds and Reptiles*, ed. D. C. Deeming and M.W.J. Ferguson, 147–173. Cambridge: Cambridge University Press.

Deerenberg, C., I. Pen, C. Dijkstra, B.-J. Arkles, G. H. Visser, and S. Daan. 1995. Parental energy expenditure in relation to manipulated brood size in the European kestrel *Falco tinnunculus*. *Zoology* 99:39–48.

de Graw, W. A., M. D. Kern, and J. R. King. 1979. Seasonal changes in the blood composition of captive and free-living White-crowned Sparrows. *Journal of Comparative Physiology* 129:151–162.

de Heij, M. E., R. Ubels, G. Henk Visser, and J. M. Tinbergen. 2008. Female great tits Parus major do not increase their daily energy expenditure when incubating enlarged clutches. *Journal of Avian Biology* 39:121–126.

de Heij, M. E., P. J. Van den Hout, and J. M. Tinbergen. 2006. Fitness cost of incubation in great tits (*Parus major*) is related to clutch size. *Proceedings of the Royal Society B: Biological Sciences* 273:2353–2361.

de Heij, M. E., A. J. van der Graaf, D. Hafner, and J. M. Tinbergen. 2007. Metabolic rate of nocturnal incubation in female great tits, *Parus major*, in relation to clutch size measured in a natural environment. *Journal of Experimental Biology* 210:2006–2012.

de Jong, G. 1993. Covariances between traits deriving from successive allocations of a resource. *Functional Ecology* 7:75–83.

De La Hera, I., J. Perez-Tris, and J. L. Tellería. 2009. Migratory behaviour affects the trade-off between feather growth rate and feather quality in a passerine bird. *Biological Journal of the Linnean Society* 97:98–105.

De Neve, L., J. A. Fargallo, P. Vergara, J. A. Lemus, M. Jaren-Galan, and I. Luaces. 2008. Effects of maternal carotenoid availability in relation to sex, parasite infection and health status of nestling kestrels (*Falco tinnunculus*). *Journal of Experimental Biology* 211:1414–1425.

Descamps, S., H. G. Gilchrist, J. Bety, E. I. Buttler, and M. R. Forbes. 2009. Costs of reproduction in a long-lived bird: Large clutch size is associated with low survival in the presence of a highly virulent disease. *Biology Letters* 5:278–281.

DesRochers, D. W., J. M. Reed, J. Awerman, J. A. Kluge, J. Wilkinson, L. I. van Griethuijsen, J. Aman, et al. 2009. Exogenous and endogenous corticosterone alter feather quality. *Comparative Biochemistry and Physiology A: Molecular and Integrative Physiology* 152:46–52.

DeSteven, D. 1980. Clutch size, breeding success, and parental survival in the Tree Swallow (*Iridoprocne bicolor*). *Evolution* 34:278–291.

de Tassigny, X.d.A., and W. H. Colledge. 2010. The role of kisspeptin signalling in reproduction. *Physiology* 25:207–217.

Deviche, P., and P. Sharp. 2001. Reproductive endocrinology of a free-living, opportunistically breeding passerine (white-winged crossbill, *Loxia leucoptera*). *General and Comparative Endocrinology* 123:268–279.

Devries, J. H., R. W. Brook, D. W. Howerter, and M. G. Anderson. 2008. Effects of spring body condition and age on reproduction in Mallards (*Anas platyrhynchos*). *Auk* 125:618–628.

Dhaka, A., V. Viswanath, and A. Patapoutian. 2006. TRP ion channel and temperature sensation. *Annual Review of Neuroscience* 29:135–161.

Dhondt, A. A. 1989. Blue tit. In *Lifetime Reproduction in Birds*, ed. I. Newton, 15–34. London: Academic Press.

Dickey, M.-H., G. Gauthier, and M.-C. Cadieux. 2008. Climatic effects on the breeding phenology and reproductive success of an arctic-nesting goose species. *Global Change Biology* 14:1973–1985.

Dijkstra, C., A. Bult, S. Bijlsma, S. Daan, T. Meijer, and M. Zijlstra. 1990. Brood size manipulations in the kestrel (*Falco tinnunculus*): Effects on offspring and parent survival. *Journal of Animal Ecology* 59:269–285.

Dijkstra, C., L. Vuursteen, S. Daan, and D. M. Masman. 1982. Clutch size and laying date in the Kestrel (*Falco tinnunculus*): The effect of supplementary food. *Ibis* 124:210–213.

Dingemanse, N. J., A. J. N. Kazem, D. Réale, and J. Wright. 2010. Behavioural reaction norms: Animal personality meets individual plasticity. *Trends in Ecology and Evolution* 25:81–89.

Dobbs, R. C., J. D. Styrsky, and C. F. Thompson. 2006. Clutch size and the costs of incubation in the house wren. *Behavioral Ecology* 17:849–856.

Doligez, B., and J. Clobert. 2003. Clutch size reduction as a response to increased nest predation rate in the Collared Flycatcher. *Ecology* 84:2582–2588.

Dor, R., and A. Lotem. 2010. Parental effort and response to nestling begging in the house sparrow: Repeatability, heritability and parent-offspring coevolution. *Journal of Evolutionary Biology* 23:1605–1612.

Dougherty, D. C., and M. M. Sanders. 2005. Estrogen action: Revitalization of the chick oviduct model. *Trends in Endocrinology and Metabolism* 16:414–419.

Dowling, D. K., and L. W. Simmons. 2009. Reactive oxygen species as universal constraints in life-history evolution. *Proceedings of the Royal Society B: Biological Sciences* 276:1737–1745.

Drent, R. 2006. The timing of birds' breeding seasons: The Perrin's hypothesis revisited especially for migrants. *Ardea* 94:305–322.

Drent, R., and S. Daan. 1980. The prudent parent: Energetic adjustments in avian breeding. *Ardea* 68:225–252.

Drobney, R. D. 1984. Effect of diet on visceral morphology of breeding Wood Ducks. *Auk* 101: 93–98.

———. 1991. Nutrient limitation of clutch size in waterfowl: Is there a universal hypothesis? *Condor* 93:1026–1028.

Dubiec, A., M. Cichon, and K. Deptuch. 2006. Sex-specific development of cell-mediated immunity under experimentally altered rearing conditions in blue tit nestlings. *Proceedings of the Royal Society B: Biological Sciences* 273:1759–1764.

Duckworth, R. A., A. V. Badyaev, and A. F. Parlow. 2003. Elaborately ornamented males avoid costly parental care in the house finch (*Carpodacus mexicanus*): A proximate perspective. *Behavioral Ecology and Sociobiology* 55:176–183.

Dufty, A. M., A. R. Goldsmith, and J. C. Wingfield. 1987. Prolactin secretion in a brood parasite, the brown-headed cowbird, *Molothrus ater*. *Journal of Zoology* 212:669–675.

Dufva, R. 1996. Blood parasites, health, reproductive success, and egg volume in female Great Tits *Parus major*. *Journal of Avian Biology* 27:83–87.

Dunn, P. 2004. Breeding dates and reproductive performance. *Advances in Ecological Research* 35:69–87.

Dunn, P. O., Winkler, D. W., Whittingham, L. A., Hannon, S. J. and Robertson, R. A. 2011. A test of the mismatch hypothesis: How is timing of reproduction related to food abundance in an aerial insectivore? *Ecology* 92:450–461.

Dunnet, G. M. 1955. The breeding of the Starling *Sturnus vulgaris* in relation to its food supply. *Ibis* 97:619–661.

Dzialowski, E. M., W. L. Reed, and P. R. Sotherland. 2009. Effects of egg size on Double-crested Cormorant (*Phalacrocorax auritus*) egg composition and hatchling phenotype. *Comparative Biochemistry and Physiology A: Molecular and Integrative Physiology* 152:262–267.

Dzus, E. H., and R. G. Clark. 1998. Brood survival and recruitment of Mallards in relation to wetland density and hatching date. *Auk* 115:311–318.

Ebbinge, B. S., and B. Spaans. 1995. The importance of body reserves accumulated in spring staging areas in the temperate zone for breeding dark-bellied Brent Geese *Brante b. bernicla* in the high Arctic. *Journal of Avian Biology* 26:105–113.

Edinger, R. S., E. Mambo, and M. I. Evans. 1997. Estrogen-dependent transcriptional activation and vitellogenin gene memory. *Molecular Endocrinology* 11:1985–1993.

Edwards, S. 2007. Genomics and ornithology. *Journal of Ornithology* 148:27–33.

Eggers, S., M. Griesser, M. Nystrand, and J. Ekman. 2006. Predation risk induces changes in nest-site selection and clutch size in the Siberian jay. *Proceedings of the Royal Society B: Biological Sciences* 273:701–706.

Eising, C. M., W. Müller, and T.G.G. Groothuis. 2006. Avian mothers create different phenotypes by hormone deposition in their eggs. *Biology Letters* 2:20–22.

Eising, C. M., D. Pavlova, T.G.G. Groothuis, M. Eens, and R. Pinxten. 2008. Maternal yolk androgens in European starlings: Social environment or individual traits of the mother? *Behaviour* 145:51–72.

Eisner, E. 1960. The relationship of hormones to the reproductive behaviour of birds, referring especially to parental behaviour: A review. *Animal Behaviour* 8:155–179.

Elf, P. K., and A. J. Fivizzani. 2002. Changes in sex steroid levels in yolks of the leghorn chicken, *Gallus domesticus*, during embryonic development. *Journal of Experimental Zoology* 293:594–600.

El Halawani, M. E., J. L. Silsby, E. J. Behnke, and S. C. Fehrer. 1986. Hormonal induction of incubation behavior in ovariectomized female turkeys (*Meleagris gallopavo*). *Biology of Reproduction* 35:59–67.

Elkin, R. G., I. MacLachlan, M. Hermann, and W. J. Schneider. 1995. Characterization of the Japanese Quail oocyte receptor for very low density lipoprotein and vitellogenin. *Journal of Nutrition* 125:1258–1266.

Ellegren, H., and B. C. Sheldon. 2008. Genetic basis of fitness differences in natural populations. *Nature* 452:169–175.

Erikstad, K. E., P. Fauchald, T. Tveraa, and H. Steen. 1998. On the cost of reproduction in long-lived birds: The influence of environmental variability. *Ecology* 79:1781–1788.

Erikstad, K. E., H. Sandvik, P. Fauchald, and T. Tveraa. 2009. Short- and long-term consequences of reproductive decisions: An experimental study in the puffin. *Ecology* 90:3197–3208.

Etches, R. J. 1996. *Reproduction in Poultry*. Wallingford, UK: CAB International.

Evans, K. L., R. P. Duncan, T. M. Blackburn, and H.Q.P. Crick. 2005. Investigating geographic variation in clutch size using a natural experiment. *Functional Ecology* 19:616–624.

Evans, M. I., R. Silva, and J. B. Burch. 1988. Isolation of chicken vitellogenin I and III cDNAs and the developmental regulation of five estrogen-responsive genes in the embryonic liver. *Genes and Development* 2:116–124.

Ewen, J. G., P. Cassey, and A. P. Møller. 2004. Facultative primary sex ratio variation: A lack of evidence in birds? *Proceedings of the Royal Society B: Biological Sciences* 271:1277–1282.

Ewert, M. A. 1992. Cold torpor, diapause, delayed hatching and aestivation in reptiles and birds. In *Egg Incubation: Its Effects on Embryonic Development in Birds and Reptiles*, ed. D. C. Deeming and M.W.J. Ferguson, 173–191. Cambridge: Cambridge University Press.

Fandrey, J. 2004. Oxygen-dependent and tissue-specific regulation of erythropoietin gene expression. *American Journal of Physiology* 286:R977–R988.

Farner, D. S., B. K. Follett, J. R. King, and M. Morton. 1966. A quantitative examination of ovarian growth in the white-crowned sparrow. *Biological Bulletin* 130:67–75.

Feare, C. 1984. *The Starling*. Oxford: Oxford University Press.

Feare, C. J., P. L. Spencer, and D.A.T. Constantine. 1982. Time of egg-laying of starlings *Sturnus vulgaris*. *Ibis* 124:174–178.

Fenoglio, S., M. Cucco, and G. Malacarne. 2003. Moorhen *Gallinula chloropus* females lay eggs of different size and beta-carotene content. *Ardea* 91:117–121.

Fernandez, F. R., and C. B. Grindem. 2006. Reticulocyte response. In *Schalm's Veterinary Hematology*, ed. B. F. Feldman, J. G. Zinkl, and N. C. Jain, 110 –116. Ames, Iowa: Blackwell.

Ferrari, R. P., R. Martinelli, and N. Saino. 2006. Differential effects of egg albumen content on barn swallow nestlings in relation to hatch order. *Journal of Evolutionary Biology* 19:981–993.

Fidler, A. E., K. van Oers, P. J. Drent, S. Kuhn, J. C. Mueller, and B. Kempanaers. 2007. *Drd4* gene polymorphisms are associated with personality variation in a passerine bird. *Proceedings of the Royal Society B: Biological Sciences* 274:1685–1691.

Finch, C. E., and M. R. Rose. 1995. Hormones and the physiological architecture of life-history evolution. *Quarterly Review of Biology* 70:1–52.

Finkel, T., and N. J. Holbrook. 2000. Oxidants, oxidative stress and the biology of ageing. *Nature* 408:239–247.

Finkler, M. S., J. B. van Orman, and P. R. Sotherland. 1998. Experimental manipulation of egg quality in chickens: Influence of albumen and yolk on the size and body composition of near-term embryos in a precocial bird. *Journal of Comparative Physiology* 168B:17–24.

Fisher, R. A. 1930. *The Genetical Theory of Natural Selection*. Oxford: Oxford University Press.

Flaws, J. A., A. DeSanti, K. I. Tilly, R. O. Javid, K. Kugu, A. L. Johnson, A. N. Hirshfield, et al. 1995. Vasoactive intestinal peptide-mediated suppression of apoptosis in the ovary: Potential mechanisms of action and evidence of a conserved antiatretogenic role through evolution. *Endocrinology* 136:4351–4359.

Flinks, H., B. Helm, and P. Rothery. 2008. Plasticity of moult and breeding schedules in migratory European Stonechats *Saxicola rubicola*. *Ibis* 150:687–697.

Flint, P. L., J. B. Grand, T. F. Fondell, and J. A. Morse. 2006. Population dynamics of greater scaup breeding on the Yukon-Kuskokwim Delta, Alaska. *Wildlife Monographs* 162:1–22.

Flint, P. L., and J. S. Sedinger. 1992. Reproductive implications of egg-size variation in the Black Brant. *Auk* 109:896–903.

Follett, B. K., and M. R. Redshaw. 1974. The physiology of vitellogenesis. In *Physiology of Amphibia*, ed. B. Lofts, 219–308. New York: Academic Press.

Fontaine, J. J., and T. E. Martin. 2006. Parent birds assess nest predation risk and adjust their reproductive strategies. *Ecology Letters* 9:428–434.

Forbes, S. 2007. Sibling symbiosis in nestling birds. *Auk* 124:1–10.

———. 2009. Portfolio theory and how parent birds manage investment risk. *Oikos* 118:1561–1569.

Forbes, S., B. Glassey, S. Thornton, and L. Earle. 2001. The secondary adjustment of clutch size in red-winged blackbirds (*Agelaius phoeniceus*). *Behavioral Ecology and Sociobiology* 50:37–44.

Forbes, S., S. Thornton, B. Glassey, M. Forbes, and N. J. Buckley. 1997. Why parent birds play favourites. *Nature* 390:351–352.

Forbes, S., and M. Wiebe. 2010. Egg size and asymmetric sibling rivalry in red-winged blackbirds. *Oecologia* 163:361–372.

Fortune, J. E. 2003. The early stages of follicular development: Activation of primordial follicles and growth of preantral follicles. *Animal Reproduction Science* 78:135–163.

Foster, M. S. 1974. A model to explain molt-breeding overlap and clutch size in some tropical birds. *Evolution* 28:182–190.

———. 1975. The overlap of molting and breeding in some tropical birds. *Condor* 77:304–314.

Freed, L. A. 1981. Loss of mass in breeding wrens: Stress or adaptation? *Ecology* 62:1179–1186.

Gallizzi, K., L. Gern, and H. Richner. 2008. A flea-induced pre-hatching maternal effect modulates tick feeding behaviour on great tit nestlings. *Functional Ecology* 22:94–99.

Garant, D., J. D. Hadfield, L. E. B. Kruuk, and B. C. Sheldon. 2007. Stability of genetic variance and covariance for reproductive characters in the face of climate change in a wild bird population. *Molecular Ecology* 17:179–188.

Garant, D., L. E. B. Kruuk, R. H. McCleery, and B. C. Sheldon. 2007. The effects of environmental heterogeneity on multivariate selection on reproductive traits in female great tits. *Evolution* 61:1546–1559.

Garcia, F., J. Sanchez, and J. Planas. 1986. Influence of laying on iron-metabolism in quail. *British Poultry Science* 27:585–592.

Gasparini, J., T. Boulinier, V. A. Gill, D. Gil, S. A. Hatch, and A. Roulin. 2007. Food availability affects the maternal transfer of androgens and antibodies into eggs of a colonial seabird. *Journal of Evolutionary Biology* 20:874–880.

Gasparini, J., K. D. McCoy, C. Haussy, T. Tveraa, and T. Boulinier. 2001. Induced maternal response to the Lyme disease spriochaete *Borrelia burgdorferi* sensu lato in a colonial seabird, the kittiwake *Rissa tridactyla*. *Proceedings of the Royal Society B: Biological Sciences* 268:647–650.

Gasparini, J., K. D. McCoy, V. Staszewski, C. Haussy, and T. Boulinier. 2006. Dynamics of anti-Borrelia antibodies in Black-legged Kittiwake (*Rissa tridactyla*) chicks suggest a maternal educational effect. *Canadian Journal of Zoology* 84:623–627.

Gasparini, J., K. D. McCoy, T. Tveraa, and T. Boulinier. 2002. Related concentrations of specific immunoglobulins against the Lyme disease agent *Borrelia burgdorferi sensu lato* in eggs, young and adults of the kittiwake (*Rissa tridactyla*). *Ecology Letters* 5:519–524.

Gaston, A. J., H. G. Gilchrist, and J. M. Hipfner. 2005. Climate change, ice conditions and reproduction in an Arctic nesting marine bird: Brunnich's Guillemot (*Uria lomvia* L.). *Journal of Animal Ecology* 74:832–841.

Gauthier, G., J. Bety, and K. A. Hobson. 2003. Are Greater Snow Geese capital breeders? New evidence from a stable-isotope model. *Ecology* 84:3250–3264.

Gayathri, K., K. Shenoy, and S. Hegde. 2004. Blood profile of pigeons (*Columba livia*) during growth and breeding. *Comparative Biochemistry and Physiology A: Molecular and Integrative Physiology* 138:187–192.

George, R., D. L. Barber, and W. J. Schneider. 1987. Characterization of the chicken oocyte receptor for low and very low density lipoproteins. *Journal of Biological Chemistry* 262:16838–16847.

Gibb, J. A. 1950. The breeding biology of the great and blue titmice. *Ibis* 92:507–539.

Gienapp, P., L. Hemerik, and M. E. Visser. 2005. A new statistical tool to predict phenology under climate change scenarios. *Global Change Biology* 11:600–606.

Gienapp, P., E. Postma, and M. E. Visser. 2006. Why breeding time has not responded to selection for earlier breeding in a songbird population. *Evolution* 60:2381–2388.

Gil, D. 2008. Hormones in avian eggs: Physiology, ecology and behavior. *Advances in the Study of Behavior* 38:337–398.

Gil, D., and J. Faure. 2007. Correlated response in yolk testosterone levels following divergent genetic selection for social behaviour in Japanese quail. *Journal of Experimental Zoology* 307A:91–94.

Gil, D., G. Leboucher, A. Lacroix, R. Cue, and M. Kreutzer. 2004. Female canaries produce eggs with greater amounts of testosterone when exposed to preferred male song. *Hormones and Behavior* 45:64–70.

Gilbert, A. B. 1979. Female genital organs. In *Form and Function in Birds*, ed. A. S. King and J. McLelland, 237–360. New York: Academic Press.

Gilbert, A. B., M. M. Perry, D. Waddington, and M. A. Hardie. 1983. Role of atresia in establishing the follicular hierarchy in the ovary of the domestic hen (*Gallus domesticus*). *Journal of Reproduction and Fertility* 69:221–227.

Gilbert, L., E. Bulmer, K. E. Arnold, and J. A. Graves. 2007. Yolk androgens and embryo sex: Maternal effects or confounding factors? *Hormones and Behavior* 51:231–238.

Gill, D. 2008. Hormones in avian eggs: Physiology, ecology and behavior. *Advances in the Study of Behavior* 38:337–398.

Gill, S. A. 2003. Timing and duration of egg laying in duetting Buff-breasted Wrens. *Journal of Field Ornithology* 74:31–36.

Gill, V. A., and S. A. Hatch. 2002. Components of productivity in black-legged kittiwakes *Rissa tridactyla*: Response to supplemental feeding. *Journal of Avian Biology* 33:113–126.
Giuliano, W. M., R. S. Lutz, and R. Patino. 1996. Reproductive responses of adult female northern bobwhite and scaled quail to nutritional stress. *Journal of Wildlife Management* 60:302–309.
Goerlich, V., C. Dijkstra, J. Boonekamp, and T. G. G. Groothuis. 2010. Change in body mass can overrule the effects of maternal testosterone on primary offspring sex ratio of first eggs in homing pigeons. *Physiological and Biochemical Zoology* 83:490–500.
Goerlich, V. C., C. Dijkstra, and T. G. G. Groothuis. 2010. No evidence for selective follicle abortion underlying primary sex ratio adjustment in pigeons. *Behavioral Ecology and Sociobiology* 64:599–606.
Goerlich, V. C., C. Dijkstra, S. M. Schaafsma, and T. G. G. Groothuis. 2009. Testosterone has a long-term effect on primary sex ratio of first eggs in pigeons: In search of a mechanism. *General and Comparative Endocrinology* 163:184–192.
Goldsmith, A. R. 1991. Prolactin and avian reproductive strategies. *Proceedings of the International Ornithological Congress* 20:2063–2071.
Goldsmith, A. R., S. Burke, and J. M. Prosser. 1984. Inverse changes in plasma prolactin and LH concentrations in female canaries after disrupting and reinitiation of incubation. *Journal of Endocrinology* 103:251–256.
Goldsmith, A. R., and D. M. Williams. 1980. Incubation in mallards (*Anas platyrhynchos*): Changes in plasma levels of prolactin and luteinizing hormone. *Journal of Endocrinology* 86:371–379.
Gorman, K. B., D. Esler, R. L. Walzem, and T. D. Williams. 2009. Plasma yolk precursor dynamics during egg production by female greater scaup (*Aythya marila*): Characterization and indices of reproductive state. *Physiological and Biochemical Zoology* 82:372–381.
Grau, C. R. 1976. Ring structure of avian egg yolk. *Poultry Science* 55:1418–1422.
Graveland, J., and A. E. Berends. 1997. Timing of the calcium intake and effect of calcium deficiency on behaviour and egg laying in captive great tits, *Parus major*. *Physiological Zoology* 70:74–84.
Graveland, J., and R. H. Drent. 1997. Calcium availability limits breeding success of passerines on poor soils. *Journal of Animal Ecology* 66:279–288.
Gray, J. M., D. Yarian, and M. Ramenofsky. 1990. Corticosterone, foraging behavior, and metabolism in dark-eyed juncos, *Junco hyemalis*. *General and Comparative Endocrinology* 79:375–384.
Green, A. K., D. M. Barnes, and W. H. Karasov. 2005. A new method to measure intestinal activity of P-glycoprotein in avian and mammalian species. *Journal of Comparative Physiology* 175B:57–66.
Green, D. J., and A. Cockburn. 2001. Post-fledging care, philopatry and recruitment in brown thornbills. *Journal of Animal Ecology* 70:505–514.
Greives, T. J., L. J. Kriegsfeld, G. E. Bentley, K. Tsutsui, and G. E. Demas. 2008. Recent advances in reproductive neuroendocrinology: A role for RFamide peptides in seasonal reproduction? *Proceedings of the Royal Society B: Biological Sciences* 275:1943–1951.
Griffin, H., and D. Hermier. 1988. Plasma lipoprotein metabolism and fattening in poultry. In *Leanness in Domestic Birds: Genetic, Metabolic and Hormonal Aspects*, ed. B. Leclercq and C. C. Whitehead, 175–201. New York: Butterworth.
Griffin, H. D., and M. M. Perry. 1985. Exclusion of plasma lipoproteins of intestinal origin from avian egg yolk because of their size. *Comparative Biochemistry and Physiology B: Biochemistry and Molecular Biology* 82:321–325.

Griffiths, R., S. Daan, and C. Dijkstra. 1996. Sex identification in birds using two CHD genes. *Proceedings of the Royal Society B: Biological Sciences* 263:1251–1256.

Griffiths, R., C. Double, K. J. Orr, and R. D. Dawson. 1998. A DNA test to sex most birds. *Molecular Ecology* 7:1071–1075.

Grindstaff, J. L. 2008. Maternal antibodies reduce costs of an immune response during development. *Journal of Experimental Biology* 211:654–660.

Grindstaff, J. L., E. D. I. Brodie, and E. D. Ketterson. 2003. Immune function across generations: Integrating mechanism and evolutionary process in maternal antibody transmission. *Proceedings of the Royal Society B: Biological Sciences* 270:2309–2319.

Grindstaff, J. L., D. Hasselquist, J.-Å. Nilsson, M. I. Sandell, H. G. Smith, and M. Stjernman. 2006. Transgenerational priming of immunity: Maternal exposure to a bacterial antigen enhances offspring humoral immunity. *Proceedings of the Royal Society B: Biological Sciences* 273:2551–1557.

Groothuis, T.G.G., C. Carere, J. Lipar, P. J. Drent, and H. Schwabl. 2008. Selection on personality in a wild songbird affects egg maternal hormone levels turned to its effect on timing of reproduction. *Biology Letters* 23:465–467.

Groothuis, T.G.G., C. M. Eising, J. D. Blount, P. F. Surai, V. Apanius, C. Dijkstra, and W. Müller. 2006. Multiple pathways of maternal effects in black-headed gull eggs: Constraint and adaptive compensatory adjustments. *Journal of Evolutionary Biology* 19:1304–1313.

Groothuis, T.G.G., C. M. Eising, C. Dijkstra, and W. Müller. 2005. Balancing between costs and benefits of maternal hormone deposition in avian eggs. *Biology Letters* 1:78–81.

Groothuis, T.G.G., W. Müller, N. von Engelhardt, C. Carere, and C. Eising. 2005. Maternal hormones as a tool to adjust offspring phenotype in avian species. *Neuroscience and Biobehavioral Reviews* 29:325–352.

Groothuis, T. G., and H. Schwabl. 2002. Determinants of within- and among-clutch variation in levels of maternal hormones in Black-Headed Gull eggs. *Functional Ecology* 16:281–289.

———. 2008. Hormone-mediated maternal effects in birds: Mechanisms matter but what do we know of them? *Philosophical Transactions of the Royal Society B* 363:1647–1661.

Groothuis, T.G.G., and N. von Engelhardt. 2005. Investigating maternal hormones in avian eggs: Measurement, manipulation, and interpretation. *Annals of the New York Academy of Science* 1046:168–180.

Groscolas, R. 1986. Changes in body mass, body temperature and plasma fuel levels during the natural breeding fast in male and female emperor penguins *Aptenodtes forsteri*. *Journal of Comparative Physiology* 156B:521–527.

Groscolas, R., and J.-P. Robin. 2001. Long-term fasting and re-feeding in penguins. *Comparative Biochemistry and Physiology A: Molecular and Integrative Physiology* 128:643–653.

Groscolas, R., F. Decrock, M. A. Thil, C. Fayolle, C. Boissery, and J. P. Robin. 2000. Refeeding signal in fasting-incubating king penguins: Changes in behavior and egg temperature. *American Journal of Physiology Regulatory Integrative and Comparative Physiology* 279:R2104–2112.

Groscolas, R., A. Lacroix, and J.-P. Robin. 2008. Spontaneous egg or chick abandonment in energy-depleted king penguins: A role for corticosterone and prolactin? *Hormones and Behavior* 53:51–60.

Gruber, M. 1972. Hormonal control of yolk protein synthesis. In *Egg Formation and Production*, ed. B. M. Freeman and P. E. Lake, 23–32. Edinburgh: British Poultry Science.

Gunnarsson, T. G., J. A. Gill, J. Newton, P. M. Potts, and W. J. Sutherland. 2005. Seasonal matching of habitat quality and fitness in a migratory bird. *Proceedings of the Royal Society B: Biological Sciences* 272:2319–2323.

Gustafsson, L. 1986. Lifetime reproductive success and heritability: Empirical support for Fisher's fundamental theorem. *American Naturalist* 128:761–764.

Gustafsson, L., D. Nordling, M. S. Andersson, B. C. Sheldon, and A. Qvarnstrom. 1994. Infectious-diseases, reproductive effort and the cost of reproduction in birds. *Philosophical Transactions of the Royal Society B* 346:323–331.

Gustafsson, L., and W. J. Sutherland. 1988. The costs of reproduction in the collared flycatcher *Ficedula albicollis*. *Nature* 335:813–815.

Gwinner, E. 2003. Circannual rhythms in birds. *Current Opinion in Neurobiology* 13:770–778.

Hackl, R., V. Bromundt, J. Daisley, and K. Kotrschal. 2003. Distribution and origin of steroid hormones in the yolk of Japanese quail (*Coturnix coturnix japonica*). *Journal of Comparative Physiology* 173B:327–331.

Haftorn, S. 1978. Egg-laying and regulation of egg temperature during incubation in the Goldcrest *Regulus regulus*. *Ornis Scandinavica* 9:2–21.

———. 1981. Incubation during the egg-laying period in relation to clutch size and other aspects of reproduction in the Great Tit (*Parus major*). *Ornis Scandinavica* 12:169–185.

———. 1985. Recent research on titmice in Norway. *Proceedings of the International Ornithological Congress* 18:137–155.

Haftorn, S., and R. E. Reinertsen. 1985. The effect of temperature and clutch size on the energetic cost of incubation in a free-living Blue Tit (*Parus caeruleus*). *Auk* 102:470–478.

Hahn, T. P. 1998. Reproductive seasonality in an opportunistic breeder, the Red Crossbill (*Loxia curvirostra*). *Ecology* 79:2365–2375.

Halford, S., S. S. Pires, M. Turton, L. Zheng, I. González-Menéndez, W. L. Davies, S. N. Peirson, et al. 2009. VA opsin-based photoreceptors in the hypothalamus of birds. *Current Biology* 19:1396–1402.

Hall, K.S.S., and T. Fransson. 2000. Lesser Whitethroats under time-constraint moult more rapidly and grow shorter wing feathers. *Journal of Avian Biology* 31:583–587.

Hall, M. E., J. D. Blount, S. Forbes, and N. J. Royle. 2010. Does oxidative stress mediate the trade-off between growth and self-maintenance in structured families? *Functional Ecology* 24:365–373.

Hall, M. E., L. Nasir, F. Daunt, E. A. Gault, J. P. Croxall, S. Wanless, and P. Monaghan. 2004. Telomere loss in relation to age and early environment in long-lived birds. *Proceedings of the Royal Society B: Biological Sciences* 271:1571–1576.

Hall, M. R., and A. R. Goldsmith. 1983. Factors affecting prolactin secretion during breeding and incubation in the domestic duck (*Anas platyrhynchos*). *General and Comparative Endocrinology* 49:270–276.

Hamdoum, A., and D. Epel. 2007. Embryo stability and vulnerability in an always changing world. *Proceedings of the National Academy of Sciences USA* 104:1745–1750.

Hamel, S., S. D. Côté, J.-M. Gaillard, and M. Festa-Bianchet. 2009. Individual variation in reproductive costs of reproduction: High-quality females always do better. *Journal of Animal Ecology* 78:143–151.

Hammond, K. A., M. A. Chappell, R. A. Cardullo, R. S. Lin, and T. S. Johnsen. 2000. The mechanistic basis of aerobic performance variation in red junglefowl. *Journal of Experimental Biology* 203:2053–2064.

Han, D., N. H. Haunerland, and T. D. Williams. 2009. Variation in yolk precursor receptor mRNA expression is a key determinant of reproductive phenotype in the zebra finch (*Taeniopygia guttata*) *Journal of Experimental Biology* 212:1277–1283.

Hanssen, S. A., K. E. Erikstad, V. Johnsen, and O. B. Jan. 2003. Differential investment and costs during avian incubation determined by individual quality: An experimental study of the common eider (*Somateria mollissima*). *Proceedings of the Royal Society B: Biological Sciences* 270:531–537.

Hanssen, S. A., D. Hasselquist, I. Folstad, and K. E. Erikstad. 2004. Costs of immunity: Immune responsiveness reduces survival in a vertebrate. *Proceedings of the Royal Society B: Biological Sciences* 271:925–930.

———. 2005. Cost of reproduction in a long-lived bird: Incubation effort reduces immune function and future reproduction. *Proceedings of the Royal Society B: Biological Sciences* 272:1039–1046.

Hargatai, R., Z. Matus, G. Hegyi, G. Michl, G. Toth, and J. Torok. 2006. Antioxidants in the egg yolk of a wild passerine: Differences between breeding seasons. *Comparative Biochemistry and Physiology B: Biochemistry and Molecular Biology* 143:145–152.

Harrison, T.J.E., J. A. Smith, G. R. Martin, D. E. Chamberlain, S. Bearhop, G. N. Robb, and S. J. Reynolds. 2010. Does food supplementation really enhance productivity of breeding birds? *Oecologia* 164:311–320.

Harrison, X. A., J. D. Blount, R. Inger, D. R. Norris, and S. Bearhop. 2010. Carry-over effects as drivers of fitness differences in animals. *Journal of Animal Ecology* 80:4–18.

Harshman, L. G., and A. J. Zera. 2007. The cost of reproduction: The devil in the details. *Trends in Ecology and Evolution* 22:80–88.

Hartley, R. C., and M. W. Kennedy. 2004. Are carotenoids a red herring in sexual display? *Trends in Ecology and Evolution* 19:353–354.

Hasselquist, D., and J.-Å. Nilsson. 2009. Maternal transfer of antibodies in vertebrates: Trans-generational effects on offspring immunity. *Philosophical Transactions of the Royal Society B: Biological Sciences* 364:51–60.

Hatchwell, B. J., M. K. Fowlie, D. J. Ross, and A. F. Russell. 1999. Incubation behavior of Long-tailed Tits: Why do males provision incubating females? *Condor* 101:681–686.

Hatchwell, B. J., and E. J. Pellatt. 1990. Intraspecific variation in egg composition and yolk formation in the common guillemot (*Uria aalge*). *Journal of Zoology* 220:279–286.

Hau, M., M. Wikelski, H. Gwinner, and E. Gwinner. 2004. Timing of reproduction in a Darwin's finch: Temporal opportunism under spatial constraints. *Oikos* 106:489–500.

Hayward, A., and J. F. Gillooly. 2011. The cost of sex: Quantifying energetic investment in gamete production by males and females. *PLoS One* 6:e16557.

Hayward, L. S., J. B. Richardson, M. N. Grogan, and J. C. Wingfield. 2006. Sex differences in the organizational effects of corticosterone in the egg yolk of quail. *General and Comparative Endocrinology* 146:144–148.

Hayward, L. S., D. G. Satterlee, and J. C. Wingfield. 2005. Japanese Quail selected for high plasma corticosterone response deposit high levels of corticosterone on their eggs. *Physiological and Biochemical Zoology* 78:1026–1031.

Haywood, S. 1993a. Role of extrinsic factors in the control of clutch-size in the Blue Tit *Parus caeruleus*. *Ibis* 135:79–84.

———. 1993b. Sensory and hormonal control of clutch size in birds. *Quarterly Review of Biology* 68:33–60.

———. 1993c. Sensory control of clutch size in the Zebra Finch (*Taeniopygia guttata*). *Auk* 110:778–786.

Heaney, V., and P. Monaghan. 1996. Optimal allocation of effort between reproductive phases: The trade-off between incubation costs and subsequent brood rearing capacity. *Proceedings of the Royal Society B: Biological Sciences* 263:1719–1724.

Heaney, V., R. Nager, and P. Monaghan. 1998. Effect of increased egg production on egg composition in the Common Tern, *Sterna hirundo*. *Ibis* 140:693–696.

Hebert, P. 2002. Incubation and the costs of reproduction. In *Avian Incubation*, ed. D. C. Deeming, 270–279. Oxford: Oxford University Press.

Hector, J.A.L., J. P. Croxall, and B. K. Follett. 1986. Reproductive endocrinology of the Wandering Albatross *Diomedea exulans* in relation to biennial breeding and deferred sexual maturity. *Ibis* 128:9–22.

Hector, J.A.L., and A. R. Goldsmith. 1985. The role of prolactin during incubation: Comparative studies of three *Diomedea* albatrosses. *General and Comparative Endocrinology* 60:236–243.

Heeb, P., I. Werner, M. Kolliker, and H. Richner. 1998. Benefits of induced host responses against an ectoparasite. *Proceedings of the Royal Society B: Biological Sciences* 265:51–56.

Hegner, R. E., and J. C. Wingfield. 1986. Behavioral and endocrine correlates of multiple brooding in the semicolonial house sparrow *Passer domesticus*. II. Females. *Hormones and Behavior* 20:313–326.

Heinsohn, R., S. Legge, and S. Barry. 1997. Extreme bias in sex allocation in *Eclectus* parrots. *Proceedings of the Royal Society B: Biological Sciences* 264:1325–1329.

Helfenstein, F., A. Berthouly, M. Tanner, F. Karadas, and H. Richner. 2008. Nestling begging intensity and parental effort in relation to prelaying carotenoid availability. *Behavioral Ecology* 19:108–115.

Helm, B., and M. E. Visser. 2010. Heritable circadian period length in a wild bird population. *Proceedings of the Royal Society B: Biological Sciences* 277:3335–3342.

Helms, C. W. 1968. Food, fat, and feathers. *American Zoologist* 8:151–167.

Hemborg, C. 1999a. Annual variation in the timing of breeding and moulting in male and female Pied Flycatchers *Ficedula hypoleuca*. *Ibis* 141:226–232.

———. 1999b. Sexual differences in moult–breeding overlap and female reproductive costs in pied flycatchers, *Ficedula hypoleuca*. *Journal of Animal Ecology* 68:429–436.

Hemborg, C., and A. Lundberg. 1998. Costs of overlapping reproduction and moult in passerine birds: An experiment with the pied flycatcher. *Behavioral Ecology and Sociobiology* 43:19–23.

Hemborg, C., and J. Merilä. 1998. A sexual conflict in collared flycatchers, *Ficedula albicollis*: Early male moult reduces female fitness. *Proceedings of the Royal Society B: Biological Sciences* 265:2003–2007.

Hiatt, E., A. R. Goldsmith, and D. S. Farner. 1987. Plasma levels of prolactin and gonadotopins during the reproductive cycle of White-crowned Sparrows (*Zonotrichia leucophrys*). *Auk* 104:208–217.

Hill, J. W., J. K. Elmquist, and C. F. Elias. 2008. Hypothalamic pathways linking energy balance and reproduction. *American Journal of Physiology Endocrinology and Metabolism* 294:E827–832.

Hill, W. L. 1993. Importance of prenatal nutrition to the development of a precocial chick. *Developmental Psychobiology* 26:237–249.

———. 1995. Intraspecific variation in egg composition. *Wilson Bulletin* 107:382–387.

Hinde, C. A., K. L. Buchanan, and R. M. Kilner. 2009. Prenatal environmental effects match offspring begging to parental provisioning. *Proceedings of the Royal Society B: Biological Sciences* 276:2787–2794.

Hipfner, J. M., and A. J. Gaston. 1999. The relationship between egg size and post-hatching development in the Thick-billed Murre. *Ecology* 80:1289–1297.

Hipfner, J. M., A. J. Gaston, and L. N. de Forest. 1997. The role of female age in determining egg size and laying date of Thick-billed Murres. *Journal of Avian Biology* 28:271–278.

Hipfner, J. M., A. J. Gaston, G. R. Herzberg, J. T. Brosnan, A. E. Storey, and D. Nettleship. 2003. Egg composition in relation to female age and relaying: Constraints on egg production in Thick-billed Murres (*Uria lomvia*). *Auk* 120:645–657.

Hiramatsu, N., R. W. Chapman, J. K. Lindzey, M. R. Haynes, and C. V. Sullivan. 2004. Molecular characterization and expression of vitellogenin receptor from white perch (*Morone americana*). *Biology of Reproduction* 70:1720–1730.

Hocking, P. M., M. Bain, C. E. Channing, R. Fleming, and S. Wilson. 2003. Genetic variation for egg production, egg quality and bone strength in selected and traditional breeds of laying fowl. *British Poultry Science* 44:365–373.

Högstedt, G. 1980. Evolution of clutch size in birds: Adaptive variation in relation to territory quality. *Science* 210:1148–1150.

———. 1981. Effect of additional food on reproductive success in the magpie (*Pica pica*). *Journal of Animal Ecology* 50:219–229.

Holberton, R. L., T. Boswell, and M. J. Hunter. 2008. Circulating prolactin and corticosterone concentrations during the development of migratory condition in the Dark-eyed Junco, *Junco hyemalis*. *General and Comparative Endocrinology* 155:641–649.

Hõrak, P., S. Jenni-Eiermann, and I. Ots. 1999. Do great tits (*Parus major*) starve to reproduce? *Oecologia* 119:293–299.

Hõrak, P., S. Jenni-Eiermann, I. Ots, and L. Tegelmann. 1998. Health and reproduction: The sex-specific clinical profile of great tits (*Parus major*) in relation to breeding. *Canadian Journal of Zoology* 76:2235–2244.

Hõrak, P., R. Mänd, and I. Ots. 1997. Identifying targets of selection: A multivariate analysis of reproductive traits in the great tit. *Oikos* 78:592–600.

Horton, B. M., and R. L. Holberton. 2009. Corticosterone manipulations alter morph-specific nestling provisioning behavior in male white-throated sparrows, *Zonotrichia albicollis*. *Hormones and Behavior* 56:510–518.

Houston, A. I., J. M. McNamara, Z. Barta, and K. C. Klasing. 2007. The effect of energy reserves and food availability on optimal immune defence. *Proceedings of the Royal Society B: Biological Sciences* 274:2835–2842.

Houston, D. C., D. Donnan, and P. J. Jones. 1995. The source of the nutrients required for egg production in zebra finches *Poephila guttata*. *Journal of Zoology* 235:469–483.

Houston, D. C., D. Donnan, P. Jones, I. Hamilton, and D. Osborne. 1995a. Changes in the muscle condition of female Zebra Finches *Poephila guttata* during egg laying and the role of protein storage in bird skeletal-muscle. *Ibis* 137:322–328.

———. 1995b. Changes in the muscle condition of female Zebra Finches *Poephila guttata* during egg laying and the role of protein storage in bird skeletal muscle. *Ibis* 137:322–328.

Hsin, H., and C. Kenyon. 1999. Signals from the reproductive system regulate the lifespan of *C. elegans*. *Nature* 399:362–266.

Hummel, S., E. G. Lynn, A. Osanger, S. Hirayama, J. Nimpf, and W. J. Schneider. 2003. Molecular characterization of the first avian LDL receptor: Role in sterol metabolism of ovarian follicular cells. *Journal of Lipid Research* 44:1633–1642.

Hunter, S. 1984. Moult of the giant petrels *Macronectes halli* and *M. giganteus* at South Georgia. *Ibis* 126:119–132.

Husby, A., L. E. B. Kruuk, and M. E. Visser. 2009. Decline in the frequency and benefits of multiple brooding in great tits as a consequence of a changing environment. *Proceedings of the Royal Society B: Biological Sciences* 276:1845–1854.

Husby, A., D. H. Nussey, M. E. Visser, A. J. Wilson, B. C. Sheldon, and L. E. B. Kruuk. 2010. Contrasting patterns of phenotypic plasticity in reproductive traits in two great tit (*Parus major*) populations. *Evolution* 64:2221–2237.

Hutchison, R. E. 1975. Effects of ovarian steroids and prolactin on the sequential development of nesting behaviour in female budgerigars *Journal of Endocrinology* 67:29–39.

Isaksson, C., A. Johansson, and S. Andersson. 2008. Egg yolk carotenoids in relation to habitat and reproductive investment in the Great Tit *Parus major*. *Physiological and Biochemical Zoology* 81:112–118.

Isaksson, C., P. McLaughlin, P. Monaghan, and S. Andersson. 2007. Carotenoid pigmentation does not reflect total non-enzymatic antioxidant activity in plasma of adult and nestling great tits, *Parus major*. *Functional Ecology* 21:1123–1129.

Isaksson, C., B. C. Sheldon, and T. Uller. 2011. The challenges of integrating oxidative stress into life-history biology. *BioScience* 61:194–202.

Jager, T. D., J. B. Hulscher, and M. Kersten. 2000. Egg size, egg composition and reproductive success in the Oystercatcher *Haematopus ostralegus*. *Ibis* 142:603–613.

Jarvinen, A., and M. Pryl. 1989. Egg dimensions of the Great Tit *Parus major* in southern Finland. *Ornis Fennica* 66:69–74.

Jawor, J. M., J. W. McGlothlin, J. M. Casto, T. J. Grieves, E. A. Snadjr, G. E. Bentley, and E. D. Ketterson. 2006. Seasonal and individual variation in response to GnRH challenge in male dark-eyed juncos (*Junco hyemalis*). *General and Comparative Endocrinology* 149:182–189.

Jelkmann, W. 2007. Erythropoietin after a century of research: Younger than ever. *European Journal of Haematology* 78:183–205.

Jenni, L., S. Muller, F. Spina, A. Kvist, and Å. Lindström. 2006. Effect of endurance flight on haematocrit in migrating birds. *Journal of Ornithology* 147:531–542.

Jenouvrier, S., J.-C. Thibault, A. Viallefont, P. Vidal, D. Ristow, J. L. Mougin, J. Brichetti, et al. 2009. Global climate patterns explain range-wide synchronicity in survival of a migratory seabird. *Global Change Biology* 15:268–279.

Jensen, F. B. 2009. The dual roles of red blood cells in tissue oxygen delivery: Oxygen carriers and regulators of local blood flow. *Journal of Experimental Biology* 212:3387–3393.

Jetz, W., C. H. Sekercioglu, and K. Böhning-Gaese. 2009. The worldwide variation in avian clutch size across species and space. *PLoS Biology* 6:e303.

John, J. L. 1994. The avian spleen: A neglected organ. *Quarterly Review of Biology* 69:327–351.

Johnsen, A., A. E. Fidler, S. Kuhn, K. L. Carter, A. Hoffmann, I. R. Barr, C. Biard, et al. 2007. Avian *Clock* gene polymorphism: Evidence for a latitudinal cline in allele frequencies. *Molecular Ecology* 16:4867–4880.

Johnson, A. L. 1990. Steroidogenesis and actions of steroids in the hen ovary. *Critical Reviews in Poultry Biology* 2: 319–346.

———. 2000. Reproduction in the female. In *Sturkie's Avian Physiology*, ed. G. C. Whittow, 569–596. San Diego: Academic Press.

Johnson, A. L., M. J. Haugen, and D. C. Woods. 2008. Role for inhibitor of differentiation/deoxyribonucleic acid-binding (Id) proteins in granulosa cell differentiation. *Endocrinology* 149:3187–3195.

Johnson, A. L., and D. C. Woods. 2007. Ovarian dynamics and follicle development. In *Reproductive Biology and Phylogeny of Birds*, ed. B.G.M. Jamieson, 243–277. Enfield, N. H.: Science Publishers.

———. 2009. Dynamics of avian ovarian follicle development: Cellular mechanisms of granulosa cell differentiation. *General and Comparative Endocrinology* 163:12–17.

Johnson, L. S., C. F. Thompson, S. K. Sakaluk, M. Neuhäuser, B.G.P. Johnson, S. S. Soukup, S. J. Forsythe, et al. 2009. Extra-pair young in house wren broods are more likely to be male than female. *Proceedings of the Royal Society B: Biological Sciences* 276:2285–2289.

Johnson, P. A., C. F. Brooks, and A. J. Davis. 2005. Pattern of secretion of immunoreactive inhibin/activin subunits by avian granulosa cells. *General and Comparative Endocrinology* 141:233–239.

Jones, J. H. 1998. Optimization of the mammalian respiratory system: Symmorphosis versus single species adaptation. *Comparative Biochemistry and Physiology B: Biochemistry and Molecular Biology* 120:125–138.

Jones, P. J. 1983. Hematocrit values of breeding red-billed queleas *Quelea quelea* (Aves, ploceidae) in relation to body condition and thymus activity. *Journal of Zoology* 201:217–222.

Jones, T., and W. Cresswell. 2010. The phenology mismatch hypothesis: Are declines of migrant birds linked to uneven global climate change? *Journal of Animal Ecology* 79:98–108.

Jouventin, P., and R. Mauget. 1996. The endocrine basis of the reproductive cycle in the king penguin (*Aptenodytes patagonicus*). *Journal of Zoology* 238:665–678.

Kacelnik, A., P. A. Cotton, L. Stirling, and J. Wright. 1995. Food allocation among nestling starlings: Sibling competition and the scope of parental choice. *Proceedings of the Royal Society B: Biological Sciences* 259:259–263.

Kalmbach, E., R. Griffiths, J. E. Crane, and R. W. Furness. 2004. Effects of experimentally increased egg production on female body condition and laying dates in the great skua *Stercorarius skua*. *Journal of Avian Biology* 35:501–514.

Kamiyoshi, M., and K. Tanaka. 1983. Endocrine control of ovulatory sequence in domestic fowl. In *Avian Endocrinology: Environmental and Ecological Perspectives*, ed. S. Mikami, K. Homma, and M. Wada, 167–177. Berlin: Springer-Verlag.

Karasov, W. H. 1996. Digestive plasticity in avian energetics and feeding ecology. In *Avian Energetics and Nutritional Ecology*, ed. C. Carey, 61–84. New York: Chapman and Hall.

Karasov, W. H., and S. R. McWilliams. 2005. Digestive constraints in mammalian and avian ecology. In *Physiological and Ecological Adaptations to Feeding in Vertebrates*, ed. J. M. Starck and T. Wang, 87–112. Enfield, N. H.: Science Publishers.

Karell, P., P. Kontiainen, H. Pietiäinen, H. Siitar, and J. E. Brommer. 2008. Maternal effects on offspring Igs and egg size in relation to natural and experimentally improved food supply. *Functional Ecology* 22:682–690.

Kennamer, R. A., S. K. Alsum, and S. V. Colwell. 1997. Composition of Wood Duck eggs in relation to egg size, laying sequence, and skipped days of laying. *Auk* 114:479–487.

Kennedy, G. Y., and H. G. Vevers. 1976. A survey of avian eggshell pigments. *Comparative Biochemistry and Physiology B: Biochemistry and Molecular Biology* 55:117–123.

Kern, M., W. Bacon, D. Long, and R. J. Cowie. 2005. Blood metabolites and corticosterone levels in breeding adult pied flycatchers. *Condor* 107:665–677.

Ketterson, E. D., and V. J. Nolan. 1999. Adaptation, exaptation, and constraint: A hormonal perspective. *American Naturalist* 153:S4–S25.

Ketterson, E. D., V. J. Nolan, M. J. Cawthorn, P. G. Parker, and C. Ziegenfus. 1996. Phenotypic engineering: Using hormones to explore the mechanistic and functional bases of phenotypic variation in nature. *Ibis* 138:70–86.

Ketterson , E. D., V. J. Nolan, L. Wolf, and A. R. Goldsmith. 1990. Effect of sex, stage of reproduction, season, and mate removal on prolactin in Dark-Eyed Juncos. *Condor* 92:922–930.

Khorrami, S., H. Tazawa, and W. Burggren. 2008. `Blood-doping' effects on hematocrit regulation and oxygen consumption in late-stage chicken embryos (*Gallus gallus*). *Journal of Experimental Biology* 211:883–889.

Kim, M. H., D. S. Seo, and Y. Ko. 2004. Relationship between egg productivity and insulin-like growth factor-I genotypes in Korean native Ogol chickens. *Poultry Science* 83:1203–1209.

King, A. S., and J. McLelland. 1984. *Birds, Their Structure and Function*. 2nd ed. London: Bailliere Tindall.

King, J. R. 1972. Energetics of reproduction in birds. In *Breeding Biology of Birds*, ed. D. S. Farner, 78–107. Washington: National Academy of Sciences.

King, J. R., B. K. Follett, D. S. Farner, and M. L. Morton. 1966. Annual gonadal cycles and pituitary gonadotropins in *Zonotrichia leucophrys gambelii*. *Condor* 68:476–487.

Kinsky, F. C. 1971. The consistent presence of paired ovaries in the Kiwi (*Apteryx*) with some discussion of this condition in other birds. *Journal of Ornithology* 112:334–357.

Kirby, J. D., J. A. Vizcarra, L. R. Berghman, J. A. Proudman, J. Yang, and S.C.G. 2005. Regulation of FSH secretion: GnRH independent? In *Functional Avian Endocrinology*, ed. A. Dawson and P. J. Sharp, 83–96. New Delhi: Narosa Publishing House.

Kirkwood, T.B.L. 1977. Evolution of ageing. *Nature* 270:301–304.

———. 2005. Understanding the odd science of aging. *Cell* 120:437–447.

Kirkwood, T.B.L., and S. N. Austad. 2000. Why do we age? *Nature* 408:233–238.

Klaassen, M., A. Lindstrom, H. Meltofte, and T. Piersma. 2001. Arctic waders are not capital breeders. *Nature* 413:794–794.

Klasing, K. C. 1998. Nutritional modulation of resistance to infectious diseases. *Poultry Science* 77:1119–1125.

Klatt, P. H., B. J. M. Stutchbury, and M. L. Evans. 2008. Incubation feeding by male Scarlet Tanagers: A mate removal experiment. *Journal of Field Ornithology* 79:1–10.

Knight, P. G., R. T. Gladwell, and T. M. Lovell. 2005. The inhibin-activin system and ovarian folliculogenesis in the chicken. In *Functional Avian Endocrinology*, ed. A. Dawson and P. J. Sharp, 323–337. New Delhi: Narosa Publishing House.

Knowles, S. C. L., S. Nakagawa, and B. C. Sheldon. 2009. Elevated reproductive effort increases blood parasitaemia and decreases immune function in birds: A meta-regression approach. *Functional Ecology* 23:405–415.

Kohler, P. O., P. M. Grimley, and B. W. O'Malley. 1969. Estrogen-induced cytodifferentiation of the ovalbumin-secreting glands of the chick oviduct. *Journal of Cell Biology* 40:8–27.

Kolliker, M., and H. Richner. 2004. Navigation in a cup: Chick provisioning in great tits, *Parus major*, nests. *Animal Behaviour* 68:941–948.

Komdeur, J. 2003. Daughters on request: About helpers and egg sexes in the Seychelles warbler. *Proceedings of the Royal Society B: Biological Sciences* 270:3–11.

Komdeur, J., S. Daan, J. Tinbergen, and C. Mateman. 1997. Extreme adaptive modification in sex ratio of the Seychelles warbler's eggs. *Nature* 385:522–525.

Komdeur, J., M. J. L. Magrath, and S. Krackow. 2002. Pre-ovulation control of hatchling sex ratio in the Seychelles warbler. *Proceedings of the Royal Society B: Biological Sciences* 269:1067–1072.

Komdeur, J., and I. Pen. 2002. Adaptive sex allocation in birds: The complexities of linking theory and practice. *Philosophical Transactions of the Royal Society B* 357:373–380.

Komdeur, J., and D. S. Richardson. 2007. Molecular ecology reveals the hidden complexities of the Seychelles warbler. *Advances in the Study of Behavior* 37:147–187.

Kordonowy, L. L., J. P. McMurtry, and T. D. Williams. 2010. Variation in plasma leptin-like immunoreactivity in free-living European starlings (*Sturnus vulgaris*). *General and Comparative Endocrinology* 166:47–53.

Korpimaki, E., and J. Wiehn. 1998. Clutch size of kestrels: Seasonal decline and experimental evidence for food limitation under fluctuating food conditions. *Oikos* 83:259–272.

Kostelecka-Myrcha, A., Z. Zukowski, and E. Oksiejczuk. 1997. Changes in the red blood indices during nestling development of the Tree Sparrow *Passer montanus* in an urban environment. *Ibis* 139:92–96.

Koutsos, E. A., A. J. Clifford, C. C. Calvert, and K. C. Klasing. 2003. Maternal carotenoid status modifies the incorporation of dietary carotenoids into immune tissues of growing chickens (*Gallus gallus domesticus*). *Journal of Nutrition* 133:1132–1138.

Kowalczyk, K., J. Daiss, J. Halpern, and R.T.F. 1985. Quantitation of maternal-fetal IgG transport in the chicken. *Immunology* 54:755–762.

Krackow, S. 1995. Potential mechanisms for sex ratio adjustment in mammals and birds. *Biological Reviews* 70:225–241.

Krist, M. 2009. Short- and long-term effects of egg size and feeding frequency on offspring quality in the collared flycatcher (*Ficedula albicollis*). *Journal of Animal Ecology* 78:907–918.

———. 2011. Egg size and offspring quality: A meta-analysis in birds. *Biological Reviews* 86: 692–716.

Kuenzel, W. J. 2003. Neurobiology of molt in avian species. *Poultry Science* 82:981–991.

Kuenzel, W. J., M. M. Beck, and R. Teruyama. 1999. Neural sites and pathways regulating food intake in birds: A comparative analysis to mammalian systems. *Journal of Experimental Zoology* 283:348–364.

Kullberg, C., D. C. Houston, and N. B. Metcalfe. 2002. Impaired flight ability—A cost of reproduction in female blue tits. *Behavioral Ecology* 13:575–579.

Kullberg, C., N. B. Metcalfe, and D. C. Houston. 2002. Impaired flight ability during incubation in the pied flycatcher. *Journal of Avian Biology* 33:179–183.

Lack, D. 1947. The significance of clutch size. *Ibis* 89:302–335.

———. 1966. *Population Studies of Birds*. Oxford: Clarendon Press.

———. 1967. The significance of clutch size in waterfowl. *Wildfowl* 18:125–128.

———. 1968. *Ecological Adaptations for Breeding in Birds*. London: Methuen.

Lambrechts, M. M., J. Blondel, M. Maistre, and P. Perret. 1997. A single response mechanism is responsible for evolutionary adaptive variation in a bird's laying date. *Proceedings of the National Academy of Sciences USA* 94:5153–5155.

Lambrechts, M. M., P. Perret, and J. Blondel. 1996. Adaptive differences in the timing of egg laying between different populations of birds result from variation in photoresponsiveness. *Proceedings of the Royal Society B: Biological Sciences* 263:19–22.

Landys, M. M., M. Ramenofsky, C. G. Guglielmo, and J. C. Wingfield. 2004. The low-affinity glucocorticoid receptor regulates feeding and lipid breakdownin the migratory Gambel's white-crowned sparrow *Zonotrichia leucophrys gambelii*. *Journal of Experimental Biology* 207:143–154.

Landys-Ciannelli, M. M., J. Jukema, and T. Piersma. 2002. Blood parameter changes during stopover in a long-distance migratory shorebird, the bar-tailed godwit *Limosa lapponica taymyrensis*. *Journal of Avian Biology* 33:451–455.

Langston, N. E., and S. Rohwer. 1996. Molt-breeding tradeoffs in albatrosses: Life history implications for big birds. *Oikos* 76:498–510.

Larsen, V. A., T. Lislevand, and I. Byrkjedal. 2003. Is clutch size limited by incubation ability in northern lapwings? *Journal of Animal Ecology* 72:784–792.

Lavelin, I., N. Meiri, M. Einat, O. Genina, and M. Pines. 2002. Mechanical strain regulation of the chicken glypican-4 gene expression in the avian eggshell gland. *American Journal of Physiology Regulatory Integrative and Comparative Physiology* 283:R853–861.

Lavelin, I., N. Meiri, O. Genina, R. Alexiev, and M. Pines. 2001. Na+-K+-ATPase gene expression in the avian eggshell gland: Distinct regulation in different cell types. *American Journal of Physiology Regulatory Integrative and Comparative Physiology* 281:R1169–1176.

Lea, R. W., A. S. M. Dods, P. J. Sharp, and A. Chadwick. 1981. The possible role of prolactin in the regulation of nesting behaviour and the secretion of luteinizing hormone in broody bantams. *Journal of Endocrinology* 91:89–97.

Lea, R. W., D. M. Vowles, and H. R. Dick. 1986. Factors affecting prolactin secretion during the breeding cycle of the ring dove (*Streptopelia risoria*) and its possible role in incubation. *Journal of Endocrinology* 110:447–458.

Lee, S. J., M. S. Witter, I. C. Cuthill, and A. R. Goldsmith. 1996. Reduction in escape performance as a cost of reproduction in gravid starlings, *Sturnus vulgaris*. *Proceedings of the Royal Society B: Biological Sciences* 263:619–624.

Leeuwenburgh, C., and J. W. Heinecke. 2001. Oxidative stress and antioxidants in exercise. *Current Medicinal Chemistry* 8:829.

Lehrman, D. S., and P. Brody. 1961. Does prolactin induce incubation behaviour in the ring dove? *Journal of Endocrinology* 22:269–275.

Le Maho, Y., H. Vu Van Kha, H. Koubi, G. Dewasmes, J. Girard, P. Ferré, and M. Cagnard. 1981. Body composition, energy expenditure, and plasma metabolites in long-term fasting geese. *American Journal of Physiology* 241:E342–E354.

Lengyel, S., B. Kiss, and C. R. Tracy. 2009. Clutch size determination in shorebirds: Revisiting incubation limitation in the pied avocet (*Recurvirostra avosetta*). *Journal of Animal Ecology* 78:396–405.

Lepage, D., G. Gauthier, and S. Menu. 2000. Reproductive consequences of egg-laying decisions in snow geese. *Journal of Animal Ecology* 69:414–427.

Lescroël, A., G. Ballard, V. Toniolo, K. J. Barton, P. R. Wilson, P.O.B. Lyver, and D. G. Ainley. 2010. Working less to gain more: When breeding quality relates to foraging efficiency. *Ecology* 91:2044–2055.

Lescroël, A., K. M. Dugger, G. Ballard, and D. G. Ainley. 2009. Effects of individual quality, reproductive success and environmental variability on survival of a long-lived seabird. *Journal of Animal Ecology* 78:798–806.

Lessells, C. M. 1986. Brood size in Canada geese: A manipulation experiment. *Journal of Animal Ecology* 55:669–689.

Lessells, C. M. 1991. The evolution of life histories. In *Behavioural Ecology: An Evolutionary Approach*, ed. J. R. Krebs and N. B. Davies, 32–68. Oxford: Blackwell Scientific.

Lessells, C. M., and M. I. Avery. 1989. Hatching asynchrony in European bee-eaters *Merops apiaster*. *Journal of Animal Ecology* 58:815–835.

Lessells, C. M., N. J. Dingemanse, and C. Both. 2002. Egg weights, egg component weights, and laying gaps in Great Tits (*Parus major*) in relation to ambient temperature. *Auk* 119:1091–1103.

Lessells, C. M., K. R. Oddie, and A. C. Mateman. 1998. Parental behaviour is unrelated to experimentally manipulated great tit brood sex ratio. *Animal Behaviour* 56:385–393.

Lewis, R. A. 1975. Reproductive biology of the White-crowned Sparrow II. Environmental control of reproductive and associated cycles. *Condor* 77:111–124.

Liedvogel, M., and B. C. Sheldon. 2010. Low variability and absence of phenotypic correlates of *Clock* gene variation in a great tit *Parus major* population. *Journal of Avian Biology* 41:543–550.

Liedvogel, M., M. Szulkin, S. C. L. Knowles, M. J. Wood, and B. C. Sheldon. 2009. Phenotypic correlates of *Clock* gene variation in a wild blue tit population: Evidence for a role in seasonal timing of reproduction. *Molecular Ecology* 18:2444–2456.

Lifjeld, J. T., and T. Slagsvold. 1986. The function of courtship feeding during incubation in the pied flycatcher *Ficidula hypoleuca*. *Animal Behaviour* 34:1441–1453.

———. 1988. Effects of energy costs on the optimal diet: An experiment with Pied Flycatchers *Ficedula hypoleuca* feeding nestlings. *Ornis Scandinavica* 19:111–118.

Lima, S. L. 1987. Clutch size in birds: A predation perspective. *Ecology* 68:1062–1070.

Lindström, A., G. H. Visser, and S. Daan. 1993. The energetic cost of feather synthesis is proportional to basal metabolic rate. *Physiological Zoology* 66:490–510.

Liu, H. K., and W. L. Bacon. 2004. Effect of chronic progesterone injection on egg production in Japanese quail. *Poultry Science* 83:2051–2058.

———. 2005. Changes in egg production rate induced by progesterone injection in broiler breeder hens. *Poultry Science* 84:321–327.

Loonen, M. J. J. E., L. W. Bruinzeel, J. M. Black, and R. H. Drent. 1999. The benefit of large broods in barnacle geese: A study using natural and experimental manipulations. *Journal of Animal Ecology* 68:753–768.

Lormee, H., P. Jouventin, O. Chastel, and R. Mauget. 1999. Endocrine correlates of parental care in an Antarctic winter breeding seabird, the emperor penguin, *Aptenodytes forsteri*. *Hormones and Behavior* 35:9–17.

Love, O. P., C. W. Breuner, F. Vézina, and T. D. Williams. 2004. Mediation of a corticosterone-induced reproductive conflict. *Hormones and Behavior* 46:59–65.

Love, O. P., E. H. Chin, K. E. Wynne-Edwards, and T. D. Williams. 2005. Stress hormones: A link between maternal condition and sex-biased reproductive investment. *American Naturalist* 166:751–766.

Love, O. P., G. Gilchrist, J. Bêty, K. E. Wynne-Edwards, L. Berzins, and T. D. Williams. 2009. Using life-histories to predict and interpret variability in yolk hormones. *General and Comparative Endocrinology* 163:169–174.

Love, O. P., K. G. Salvante, J. Dale, and T. D. Williams. 2008. Sex-specific variability in the immune system across life-history stages. *American Naturalist* 172:E99–E112.

Love, O. P., and T. D. Williams. 2008a. The adaptive value of stress-induced phenotypes: Effects of maternally derived corticosterone on sex-biased investment, cost of reproduction, and maternal fitness. *American Naturalist* 172:E135–E149.

———. 2008b. Plasticity in the adrenocortical response of a free-living vertebrate: The role of pre- and post-natal developmental stress. *Hormones and Behavior* 54:496–505.

———. 2011. Manipulation of developmental stress reveals sex-specific effects of egg size on offspring phenotype. *Journal of Evolutionary Biology* 24: 1497–1504.

Love, O. P., K. E. Wynne-Edwards, L. Bond, and W. T.D. 2008. Determinants of within- and among-clutch variation in yolk corticosterone in the European starling. *Hormones and Behavior* 53:104–111.

Lovern, M. B., and J. Wade. 2003. Sex steroids in green anoles (*Anolis carolinensis*): Uncoupled maternal plasma and yolking follicle concentrations, potential embryonic steroidogenesis, and evolutionary implications. *General and Comparative Endocrinology* 134:109–115.

Lundberg, A., and R. V. Alatalo. 1992. *The Pied Flycatcher*. London: T. and A. D. Poyser.

MacColl, A.D.C., and B. J. Hatchwell. 2003. Heritability of parental effort in a passerine bird. *Evolution* 57:2191–2195.

———. 2004. Determinants of lifetime fitness in a cooperative breeder, the long-tailed tit *Aegithalos caudatus*. *Journal of Animal Ecology* 73:1137–1148.

Mackley, E. K., R. A. Phillips, J. R. D. Silk, E. D. Wakefield, V. Afanasyev, J. W. Fox, and R. W. Furness. 2010. Free as a bird? Activity patterns of albatrosses during the nonbreeding period. *Marine Ecology-Progress Series* 406:291–303.

MacLachlan, I., J. Nimpf, and W. J. Schneider. 1994. Avian riboflavin binding protein binds to lipoprotein receptors in association with vitellogenin. *Journal of Biological Chemistry* 269:24127–24132.

Maddox, J. D., and P. J. Weatherhead. 2009. Seasonal sex allocation by Common Grackles? Revisiting a foundational study. *Ecology* 90:3190–3196.

Madsen, J. 2001. Spring migration strategies in Pink-footed Geese *Anser brachyrhynchus* and consequences for spring fattening and fecundity. *Ardea* 89:43–55.

Magrath, R. D. 1990. Hatching asynchrony in altricial birds. *Biological Reviews* 95:587–622.

———. 1991. Nestling weight and juvenile survival in the blackbird *Turdus merula*. *Journal of Animal Ecology* 59:224–240.

———. 1992. Seasonal changes in egg-mass within and among clutches of birds: General explanations and a field study of the Blackbird *Turdus merula*. *Ibis* 134:171–179.

Mahon, M. G., K. A. Lindstedt, M. Hermann, J. Nimpf, and W. J. Schneider. 1999. Multiple involvement of clusterin in chicken ovarian follicle development. *Journal of Biological Chemistry* 274:4036–4044.

Maigret, J. L., and M. T. Murphy. 1997. Costs and benefits of parental care in eastern kingbirds. *Behavioral Ecology* 8:250–259.

Mandeles, S., and E. D. Ducay. 1962. Site of egg white protein formation. *Journal of Biological Chemistry* 237:3196–3199.

Mantei, K. E., S. Ramakrishnan, P. J. Sharp, and J. D. Buntin. 2008. Courtship interactions stimulate rapid changes in GnRH synthesis in male ring doves. *Hormones and Behavior* 54:669–675.

Marra, P. P., K. A. Hobson, and R. T. Holmes. 1998. Linking winter and summer events in a migratory bird by using stable-carbon isotopes. *Science* 282:1884–1886.

Marshall, D. J., and T. Uller. 2007. When is a maternal effect adaptive? *Oikos* 116:1957–1963.

Martin, L. B. 2009. Stress and immunity in wild vertebrates: Timing is everything. *General and Comparative Endocrinology* 163:70–76.

Martin, L. B., A. Scheuerlein, and M. Wikelski. 2003. Immune activity elevates energy expenditure of house sparrows: A link between direct and indirect costs? *Proceedings of the Royal Society B: Biological Sciences* 270:153–158.

Martin, T. E. 1995. Avian life-history evolution in relation to nest sites, nest predation, and food. *Ecological Monographs* 65:101–127.

Martin, T. E., R. D. Bassar, S. K. Bassar, J. J. Fontaine, P. Lloyd, H. A. Mathewson, A. M. Niklison, et al. 2006. Life-history and ecological correlates of geographical variation in egg and clutch mass among passerine species. *Evolution* 60:390–398.

Martin, T. E., P. R. Martin, C. R. Olson, B. J. Heidinger, and J. J. Fontaine. 2000. Parental care and clutch sizes in North and South American birds. *Science* 287:1482–1485.

Martins, T.L.F., and J. Wright. 1993. Cost of reproduction and allocation of food between parent and young in the swift (*Apus apus*). *Behavioral Ecology* 4:213–223.

Massaro, M., L. S. Davis, and R. S. Davidson. 2006. Plasticity of brood patch development and its influence on incubation periods in the yellow-eyed penguin *Megadyptes antipodes*: An experimental approach. *Journal of Avian Biology* 37:497–506.

Matthysen, E., F. Adriaensen, and A. A. Dhondt. 2010. Multiple responses to increasing spring temperatures in the breeding cycle of blue and great tits (*Cyanistes caeruleus, Parus major*). *Global Change Biology* 17:1–16.

Maurer, B. 1996. Energetics of avian foraging. In *Avian Energetics and Nutritional Ecology*, ed. C. Carey, 250–279. New York: Chapman and Hall.

Mazuc, J., C. Bonneaud, O. Chastel, and G. Sorci. 2003. Social environment affects female and egg testosterone in the house sparrow (*Passer domesticus*). *Ecology Letters* 6:1084–1090.

McCleery, R. H., C. M. Perrins, D. Wheeler, and S. Groves. 2007. The effect of breeding status on the timing of moult in Mute Swans *Cygnus olor*. *Ibis* 149:86–90.

McCleery, R. H., R. A. Pettifor, P. Armbruster, K. Meyer, B. C. Sheldon, and C. M. Perrins. 2004. Components of variance underlying fitness in a natural population of great tits *Parus major*. *American Naturalist* 164:E62–E72.

McEwen, B. S., and J. C. Wingfield. 2003. The concept of allostasis in biology and biomedicine. *Hormones and Behavior* 43:2–15.

———. 2010. What is in a name? Integrating homeostasis, allostasis and stress. *Hormones and Behavior* 57:105–111.

McGinley, M. A., D. H. Temme, and M. A. Geber. 1987. Parental investment in offspring in variable environments: Theoretical and empirical considerations. *American Naturalist* 130:370–398.

McGraw, K. J., E. Adkins-Regan, and R. S. Parker. 2005. Maternally derived carotenoid pigments affect offspring survival, sex ratio, and sexual attractiveness in a colorful songbird. *Naturwissenschaften* 92:375–380.

McMaster, D. G., S. G. Sealy, S. A. Gill, and D. L. Neudorf. 1999. Timing of egg laying in Yellow Warblers. *Auk* 116:236–240.

McNabb, F.M.A., and C. A. Wilson. 1997. Thyroid hormone deposition in avian eggs and effects on embryonic development. *American Zoologist* 37:553–560.

McWhorter, T. J., E. Caviedes-Vidal, and W. H. Karasov. 2009. The integration of digestion and osmoregulation in the avian gut. *Biological Reviews* 84:533–565.

Meathrel, C. E. 1991. Variation in eggs and the period of rapid yolk deposition of the silver gull *Larus novaehollandiae* during a protracted laying season. *Journal of Zoology* 223:501–508.

Meijer, T. 1991. The effect of a period of food restriction on gonad size and moult of male and female Starlings *Sturnus vulgaris* under constant photoperiod. *Ibis* 133:80–84.

Meijer, T., S. Daan, and M. Hall. 1990. Family planning in the Kestrel (*Falco tinnunculus*): The proximate control of covariation of laying date and clutch size. *Behavior* 114:117–136.

Meijer, T., C. Deerenberg, S. Daan, and C. Dijkstra. 1992. Egg-laying and photorefractoriness in the European Kestrel *Falco finnunculus*. *Ornis Scandinavica* 23:405–410.

Meijer, T., and R. Drent. 1999. Re-examination of the capital and income dichotomy in breeding birds. *Ibis* 141:399–414.

Meijer, T., and U. Langer. 1995. Food availability and egg-laying of captive European Starlings. *Condor* 97:718–728.

Meijer, T., U. Nienaber, U. Langer, and F. Trillmich. 1999. Temperature and timing of egg-laying of European Starlings. *Condor* 101:124–132.

Merilä, J., and B. C. Sheldon. 2000. Lifetime reproductive success and heritability in nature. *American Naturalist* 155:301–310.

Merilä, J., B. C. Sheldon, and L. E. B. Kruuk. 2001. Explaining stasis: Microevolutionary studies in natural populations. *Genetica* 112–113:199–222.

Merilä, J., and D. A. Wiggins. 1997. Mass loss in breeding blue tits: The role of energetic stress. *Journal of Animal Ecology* 66:452–460.

Metcalfe, N. B., and C. Alonso-Alvarez. 2010. Oxidative stress as a life-history constraint: The role of reactive oxygen species in shaping phenotypes from conception to death. *Functional Ecology* 24:984–996.

Millam, J. R., C. B. Craig-Veit, M. E. Batchelder, M. R. Viant, T. M. Herbeck, and L. W. Woods. 2002. An avian bioassay for environmental estrogens: The growth response of zebra finch (*Taeniopygia guttata*) chick oviduct to oral estrogens. *Environmental Toxicology and Chemistry* 21:2663–2668.

Millam, J. R., B. Zhang, and M. E. El Halawani. 1996. Egg production of cockatiels (*Nymphicus hollandicus*) is influenced by number of eggs in nest after incubation begins. *General and Comparative Endocrinology* 101:205–210.

Miller, D. A., C. M. Vleck, and D. L. Otis. 2009. Individual variation in baseline and stress-induced corticosterone and prolactin levels predicts parental effort by nesting mourning doves. *Hormones and Behavior* 56:457–464.

Mitchell, M. A., and A. J. Carlisle. 1991. Plasma zinc as an index of vitellogenin production and reproductive status in the domestic fowl. *Comparative Biochemistry and Physiology A: Molecular and Integrative Physiology* 100:719–724.

Mock, D. W., and G. A. Parker. 1997. *The Evolution of Sibling Rivalry*. Oxford: Oxford University Press.

Mock, D. W., P. L. Schwagmeyer, and M. B. Dugas. 2009. Parental provisioning and nestling mortality in house sparrows. *Animal Behaviour* 78:677–684.

Møller, A. P. 2008. Climate change and micro-geographic variation in laying date. *Oecologia* 155:845–857.

Møller, A. P., E. Flensted-Jensen, K. Klarborg, W. Mardal, and J. T. Nielsen. 2010. Climate change affects the duration of the reproductive season in birds. *Journal of Animal Ecology* 79:777–784.

Monaghan, P. 2010. Telomeres and life histories: The long and the short of it. *Annals of the New York Academy of Sciences* 1206:130–142.

Monaghan, P., M. Bolton, and D. C. Houston. 1995. Egg production constraints and the evolution of avian clutch size. *Proceedings of the Royal Society B: Biological Sciences* 259:189–191.

Monaghan, P., A. Charmantier, D. H. Nussey, and R. E. Ricklefs. 2008. The evolutionary ecology of senescence. *Functional Ecology* 22:371–378.

Monaghan, P., N. B. Metcalfe, and R. Torres. 2009. Oxidative stress as a mediator of life history trade-offs: Mechanisms, measurements and interpretation. *Ecology Letters* 12:75–92.

Monaghan, P., and R. G. Nager. 1997. Why don't birds lay more eggs? *Trends in Ecology and Evolution* 12:270–274.

Monaghan, P., R. G. Nager, and D. C. Houston. 1998. The price of eggs: Increased investment in egg production reduces the offspring rearing capacity of parents. *Proceedings of the Royal Society B: Biological Sciences* 265:1731–1735.

Monticelli, D., J. A. Ramos, and G. D. Quartly. 2007. Effects of annual changes in primary productivity and ocean indices on breeding performance of tropical roseate terns in the western Indian Ocean. *Marine Ecology Progress Series* 351:273–286.

Moore, M. C., and G.I.H. Johnston. 2008. Toward a dynamic model of deposition and utilization of yolk steroids. *Integrative and Comparative Biology* 48:411–418.

Morales, J., J. Moreno, S. Merino, J. J. Sanz, G. Tomas, E. Arriero, E. Lobato, et al. 2007. Early moult improves local survival and reduces reproductive output in female pied flycatchers. *Ecoscience* 14:31–39.

Morales, J., A. Velando, and J. Moreno. 2008. Pigment allocation to eggs decreases plasma antioxidants in a songbird. *Behavioral Ecology and Sociobiology* 63:227–233.

Moreno, J., R. J. Cowie, J. J. Sanz, and R.S.R. Williams. 1995. Differential response by males and females to brood manipulations in the pied flycatcher: Energy expenditure and nestling diet. *Journal of Animal Ecology* 64:721–732.

Moreno, J., L. Gustafsson, A. Carlson, and T. Part. 1991. The cost of incubation in relation to clutch-size in the Collared Flycatcher *Ficedula albicollis*. *Ibis* 133:186–193.

Moreno, J., J. Potti, and S. Merino. 1997. Parental energy expenditure and offspring size in the pied flycatcher *Ficedula hypoleuca*. *Oikos* 79:559–567.

Moreno, J., and J. J. Sanz. 1994. The relationship between the energy expenditure during incubation and clutch size in the Pied Flycatcher *Ficedula hypoleuca*. *Journal of Avian Biology* 25:125–130.

Moreno-Rueda, G. 2010. Experimental test of a trade-off between moult and immune response in house sparrows *Passer domesticus*. *Journal of Evolutionary Biology* 23:2229–2237.

Morrison, E. S., D. R. Ardia, and E. D. Clotfelter. 2009. Cross-fostering reveals sources of variation in innate immunity and hematocrit in nestling tree swallows *Tachycineta bicolor*. *Journal of Avian Biology* 40:573–578.

Morton, G. A., and M. L. Morton. 1990. Dynamics of postnuptial molt in free-living mountain White-crowned Sparrows. *Condor* 92:813–828.

Morton, M. L. 1994. Hematocrits in montane sparrows in relation to reproductive schedule. *Auk* 96:119–126.

Morton, M. L., M. E. Pereya, and L. Baptista. 1985. Photoperiodically-induced ovarian growth in the white-crowned sparrow (*Zonotrichia leucophrys gambelii*) and its augmentation by song. *Comparative Biochemistry and Physiology A: Molecular and Integrative Physiology* 80:93–97.

Mousseau, T. A., and C. W. Fox. 1998. *Maternal Effects as Adaptations*. Oxford: Oxford University Press.

Müller, W., and M. Eens. 2009. Elevated yolk androgen levels and the expression of multiple sexually selected male characters. *Hormones and Behavior* 55:175–181.

Müller, W., C. M. Eising, C. Dijkstra, and T. G. G. Groothuis. 2004. Within-clutch patterns of yolk testosterone vary with the onset of incubation in black-headed gulls. *Behavioral Ecology* 15:893–397.

Müller, W., C. M. Lessells, P. Korsten, and N. von Engelhardt. 2007. Manipulative signals in family conflict? On the function of maternal yolk hormones in birds. *American Naturalist* 169:E84–E96.

Müller, W., J. Vergauwen, and M. Eens. 2009. Long-lasting effects of elevated yolk testosterone levels on female reproduction. *Behavioral Ecology and Sociobiology* 63:809–816.

Murphy, E. C., and E. Haukioja. 1986. Clutch size in nidicolous birds. *Current Ornithology* 4:141–180.

Murphy, K. G., and S. R. Bloom. 2006. Gut hormones and the regulation of energy homeostasis. *Nature* 444:854–859.

Murphy, M. E. 1996. Energetics and nutrition of molt. In *Avian Energetics and Nutritional Ecology*, ed. C. Carey, 158–198. New York: Chapman and Hall.

Murphy, M. E., and J. R. King. 1984. Sulfur amino acid nutrition during molt in the White-crowned Sparrow. I. Does dietary sulfur amino acid concentration affect the energetics of molt as assayed by metabolized energy? *Condor* 86:314–332.

Murphy, M. E., and J. R. King. 1992. Energy and nutrient use during moult by White-Crowned Sparrows *Zonotrichia leucophrys gambelii. Ornis Scandinavica* 23:304–313.

Murphy, M. T., B. Armbrecth, E. Vlamis, and A. Pierce. 2000. Is reproduction by Tree Swallows cost free? *Auk* 117:902–912.

Murton, R. K., and N. J. Westwood. 1977. *Avian Breeding Cycles*. Oxford: Clarendon Press.

Myers, B. R., Y. M. Sigal, and D. Julius. 2009. Evolution of thermal response properties in a cold-activated TRP channel. *PLoS One* 4:e5741.

Nagaraja, S. C., S. E. Aggrey, J. Yao, D. Zadworny, R. W. Fairfull, and U. Kuhnlein. 2000. Trait association of a genetic marker near the IGF-I gene in egg-laying chickens. *Journal of Heredity* 91:150–156.

Nager, R. G. 2006. The challenges of making eggs. *Ardea* 94:323–346.

Nager, R. G., P. Monaghan, and D. C. Houston. 2000. Within-clutch trade-offs between the number and quality of eggs: Experimental manipulation in gulls. *Ecology* 81:1339–1350.

———. 2001. The cost of egg production: Increased egg production reduces future fitness in gulls. *Journal of Avian Biology* 32:159–166.

Nager, R. G., C. Rüegger, and A. J. V. Noordwijk. 1997. Nutrient or energy limitation on egg formation: A feeding experiment in Great Tits. *Journal of Animal Ecology* 66:495–507.

Nager, R. G., and A. J. van Noordwijk. 1992. Energetic limitation in the egg laying period of great tits. *Proceedings of the Royal Society B: Biological Sciences* 249:259–263.

Nager, R. G., and H. S. Zandt. 1994. Variation in egg size in Great Tits. *Ardea* 82:315–328.

Nagy, L. R., and R. T. Holmes. 2005a. Food limits annual fecundity of a migratory songbird: An experimental study. *Ecology* 86:675–681.

———. 2005b. To double brood or not? Individual variation in the reproductive effort in Black-throated Blue Warblers (*Dendroica caerulescens*). *Auk* 122:902–914.

Nakanea, Y., K. Ikegami, H. Ono, N. Yamamoto, S. Yoshida, K. Hirunagi, S. Ebihara, et al. 2010. A mammalian neural tissue opsin (Opsin 5) is a deep brain photoreceptor in birds. *Proceeding of the National Academy of Sciences USA* 107:15264–15268.

Nakao, N., S. Yasuo, A. Nishimura, T. Yamamura, T. Watanabe, T. Anraku, T. Okano, et al. 2007. Circadian clock gene regulation of steroidogenic acute regulatory protein gene expression in preovulatory ovarian follicles. *Endocrinology* 148:3031–3038.

Navara, K. J., G. E. Hill, and M. T. Mendonça. 2006. Yolk testosterone stimulates growth and immunity in house finch chicks. *Physiological and Biochemical Zoology* 79:550–555.

Navara, K. J., and M. T. Mendonça. 2008. Yolk androgens as pleiotropic mediators of physiological processes: A mechanistic review. *Comparative Biochemistry and Physiology A: Molecular and Integrative Physiology* 150:378–386.

Navarro, V. M., J. M. Castellano, D. García-Galiano, and M. Tena-Sempere. 2007. Neuroendocrine factors in the initiation of puberty: The emergent role of kisspeptin. *Reviews in Endocrine and Metabolic Disorders* 8:11–20.

Nemoto, S., and T. Finkel. 2004. Ageing and the mystery at Arles. *Nature* 429:149–152.

Neto, J. M., and A. G. Gosler. 2006. Post-juvenile and post-breeding moult of Savi's Warblers *Locustella luscinioides* in Portugal. *Ibis* 148:39–49.

Nevoux, M., J. Forcada, C. Barbraud, J. Croxall, and H. Weimerskirch. 2010. Bet-hedging response to environmental variability, an intraspecific comparison. *Ecology* 91:2416–2427.

Newbrey, J. L., W. L. Reed, S. P. Foster, and G. L. Zander. 2008. Laying-sequence variation in yolk carotenoid concentrations in eggs of Yellow-Headed Blackbirds (*Xanthocephalus xanthocephalus*). *Auk* 125:124–130.

Newton, I., ed. 1989a. *Lifetime Reproduction in Birds*. London: Academic Press.

———. 1989b. Sparrowhawk. In *Lifetime Reproduction in Birds*, ed. I. Newton, 279–296. London: Academic Press.

Newton, I., and M. Marquiss. 1981. Effect of additional food on laying dates and clutch sizes of Sparrowhawks. *Ornis Scandinavica* 12:224–229.

Nicholls, T. J., A. R. Goldsmith, and A. Dawson. 1988. Photorefractoriness in birds and comparison with mammals. *Physiological Reviews* 68:133–176.

Nicolaus, M., C. Both, R. Ubels, P. Edelaar, and J. M. Tinbergen. 2009. No experimental evidence for local competition in the nestling phase as a driving force for density-dependent avian clutch size. *Journal of Animal Ecology* 78:828–838.

Nilsson, J.-A. 1991. Clutch size determination in the Marsh Tit (*Parus palustris*). *Ecology* 72:1757–1762.

———. 1994. Energetic bottle-necks during breeding and the reproductive cost of being too early. *Journal of Animal Ecology* 63:200–208.

Nilsson, J.-A., and H. Källander. 2006. Leafing phenology and timing of egg laying in great tits *Parus major* and blue tits *P. caeruleus*. *Journal of Avian Biology* 37:357–363.

Nilsson, J.-A., and L. Raberg. 2001. The resting metabolic cost of egg laying and nestling feeding in great tits. *Oecologia* 128:187–192.

Nilsson, J.-A., and E. Svensson. 1993. The frequency and timing of laying gaps. *Ornis Scandinavica* 24:122–126.

———. 1996. The cost of reproduction: A new link between current reproductive effort and future reproductive success. *Proceedings of the Royal Society B: Biological Sciences* 263:711–714.

Nisbet, I.C.T. 1973. Courtship feeding, egg size and breeding success in common terns. *Nature* 241:141–142.

———. 1978. Dependence of fledging success on egg size, parental performance and egg composition among Common and Roseate Terns, *Sterna hirundo* and *S. dougallii*. *Ibis* 120:207–215.

Nordling, D., M. Andersson, S. Zohari, and G. Lars. 1998. Reproductive effort reduces specific immune response and parasite resistance. *Proceedings of the Royal Society B: Biological Sciences* 265:1291–1298.

Norris, D. R., and P. P. Marra. 2007. Seasonal interactions, habitat quality, and population dynamics in migratory birds. *Condor* 109:535–547.

Norris, D. R., P. P. Marra, T. K. Kyser, T. W. Sherry, and L. M. Ratcliffe. 2004. Tropical winter habitat limits reproductive success on the temperate breeding grounds in a migratory bird. *Proceedings of the Royal Society B: Biological Sciences* 271:59–64.

Norris, D. R., P. P. Marra, R. Montgomerie, T. K. Kyser, and L. M. Ratcliffe. 2004. Reproductive effort molting latitude, and feather color in a migratory songbird. *Science* 306:2249–2250.

Norte, A. C., J. A. Ramos, J. P. Sampaio, J. P. Sousa, and B. C. Sheldon. 2010. Physiological condition and breeding performance of the Great Tit. *Condor* 112:79–86.

Norte, A. C., B. C. Sheldon, J. P. Sousa, and J. A. Ramos. 2009. Physiological condition and breeding performance of the Great Tit. *Journal of Avian Biology* 40:157–165.

Nuñez Rodriguez, J., E. Bon, and F. Le Menn. 1996. Vitellogenin receptors during vitellogenesis in the rainbow trout *Oncorhynchus mykiss*. *Journal of Experimental Zoology* 274:163–170.

Nur, N. 1984a. The consequences of brood size for breeding Blue Tits I. Adult survival, weight change and the cost of reproduction. *Journal of Animal Ecology* 53:479–496.

———. 1984b. The consequences of brood size for breeding blue tits II. Nestling weight, offspring survival and optimal brood size. *Journal of Animal Ecology* 53:497–517.

———. 1984c. Feeding frequencies of nestling blue tits (*Parus caeruleus*): Costs, benefits and a model of optimal feeding frequency. *Oecologia* 65:125–137.

Nussey, D. H., E. Postma, P. Gienapp, and M. E. Visser. 2005. Selection on heritable phenotypic plasticity in a wild bird population. *Science* 310:304–306.

Nussey, D. H., A. J. Wilson, and J. E. Brommer. 2007. The evolutionary ecology of individual phenotypic plasticity in wild populations. *Journal of Evolutionary Biology* 20:831–844.

O'Brien, S., and M. Hau. 2005. Food cues and gonadal development in neotropical spotted antbirds (*Hylophylax naevioides*). *Journal of Ornithology* 146:332–337.

Ogden, L.J.E., and B.J.M. Stutchbury. 1996. Constraints on double brooding in a neotropical migrant, the Hooded Warbler. *Condor* 98:736–744.

Ojanen, M. 1983a. Effects of laying sequence and ambient temperature on the composition of eggs of the great tit *Parus major* and the pied flycatcher *Ficedula hypoleuca*. *Annales Zoologici Fennici* 20:65–71.

———. 1983b. Egg development and the related nutrient reserve depletionin the pied flycatcher, *Ficedula hypoleuca*. *Annals Zoologica Fennica* 58:93–108.

Olsen, N. J., and W. J. Kovacs. 1996. Gonadal steroids and immunity. *Endocrine Reviews* 17:369–384.

Olson, V. A., and I.P.F. Owens. 1998. Costly sexual signals: Are carotenoids rare, risky or required? *Trends in Ecology and Evolution* 13:510–514.

Olsson, O. 1997. Clutch abandonment: A state-dependent decision in King Penguins. *Journal of Avian Biology* 28:264–267.

Onagbesan, O. M., V. Bruggeman, and E. Decuypere. 2009. Intra-ovarian growth factors regulating ovarian function in avian species: A review. *Animal Reproduction Science* 111:121–140.

Onagbesan, O. M., M. Safi, E. Decuypere, and V. Bruggeman. 2004. Developmental changes in inhibin α and inhibin/activin βA and βB mRNA levels in the gonads during post-hatch prepubertal development of male and female chickens. *Molecular Reproduction and Development* 68:319–326.

Onagbesan, O. M., B. Vleugels, N. Buys, V. Bruggeman, M. Safi, and E. Decuypere. 1999. Insulin-like growth factors in the regulation of avian ovarian functions. *Domestic Animal Endocrinology* 17:299–313.

Oppenheimer, S. D., M. E. Pereyra, and M. L. Morton. 1996. Egg laying in Dusky Flycatchers and White-crowned Sparrows. *Condor* 98:428–430.

Opresko, L. K., and H. S. Wiley. 1987. Receptor-mediated endocytosis in *Xenopus* oocytes II. Evidence for two novel mechanisms of hormonal regulation. *Journal of Biological Chemistry* 262:4116–4123.

Orell, M., and M. Ojanen. 1983. Timing and length of the breeding season of the Great Tit *Parus major* and the Willow Tit *Parus montanus* near Oulu, northern Finland. *Ardea* 71:183–198.

Ots, I., A. Murumägi, and P. Hõrak. 1998. Haematological health state indices of reproducing Great Tits: Methodology and sources of natural variation. *Functional Ecology* 12:700–707.

Ouyang, J. Q., P. J. Sharp, A. Dawson, M. Quetting, and M. Hau. 2011. Hormone levels predict individual differences in reproductive success in a passerine bird. *Proceedings of the Royal Society B: Biological Sciences* 278:2537–2545.

Painter, D., D. H. Jennings, and M. C. Moore. 2002. Placental buffering ofmaternal steroid hormone effects on fetal and yolk hormone levels: A comparative study of a viviparous lizard, *Sceloporus jarrovi*, and an oviparous lizard, *Sceloporus graciosus*. *General and Comparative Endocrinology* 127:105–116.

Paitz, R. T., and R. M. Bowden. 2008. A proposed role of the sulfotransferase/sulfatase pathway in modulating yolk steroid effects. *Integrative and Comparative Biology* 48:419–427.

———. 2009. Rapid decline in the concentrations of three yolk steroids during development: Is it embryonic regulation? *General and Comparative Endocrinology* 161:246–251.

Paitz, R. T., R. M. Bowden, and J. M. Casto. 2010. Embryonic modulation of maternal steroids in European starlings (*Sturnus vulgaris*). *Proceedings of the Royal Society B: Biological Sciences* 278:99-106.

Palmer, B. D., and L. J. Guilletter. 1991. Oviductal proteins and their influence on embryonic development in birds and reptiles. In *Egg Incubation: Its Effects on Embryonic Development in Birds and Reptiles*, ed. D. C. Deeming and M.W.J. Ferguson, 29–46. Cambridge: Cambridge University Press.

Palmer, S. S., and J. M. Bahr. 1992. Follicle stimulating hormone increases serum oestradiol-17β concentrations, number of growing follicles and yolk deposition in aging hens (*Gallus domesticus domesticus*) with decreased egg production. *British Poultry Science* 33:403–414.

Parejo, D., and E. Danchin. 2006. Brood size manipulation affects frequency of second clutches in the blue tit. *Behavioral Ecology and Sociobiology* 60:184–194.

Parker, H., and H. Holm. 1990. Patterns of nutrient and energy expenditure in female Common Eiders nesting in the high arctic. *Auk* 107:660–668.

Parmesan, C. 2006. Ecological and evolutionary responses to recent climate change. *Annual Review of Ecology, Evolution and Systematics* 37:637–669.

Partecke, J., and H. Schwabl. 2008. Organizational effects of maternal testosterone on reproductive behavior of adult house sparrows. *Developmental Neurobiology* 68:1538–1548.

Partecke, J., T. Van't Hof, and E. Gwinner. 2005. Underlying physiological control of reproduction in urban and forest-dwelling European blackbirds *Turdus merula*. *Journal of Avian Biology* 36:295–305.

Partridge, L., D. Gems, and D. J. Withers. 2005. Sex and death: What is the connection? *Cell* 120:461–472.

Patten, M. A. 2007. Geographic variation in calcium and clutch size. *Journal of Avian Biology* 38:637–643.

Paul, M. J., L. M. Pyter, D. A. Freeman, J. Galang, and B. J. Prendergast. 2009. Photic and nonphotic seasonal cues differentially engage hypothalamic kisspeptin and RFamide-related peptide mRNA expression in Siberian hamsters. *Journal of Neuroendocrinology* 21:1007–1014.

Pedersen, H. C. 1989. Effects of exogenous prolactin on parental behaviour in free-living female willow ptarmigan *Lagopus l. lagopus*. *Animal Behaviour* 38:926–934.

Perazzolo, L. M., K. Coward, B. Davail, E. Normand, C. R. Tyler, F. Pakdel, W. J. Schneider, et al. 1999. Expression and localization of messenger ribonucleic acid for the vitellogenin receptor in ovarian follicles throughout oogenesis in the rainbow trout, *Oncorhynchus mykiss*. *Biology of Reproduction* 60:1057–1068.

Perfito, N., J. M. Y. Kwong, G. E. Bentley, and M. Hau. 2008. Cue hierarchies and testicular development: Is food a more potent stimulus than day length in an opportunistic breeder (*Taeniopygia g. guttata*)? *Hormones and Behavior* 53:567–572.

Perfito, N., S. L. Meddle, A. D. Tramontin, P. J. Sharp, and J. C. Wingfield. 2005. Seasonal gonadal recrudescence in song sparrows: Response to temperature cues. *General and Comparative Endocrinology* 143:121–128.

Perfito, N., R. A. Zann, G. E. Bentley, and M. Hau. 2007. Opportunism at work: Habitat predictability affects reproductive readiness in free-living zebra finches. *Functional Ecology* 21:291–301.

Perrins, C. M. 1965. Population fluctuations and clutch-size in the great tit *Parus major*. *Journal of Animal Ecology* 34:601–647.

———. 1970. The timing of bird's breeding seasons. *Ibis* 112:242–255.

———. 1979. *British Tits*. London: Collins.

———. 1991. Tits and their caterpillar food supply. *Ibis* 133:S49–S54.

———. 1996. Eggs, egg formation and the timing of breeding. *Ibis* 138:2–15.

Perrins, C. M., and R. H. McCleery. 1989. Laying dates and clutch size in the Great Tit. *Wilson Bulletin* 101:236–253.

Perrins, C. M., and D. Moss. 1975. Reproductive rates in the great tit. *Journal of Animal Ecology* 44:695–706.

Perry, M. M., and A. B. Gilbert. 1979. Yolk transport in the ovarian follicle of the hen (*Gallus domesticus*): Lipoprotein-like particles at the periphery of the oocyte in the rapid growth phase. *Journal of Cell Science* 39:257–272.

Peterson, C. C., K. A. Nagy, and J. Diamond. 1990. Sustained metabolic scope. *Proceedings of the National Academy of Sciences USA* 87:2324–2328.

Petitte, J. N., and R. J. Etches. 1991. Daily infusion of corticosterone and reproductive function in the domestic hen (*Gallus domesticus*). *General and Comparative Endocrinology* 83:397–405.

Petrie, M., H. Schwabl, N. Brande-Lavridsen, and T. Burke. 2001. Sex differences in avian yolk hormone levels. *Nature* 412:489.

Pettifor, R. A., C. M. Perrins, and R. H. McCleery. 1988. Individual optimization of clutch size in great tits. *Nature* 336:160–162.

———. 2001. The individual optimization of fitness: Variation in reproductive output, including clutch size, mean nestling mass and offspring recruitment, in manipulated broods of great tits *Parus major*. *Journal of Animal Ecology* 70:62–79.

Piersma, T., J. M. Everaarts, and J. Jukema. 1996. Build-up of red blood cells in refuelling Bar-Tailed Godwits in relation to individual migratory quality. *Condor* 98:363–370.

Pihlaja, M., H. Siitari, and R. V. Alatalo. 2006. Maternal antibodies in a wild altricial bird: Effects on offspring immunity, growth and survival. *Journal of Animal Ecology* 75:1154–1164.

Pike, T. W., and M. Petrie. 2003. Potential mechanisms of avian sex manipulation. *Biological Reviews* 78:553–574.

———. 2006. Experimental evidence that corticosterone affects offspring sex ratios in quail. *Proceedings of the Royal Society B: Biological Sciences* 273:1093–1098.

Pilz, K. M., M. Quiroga, H. Schwabl, and E. Adkins-Regan. 2004. European starling chicks benefit from high yolk testosterone levels during a drought year. *Hormones and Behavior* 46:179–192.

Pilz, K. M., H. G. Smith, M. I. Sandell, and H. Schwabl. 2003. Interfemale variation in egg yolk androgen allocation in the European starling: Do high-quality females invest more? *Animal Behaviour* 65:841–850.

Pitala, N., S. Ruuskanen, T. Laaksonen, B. Doligez, T. B., and L. Gustafsson. 2009. The effects of experimentally manipulated yolk androgens on growth and immune function of male and female nestling collared flycatchers *Ficedula albicollis*. *Journal of Avian Biology* 40:225–230.

Portner, H., and A. P. Farrell. 2008. Physiology and climate change. *Science* 322:690–692.

Portner, H. O., and R. Knust. 2007. Climate change affects marine fishes through the oxygen limitation of thermal tolerance. *Science* 315:95–97.

Portugal, S. J., J. A. Green, and P. J. Butler. 2007. Annual changes in body mass and resting metabolism in captive barnacle geese (*Branta leucopsis*): The importance of wing moult. *Journal of Experimental Biology* 210:1391–1397.

Postma, E., F. Heinrich, U. Koller, R. J. Sardell, J. M. Reid, P. Arcese, and L. F. Keller. 2011. Disentangling the effect of genes, the environment and chance on sex ratio variation in a wild bird population. *Proceedings of the Royal Society B: Biological Sciences* 278:2996–3002.

Potti, J. 1998. Variation in the onset of incubation in the pied flycatcher (*Ficedula hypoleuca*): Fitness consequences and constraints. *Journal of Zoology* 245:335–344.

———. 2007. Variation in the hematocrit of a passerine bird across life stages is mainly of environmental origin. *Journal of Avian Biology* 38:726–730.

Potti, J., J. Moreno, S. Merino, O. Frías, and R. Rodríguez. 1999. Environmental and genetic variation in the haematocrit of fledgling pied flycatchers *Ficedula hypoleuca*. *Oecologia* 120:1–8.

Prats, M. T., L. Palacios, S. Gallego, and M. Riera. 1996. Blood oxygen transport properties during migration to higher altitude of wild quail, *Coturnix coturnix coturnix*. *Physiological Zoology* 69:912–929.

Pravosudov, V. V. 2003. Long-term moderate elevation of corticosterone facilitates avian food-caching behaviour and enhances spatial memory. *Proceedings of the Royal Society B: Biological Sciences* 270:2599–2604.

Price, T., M. Kirkpatrick, and S. J. Arnold. 1988. Directional selection and the evolution of breeding date in birds. *Science* 240:798–799.

Price, T., and L. Liou. 1989. Selection on clutch size in birds. *American Naturalist* 134:950–959.

Prop, J., J. M. Black, and P. Shimmings. 2003. Travel schedules to the high arctic: Barnacle geese trade-off the timing of migration with accumulation of fat deposits. *Oikos* 103:403–414.

Quillfeldt, P., N. Everaert, J. Buyse, J. F. Masello, and S. Dridi. 2009. Relationship between plasma leptin-like protein levels, begging and provisioning in nestling thin-billed prions *Pachyptila belcheri*. *General and Comparative Endocrinology* 161:171–178.

Råberg, L., M. Grahn, D. Hasselquist, and E. Svensson. 1998. On the adaptive significance of stress-induced immunosuppression. *Proceedings of the Royal Society B: Biological Sciences* 265:1637–1641.

Ramakrishnan, S., A. D. Strader, B. Wimpee, P. Chen, M. S. Smith, and J. D. Buntin. 2007. Evidence for increased neuropeptide Y synthesis in mediobasal hypothalamus in relation to parental hyperphagia and gonadal activation in breeding ring doves. *Journal of Neuroendocrinology* 19:163–171.

Ramsay, S. L., and D. C. Houston. 1997. Nutritional constraints on egg production in the blue tit: A supplementary feeding study. *Journal of Animal Ecology* 66:649–657.

———. 1998. The effect of dietary amino acid composition on egg production in the blue tit. *Proceedings of the Royal Society B: Biological Sciences* 265:1401–1405.

Reale, D., N. J. Dingemanse, A.J.N. Kazem, and J. Wright. 2010. Evolutionary and ecological approaches to the study of personality. *Philosophical Transactions of the Royal Society B: Biological Sciences* 365:3937–3946.

Reddy, I. J., C. G. David, and S. S. Raju. 2006. Chemical control of prolactin secretion and its effects on pause days, egg production and steroid hormone concentration in girirani birds. *International Journal of Poultry Science* 5:685–692.

———. 2007. Effect of suppression of plasma prolactin on luteinizing hormone concentration, intersequence pause days and egg production in domestic hen. *Domestic Animal Endocrinology* 33:167–175.

Redshaw, M. R., and B. K. Follett. 1976. Physiology of egg yolk production by the fowl: The measurement of circulating levels of vitellogenin employing a specific radio-immunoassay. *Comparative Biochemistry and Physiology A: Molecular and Integrative Physiology* 55:399–405.

Reed, T. E., S. Wanless, M. P. Harris, M. Frederiksen, L.E.B. Kruuk, and E.J.A. Cunningham. 2006. Responding to environmental change: Plastic responses vary little in a synchronous breeder. *Proceedings of the Royal Society B: Biological Sciences* 273:2713–2719.

Reed, T. E., P. Warzybok, A. J. Wilson, R. W. Bradley, S. Wanless, and W. J. Sydeman. 2009. Timing is everything: Flexible phenology and shifting selection in a colonial seabird. *Journal of Animal Ecology* 78:376–387.

Reed, W. L., A. M. Turner, and P. R. Sotherland. 1999. Consequences of egg-size variation in the Red-winged Blackbird. *Auk* 116:549–552.

Reid, J. M., P. Arcese, L. F. Keller, and D. Hasselquist. 2006. Long-term maternal effect on offspring immune response in song sparrows *Melospiza melodia*. *Biology Letters* 2:573–576.

Reid, J. M., P. Monaghan, and R. G. Nager. 2002. Incubation and the costs of reproduction. In *Avian Incubation*, ed D. C. Deeming, 314–325. Oxford: Oxford University Press.

Reid, J. M., P. Monaghan, and G. D. Ruxton. 2000a. The consequences of clutch size for incubation conditions and hatching success in starlings. *Functional Ecology* 14:560–565.

———. 2000b. Resource allocation between reproductive phases: The importance of thermal conditions in determining the cost of incubation. *Proceedings of the Royal Society B: Biological Sciences* 267:37–41.

Reid, J. M., G. D. Ruxton, P. Monaghan, and G. M. Hilton. 2002. Energetic consequences of clutch temperature and clutch size for a uniparental intermittent incubator: The Starling. *Auk* 119:54–61.

Reid, W. V. 1987. The cost of reproduction in the glaucous-winged gull. *Oecologia* 74:458–467.
Reinert, B. D., and F. E. Wilson. 1997. The effects of thyroxine (T4) or triiodothyronine (T3) replacement therapy on the programming of seasonal reproduction and postnuptial molt in thyroidectomized male American tree sparrows (*Spizella arborea*) exposed to long days. *Journal of Experimental Zoology* 279:367–376.
Remes, V., M. Krist, V. Bertacche, and R. Stradi. 2007. Maternal carotenoid supplementation does not affect breeding performance in the Great Tit (*Parus major*). *Functional Ecology* 21:776–783.
Revel, F. G., L. Ansel, P. Klosen, M. Saboureau, P. Pévet, J. K. Mikkelsen, and V. Simonneaux. 2007. Kisspeptin: A key link to seasonal breeding. *Reviews in Endocrine and Metabolic Disorders* 8:57–65.
Reynolds, S. J., and S. Waldron. 1999. Body water dynamics at the onset of egg-laying in the zebra finch *Taeniopygia guttata*. *Journal of Avian Biology* 30:1–6.
Reznick, D. 1985. Costs of reproduction: An evaluation of the empirical evidence. *Oikos* 44:257–267.
Reznick, D., L. Nunney, and A. Tessier. 2000. Big houses, big cars, superfleas and the costs of reproduction. *Trends in Ecology and Evolution* 15:421–425.
Richards, M. P. 2003. Genetic regulation of feed intake and energy balance in poultry. *Poultry Science* 82:907–916.
Richter, T., and T. v. Zglinicki. 2007. A continuous correlation between oxidative stress and telomere shortening in fibroblasts. *Experimental Gerontology* 42:1039–1042.
Ricklefs, R. E. 2000. Lack, Skutch, and Moreau: The early development of life-history thinking. *Condor* 102:3–8.
Risch, T. S., and F. C. Rohwer. 2000. Effects of parental quality and egg size on growth and survival of herring gull chicks. *Canadian Journal of Zoology* 78:967–973.
Rivera, J.H.V., C. A. Haas, J. H. Rappole, and W. J. McShea. 2000. Parental care of fledgling Wood Thrushes. *Wilson Bulletin* 112:233–237.
Robb, G. N., R. A. McDonald, D. E. Chamberlain, S. J. Reynolds, T.J.E. Harrison, and S. Bearhop. 2008. Winter feeding of birds increases productivity in the subsequent breeding season. *Biology Letters* 4:220–223.
Roberts, M. L., K. L. Buchanan, D. Hasselquist, A.T.D. Bennett, and M. R. Evans. 2007. Physiological, morphological and behavioral effects of selecting zebra finches for divergent levels of corticosterone. *Journal of Experimental Biology* 210:4368–4378.
Robertson, G. J. 1995. Annual variation in common eider egg size: Effects of temperature, clutch size, laying date, and laying sequence. *Canadian Journal of Zoology* 73:1579–1587.
Robin, J.-P., L. Boucontet, P. Chillet, and R. Groscolas. 1998. Behavioral changes in fasting emperor penguins: Evidence for a "refeeding signal" linked to a metabolic shift. *American Journal of Physiology Regulatory Integrative and Comparative Physiology* 274:R746–753.
Robinson, W. D., J. D. Styrsky, B. J. Payne, R. G. Harper, and C. F. Thompson. 2008. Why are incubation periods longer in the tropics? A common-garden experiment with house wrens reveals it is all in the egg. *American Naturalist* 171:532–535.
Rockwell, R. F., C. S. Findlay, and F. Cooke. 1987. Is there an optimal clutch size in snow geese? *American Naturalist* 130:839–863.
Rodenhouse, N. L., and R. T. Holmes. 1992. Results of experimental and natural food reductions for breeding Black-throated Blue Warblers. *Ecology* 73:357–372.
Roff, D. A. 2002. *Life History Evolution*. Sunderland, MA: Sinauer.

Rohwer, F. C. 1988. Inter- and intraspecific relationships between egg size and clutch size in waterfowl. *Auk* 105: 161–176.

Rohwer, S., A. Viggiano, and J. M. Marzluff. 2011. Reciprocal tradeoffs between molt and breeding in albatrosses. *Condor* 113:61–70.

Romano, M., M. Caprioli, R. Ambrosini, D. Rubolini, M. Fasola, and N. Saino. 2008. Maternal allocation strategies and differential effects of yolk carotenoids on the phenotype and viability of yellow-legged gull (*Larus michahellis*) chicks in relation to sex and laying order. *Journal of Evolutionary Biology* 21:1626–1640.

Romero, L. M. 2002. Seasonal changes in plasma glucocorticoid concentrations in free-living vertebrates. *General and Comparative Endocrinology* 128:1–24.

———. 2004. Physiological stress in ecology: Lessons from biomedical research. *Trends in Ecology and Evolution* 19:249–255.

Romero, L. M., M. J. Dickens, and N. E. Cyr. 2009. The reactive scope model: A new model integrating homeostasis, allostasis, and stress. *Hormones and Behavior* 55:375–389.

Roskaft, E. 1985. The effect of enlarged brood size on the future reproductive potential of the rook. *Journal of Animal Ecology* 54:255–260.

Rowan, W. 1926. On photoperiodism, reproductive periodicity and the annual migrations of birds and certain fishes. *Proceedings of the Boston Society of Natural History* 38:147–189.

Rowe, L., D. Ludwig, and D. Schluter. 1994. Time, condition and the seasonal decline of avian clutch size. *American Naturalist* 143:698–722.

Rowland, E., O. P. Love, J. J. Verspoor, L. Sheldon, and T. D. Williams. 2007. Manipulating rearing conditions reveals developmental sensitivity of the smaller sex in a passerine bird, the European starling *Sturnus vulgaris*. *Journal of Avian Biology* 38:612–618.

Royama, T. 1966. Factors governing feeding rate, food requirements and brood size of nestling Great Tits *Parus major*. *Ibis* 108:313–347.

Royle, N. J., P. F. Surai, and I. R. Hartley. 2001. Maternally derived androgens and antioxidants in bird eggs: Complementary but opposing effects. *Behavioral Ecology* 12:381–385.

———. 2003. The effect of variation in dietary intake on maternal deposition of antioxidants in zebra finch eggs. *Functional Ecology* 17:472–481.

Royle, N. J., P. F. Surai, R. J. McCartney, and B. K. Speake. 1999. Parental investment and egg yolk lipid composition in gulls. *Functional Ecology* 13:298–306.

Rozenboim, I., N. Mobarky, R. Heiblum, Y. Chaiseha, S. W. Kang, I. Biran, A. Rosenstrauch, et al. 2004. The role of prolactin in reproductive failure associated with heat stress in the domestic turkey. *Biology of Reproduction* 71:1208–1213.

Rubenstein, D. R., and K. A. Hobson. 2004. From birds to butterflies: Animal movement patterns and stable isotopes. *Trends in Ecology and Evolution* 19:256–263.

Rubin, C.-J., M. C. Zody, J. Eriksson, J.R.S. Meadows, E. Sherwood, M. T. Webster, L. Jiang, et al. 2010. Whole-genome resequencing reveals loci under selection during chicken domestication. *Nature* 464:587–591.

Rubolini, D., M. Romano, R. Martinelli, B. Leoni, and N. Saino. 2006. Effects of prenatal yolk androgens on armaments and ornaments of the ring-necked pheasant. *Behavioral Ecology and Sociobiology* 59:549–560.

Rubolini, D., M. Romano, R. Martinelli, and N. Saino. 2006. Effects of elevated yolk testosterone levels on survival, growth and immunity of male and female yellow-legged gull chicks. *Behavioral Ecology and Sociobiology* 59:344–352.

Ruiz, G., M. Rosenmann, and F. F. Novoa. 1995. Seasonal changes of blood values in Rufous-collared sparrows from high and low altitude. *International Journal of Biometeorology* 39:103–107.

Rutkowska, J., and A. V. Badyaev. 2008. Meiotic drive and sex determination: Molecular and cytological mechanisms of sex ratio adjustment in birds. *Philosophical Transactions of the Royal Society B* 363:1675–1686.

Rutkowska, J., and M. Cichon. 2006. Maternal testerone affects the primary sex ratio and offspring survival in zebra finches. *Animal Behaviour* 71:1283–1288.

Rutkowska, J., T. Wilk, and M. Cichon. 2007. Androgen-dependent maternal effects on offspring fitness in zebra finches. *Behavioral Ecology and Sociobiology* 61:1211–1217.

Safran, R. J., K. J. McGraw, K. M. Pilz, and S. M. Correa. 2010. Egg-yolk androgen and carotenoid deposition as a function of maternal social environment in barn swallows *Hirundo rustica*. *Journal of Avian Biology* 41:470–478.

Safran, R. J., K. M. Pilz, K. J. McGraw, S. M. Correa, and H. Schwabl. 2008. Are yolk androgens and carotenoids in barn swallow eggs related to parental quality? *Behavioral Ecology and Sociobiology* 62:427–438.

Safriel, U. N. 1975. On the significance of clutch size in nidifugous birds. *Ecology* 56:703–708.

Saino, N., V. Bertacche, R. P. Ferrari, R. Martinelli, A. P. Møller, and R. Stradi. 2002. Carotenoid concentration in barn swallow eggs is influenced by laying order, maternal infection, and paternal ornamentation. *Proceedings of the Royal Society B: Biological Sciences* 269:1729–1733.

Saino, N., P. Dall'ara, R. Martinelli, and A. P. Møller. 2002. Early maternal effects and antibacterial immune factors in the eggs, nestlings and adults of the barn swallow. *Journal of Evolutionary Biology* 15:735–743.

Saino, N., P. Dall'ara, and A. P. Møller. 2001. Immunoglobulin plasma concentration in relation to egg laying and mate ornamentation of female barn swallows (*Hirundo rustica*). *Journal of Evolutionary Biology* 14:95–109.

Saino, N., R. P. Ferrari, M. Romano, R. Martinelli, and A. P. Møller. 2003. Experimental manipulation of egg carotenoids affects immunity of barn swallow nestlings. *Proceedings of the Royal Society B: Biological Sciences* 270:2485–2489.

Saino, N., M. Romano, R. Ambrosini, R. P. Ferrari, and A. P. Møller. 2004. Timing of reproduction and egg quality covary with temperature in the insectivorous Barn Swallow, *Hirundo rustica*. *Functional Ecology* 18:50–57.

Saino, N., M. Romano, R. P. Ferrari, R. Martinelli, and A. P. Møller. 2005. Stressed mothers lay eggs with high corticosterone levels which produce low-quality offspring. *Journal of Experimental Zoology* 303A:998–1006.

Salomons, H. M., G. A. Mulder, L. van de Zande, M. F. Haussmann, M.H.K. Linskens, and S. Verhulst. 2009. Telomere shortening and survival in free-living corvids. *Proceedings of the Royal Society B: Biological Sciences* 276:3157–3165.

Salvante, K. G., G. Lin, R. L. Walzem, and T. D. Williams. 2007a. Characterization of VLDL particle size diameter dynamics in relation to egg production in a passerine bird. *Journal of Experimental Biology* 210:1064–1074.

———. 2007b. What comes first, the zebra finch or the egg? Temperature-dependent reproductive, physiological and behavioural plasticity in egg-laying zebra finches. *Journal of Experimental Biology* 210:1325–1334.

Salvante, K. G., and T. D. Williams. 2002. Vitellogenin dynamics during egg-laying: Daily variation, repeatability and relationship with reproductive output. *Journal of Avian Biology* 33:391–398.

———. 2003. Effects of corticosterone on the proportion of breeding females, reproductive output and yolk precursor levels. *General and Comparative Endocrinology* 130:205–214.

Sanchez-Lafuente, A. 2004. Trade-off between clutch size and egg mass, and their effects on hatchability and chick mass in semi-precocial Purple Swamphen. *Ardeola* 51:319–330.

Sandercock, B. K. 1997. Incubation capacity and clutch size determination in two calidrine sandpipers: A test of the four-egg threshold. *Oecologia* 110:50–59.

Sanz, J. J. 1997. Clutch size manipulation in the Pied Flycatcher: Effects on nestling growth, parental care and moult. *Journal of Avian Biology* 28:157–162.

———. 1998. Effects of geographic location and habitat on breeding parameters of Great Tits. *Auk* 115:1034–1051.

Sanz, J. J., and J. Moreno. 1995. Experimentally induced clutch size enlargements affect reproductive success in the Pied Flycatcher. *Oecologia* 103:358–364.

Sanz, J. J., and J. M. Tinbergen. 1999. Energy expenditure, nestling age, and brood size: An experimental study of parental behavior in the great tit *Parus major*. *Behavioral Ecology* 10:598–606.

Sanz, J. J., J. M. Tinbergen, M. Orell, and S. Rytkonen. 1998. Daily energy expenditure during brood rearing of Great Tits *Parus major* in northern Finland. *Ardea* 86:101–107.

Sanz-Aguilar, A., G. Tavecchia, R. Pradel, E. Minguez, and D. Oro. 2008. The cost of reproduction and experience-dependent vital rates in a small petrel. *Ecology* 89:3195–3203.

Sapolsky, R. M., L. M. Romero, and A. U. Munck. 2000. How do glucocorticoids influence stress-responses? Integrating permissive, suppressive, stimulatory, and adaptive actions. *Endocrine Reviews* 21:55–89.

Scanes, C. G., J. A. Proudman, and S. V. Radecki. 1999. Influence of continuous growth hormone or insulin-like growth factor I administration in adult female chickens. *General and Comparative Endocrinology* 114:315–323.

Schäfer, K., R. Necker, and H. A. Braun. 1989. Analysis of avian cold receptor function. *Brain Research* 501:66–72.

Schamel, D., and D. M. Tracy. 1987. Latitudinal trends in breeding Red Phalaropes (*Phalaropus fulicaria*). *Journal of Field Ornithology* 58:126–134.

Schifferli, L. 1979. Warum legen singvögel (Passeres) iher eier am frühen morgen? *Ornithologische Beobachter* 76:33–36.

Schimke, R. T., G. S. McKnight, and D. J. Shapiro. 1975. Nucleic acid probes and analysis of hormone action in oviduct. In *Biochemical Actions of Hormones*, ed. G. Litwack, 245–269. New York: Academic Press.

Schneider, W. J. 2007. Low density lipoprotein receptor relatives in chicken ovarian follicle and oocyte development. *Cytogenetic and Genome Research* 117:248–255.

———. 2009. Receptor-mediated mechanisms in ovarian follicle and oocyte development. *General and Comparative Endocrinology* 163:18–23.

Schneider, W. J., R. Carroll, D. L. Severson, and J. Nimpf. 1990. Apolipoprotein VLDL-II inhibits lipolysis of triacylglyceride-rich lipoproteins in the laying hen. *Journal of Lipid Research* 31:507–513.

Schoech, S. J., R. L. Mumme, and J. C. Wingfield. 1996. Prolactin and helping behaviour in the cooperatively breeding Florida scrub-jay, *Apheloma e. coerulesens*. *Animal Behaviour* 52:445–456.

Schonbaum, C. P., J. J. Perrino, and A. P. Mahowald. 2000. Regulation of the vitellogenin receptor during *Drosophila melanogaster* oogenesis. *Molecular Biology of the Cell* 11:511–521.

Schuler, B., M. Arras, S. Keller, A. Rettich, C. Lundby, J. Vogel, and M. Gassmann. 2010. Optimal hematocrit for maximal exercise performance in acute and chronic erythropoietin-treated mice. *Proceedings of the National Academy of Sciences USA* 107:419–423.

Schulkin, J. 2003. Allostasis: A neural behavioral perspective. *Hormones and Behavior* 43:21–27.

Schultz, I. R., G. Orner, J. L. Merdink, and A. Skilman. 2001. Dose–response relationships and pharmacokinetics of vitellogenin in rainbow trout after intravascular administration of 17α-ethynylestradiol. *Aquatic Toxicology* 51:305–318.

Schwabl, H. 1993. Yolk is a source of maternal testosterone for developing birds. *Proceedings of the National Academy of Sciences USA* 90:11446–11450.

Schwagmeyer, P. L., and D. W. Mock. 2008. Parental provisioning and offspring fitness: Size matters. *Animal Behaviour* 75:291–298.

Serra, L., M. Griggio, D. Licheri, and A. Pilastro. 2007. Moult speed constrains the expression of a carotenoid-based sexual ornament. *Journal of Evolutionary Biology* 20:2028–2034.

Sharp, P. J., and N. Ciccone. 2005. The gonadotropin releasing hormone neurone: Key to avian reproductive function. In *Functional Avian Endocrinology*, ed. A. Dawson and P. J. Sharp, 59–72. New Delhi: Narosa Publishing House.

Sharp, P. J., I. C. Dunn, and D. Waddington. 2008. Chicken leptin. *General and Comparative Endocrinology* 158:2–4.

Sharp, P. J., M. C. Macnamee, R. J. Sterling, R. W. Lea, and H. C. Pedersen. 1988. Relationships between prolactin, LH and broody behaviour in bantam hens. *Journal of Endocrinology* 118:279–286.

Sharp, P. J., and R. Moss. 1981. A comparison of the responses of captive willow ptarmigan (*Lagopus lagopus lagopus*), red grouse (*Lagopus lagopus scoticus*), and hybrids to increasing daylengths with observations on the modifying effects of nutrition and crowding in red grouse. *General and Comparative Endocrinology* 45:181–188.

Sharp, P. J., and K. P. Sreekumar. 2001. Photoperiodic control of prolactin secretion. In *Avian Endocrinology*, ed. A. Dawson and C. M. Chaturvedi, 245–255. New Delhi: Narosa Publishing House.

Sharp, P. J., R. J. Sterling, R. T. Talbot, and N. S. Huskisson. 1989. The role of hypothalamic vasoactive intestinal polypeptide in the maintenance of prolactin secretion in incubating bantam hens: Observations using passive immunization, radioimmunoassay and immunohistochemistry. *Journal of Endocrinology* 122:5–13.

Shawkey, M. D., M. K. Firestone, E. L. Brodie, and S. R. Beissinger. 2009. Avian incubation inhibits growth and diversification of bacterial assemblages on eggs. *PLoS One* 4:e4522.

Shawkey, M. D., K. L. Kosciuch, M. Liu, F. C. Rohwer, E. R. Loos, J. M. Wang, and S. R. Beissinger. 2008. Do birds differentially distribute antimicrobial proteins within clutches of eggs? *Behavioral Ecology* 19:920–927.

Sheldon, B. C., L. E. B. Kruuk, and J. Merilä. 2003. Natural selection and inheritance of breeding time and clutch size in the collared flycatcher. *Evolution* 57:406–420.

Shen, X., E. Steyrer, H. Retzek, E. J. Sanders, and W. J. Schneider. 1993. Chicken oocyte growth: Receptor-mediated yolk deposition. *Cell and Tissue Research* 272:459–471.

Shlosberg, A., M. Bellaiche, E. Berman, S. Perk, N. Deeb, E. Neumark, and A. Cahaner. 1998. Relationship between broiler chicken hematocrit-selected parents and their progeny, with regard to hematocrit, mortality from ascites and bodyweight. *Research in Veterinary Science* 64:105–109.

Shuman, T. W., R. J. Robel, J. L. Zimmerman, and K. E. Kemp. 1989. Variance in digestive efficiencies of four sympatric avian granivores. *Auk* 106:324–326.

Shutler, D., R. G. Clark, C. Fehr, and A. W. Diamond. 2006. Time and recruitment costs as currencies in manipulation studies on the costs of reproduction. *Ecology* 87:2938–2946.

Siikamaki, P. 1998. Limitation of reproductive success by food availability and breeding time in Pied Flycatchers. *Ecology* 79:1789–1796.

Silverin, B. 1986. Corticosterone-binding proteins and behavioural effects of high plasma levels of corticosterone during the breeding period. *General and Comparative Endocrinology* 64:67–74.

———. 1991. Annual changes in plasma levels of LH, and prolactin in free-living female great tits (*Parus major*). *General and Comparative Endocrinology* 83:425–431.

———. 1998. Behavioral and hormonal responses of the pied flycatcher to environmental stressors. *Animal Behaviour* 54:1411–1420.

Silverin, B., and A. Goldsmith. 1984. The effects of modifying incubation on prolactin secretion in free-living pied flycatchers. *General and Comparative Endocrinology* 55:239–244.

Silverin, B., R. Massa, and K. A. Stokkan. 1993. Photoperiodic adaptation to breeding at different latitudes in Great Tits. *General and Comparative Endocrinology* 90:14–22.

Silverin, B., and J. Westin. 1995. Influence of the opposite sex on photoperiodically induced LH and gonadal cycles in the willow tit (*Parus montanus*). *Hormones and Behavior* 29:207–215.

Silverin, B., and J. C. Wingfield. 1982. Patterns of breeding behaviour and plasma levels of hormones in a free-living population of Pied Flycatchers, *Ficedula hypoleuca*. *Journal of Zoology, London* 198:117–129.

Silverin, B., J. Wingfield, K.-A. Stokkan, R. Massa, A. Järvinen, N.-Å. Andersson, M. Lambrechts, et al. 2008. Ambient temperature effects on photo induced gonadal cycles and hormonal secretion patterns in Great Tits from three different breeding latitudes. *Hormones and Behavior* 54:60–68.

Simon, A., D. W. Thomas, P. Bourgault, J. Blondel, P. Perret, and M. M. Lambrechts. 2005. Between-population differences in nestling size and hematocrit level in blue tits (*Parus caeruleus*): A cross-fostering test for genetic and environmental effects. *Canadian Journal of Zoology* 83:694–701.

Simons, L. S., and T. E. Martin. 1990. Food limitation of avian reproduction: An experiment with the Cactus Wren. *Ecology* 71:869–876.

Sinervo, B. 1999. Mechanistic analysis of natural selection and a refinement of Lack's and Williams's principles. *American Naturalist* 154:S26–S42.

Sinervo, B., and D. F. DeNardo. 1996. Costs of reproduction in the wild: Path analysis of natural selection and experimental tests of causation. *Evolution* 50:1299–1313.

Sinervo, B., and P. Licht. 1991. Hormonal and physiological control of clutch size, egg size, and egg shape in side-blotched lizards (*Uta stansburiana*): Constraints on the evolution of lizard life histories. *Journal of Experimental Zoology* 257:252–264.

Sinervo, B., F. Mendez-de-la-Cruz, D. B. Miles, B. Heulin, E. Bastiaans, M. Villagran-Santa Cruz, R. Lara-Resendiz, et al. 2010. Erosion of lizard diversity by climate change and altered thermal niches. *Science* 328:894–899.

Skinner, M. K. 2005. Regulation of primordial follicle assembly and development. *Human Reproduction Update* 11:461–471.

Slagsvold, T. 1976. Annual and geographical variation in the time of breeding of the Great Tit *Parus major* and the Pied Flycatcher *Ficedula hypoleuca* in rela-

tion to environmental phenology and spring temperature. *Ornis Scandinavica* 7:127–145.
———. 1984. Clutch size variation of birds in relation to nest predation: On the cost of reproduction. *Journal of Animal Ecology* 53:945–953.
Slagsvold, T., and J. T. Lifjeld. 1988. Ultimate adjustment of clutch size to parental feeding capacity in a passerine bird. *Ecology* 69:1918–1922.
———. 1990. Influence of male and female quality on clutch size in tits (*Parus* spp.). *Ecology* 71:1258–1266.
Slagsvold, T., J. Sandvik, G. Rofstad, O. Lorentsen, and M. Husby. 1984. On the adaptive value of intraclutch egg size variation in birds. *Auk* 101:685–697.
Smith, C. C., and S. D. Fretwell. 1974. The optimal balance between size and number of offspring. *American Naturalist* 108:499–506.
Smith, H. G. 1993. Seasonal decline in clutch size of the Marsh Tit (*Parus palustris*) in relation to date-specific survival of offspring. *Auk* 110:889–899.
———. 2004. Selection for synchronous breeding in the European starling. *Oikos* 105:301–311.
Smith, H. G., and M. Bruun. 1998. The effect of egg size and habitat on starling nestling growth and survival. *Oecologia* 115:59–63.
Smith, H. G., H. Kallander, and J. A. Nilsson. 1987. Effects of experimentally altered brood size on frequency and timing of second clutches in the Great Tit. *Auk* 104:700–706.
Smith, H. G., T. Ohlsson, and K.-J. Wettermark. 1995. Adaptive significance of egg size in the European Starling: Experimental tests. *Ecology* 76:1–7.
Smith, R. J., and F. R. Moore. 2003. Arrival fat and reproductive performance in a long-distance passerine migrant. *Oecologia* 134:325–331.
Snow, D. W. 1958. The breeding of the Blackbird *Turdus mer*ula at Oxford. *Ibis* 100:1–30.
Sockman, K. W., and H. Schwabl. 1999. Daily estradiol and progesterone levels relative to laying and onset of incubation in canaries. *General and Comparative Endocrinology* 114:257–268.
———. 2000. Yolk androgens reduce offspring survival. *Proceedings of the Royal Society B: Biological Sciences* 267:1451–1456.
———. 2001. Covariation of clutch size, laying date, and incubation tendency in the American Kestrel. *Condor* 103:570–578.
Sockman, K. W., H. Schwabl, and P. J. Sharp. 2000. The role of prolactin in the regulation of clutch size and onset of incubation behavior in the American kestrel. *Hormones and Behavior* 38:168–176.
Sockman, K. W., P. J. Sharp, and H. Schwabl. 2006. Orchestration of avian reproductive effort: An integration of the ultimate and proximate basis of flexibility in clutch size, incubation behaviour, and yolk androgen. *Biological Reviews* 81:629–666.
Sockman, K. W., J. Weiss, M. S. Webster, V. Talbott, and H. Schwabl. 2008. Sex-specific effects of yolk-androgens on growth of nestling American kestrels. *Behavioral Ecology and Sociobiology* 62:617–625.
Sockman, K. W., T. D. Williams, A. S. Dawson, and G. F. Ball. 2004. Prior experience with photostimulation enhances photo-induced reproductive development in female European starlings: A possible basis for the age-related increase in avian reproductive performance. *Biology of Reproduction* 71:979–986.
Soler, M., and J. J. Soler. 1996. Effects of experimental food provisioning on reproduction in the Jackdaw *Corvus monedula*, a semi-colonial species. *Ibis* 138:377–383.

Solomon, S. E. 1983. Oviduct. In *Physiology and Biochemistry of the Domestic Fowl*, ed. B. M. Freeman, 379–419. London: Academic Press.

Sorensen, M. C., J. M. Hipfner, T. K. Kyser, and D. R. Norris. 2009. Carry-over effects in a Pacific seabird: Stable isotope evidence that pre-breeding diet quality influences reproductive success. *Journal of Animal Ecology* 78:460–467.

Spaans, B., W. V. D. Veer, and B. S. Ebbinge. 1999. Cost of incubation in a Greater White-Fronted Goose. *Waterbirds: The International Journal of Waterbird Biology* 22:151–155.

Speakman, J. R. 1997. *Doubly Labelled Water: Theory and Practice*. Kluwer Academic, New York.

Spear, L., and N. Nur. 1994. Brood size, hatching order and hatching date: Effects on four life-history stages from hatching to recruitment in western gulls. *Journal of Animal Ecology* 63:283–298.

Stadtman, E. R. 2002. Importance of individuality in oxidative stress and aging. *Free Radical Biology and Medicine* 33:597–604.

Stanley, S., K. Wynne, B. McGowan, and S. Bloom. 2005. Hormonal regulation of food intake. *Physiological Reviews* 86:1131–1158.

Starck, J. M. 1998. Structural variants and invariants in avian embryonic and postnatal development. In *Avian Growth and Development*, ed. J. M. Starck and R. E. Ricklefs, 59–88. New York: Oxford University Press.

Starck, J. M., and R. E. Ricklefs. 1998. *Avian Growth and Development*. New York: Oxford University Press.

Staszewski, V., J. Gasparini, K. D. McCoy, T. Tveraa, and T. Boulinier. 2007. Evidence of an interannual effect of maternal immunization on the immune response of juveniles in a long-lived colonial bird. *Journal of Animal Ecology* 76:1215–1223.

Stearns, S. C. 1992. *The Evolution of Life Histories*. Oxford: Oxford University Press.

Stetson, M. H., R. A. Lewis, and D. S. Farner. 1973. Some effects of exogenous gonadotropins and prolactin on photostimulated and photorefractory white-crowned sparrows. *General and Comparative Endocrinology* 21:424–430.

Stevenson, I. R., and D. M. Bryant. 2000. Climate change and constraints on breeding. *Nature* 406:366–367.

Stevenson, T. J., G. E. Bentley, T. Ubuka, L. Arckens, E. Hampson, and S. A. MacDougall-Shackleton. 2008. Effects of social cues on GnRH-I, GnRH-II, and reproductive physiology in female house sparrows (*Passer domesticus*). *General and Comparative Endocrinology* 156:385–394.

Stifani, S., D. Barber, R. Aebersold, E. Steyrer, X. Shen, J. Nimpf, and W. J. Schneider. 1991. The laying hen expresses two different low density lipoprotein receptor-related proteins. *Journal of Biological Chemistry* 266:19079–19087.

Stifani, S., D. L. Barber, J. Nimpf, and W. J. Schneider. 1990. A single chicken oocyte plasma membrane protein mediates uptake of very low density lipoprotein and vitellogenin. *Proceedings of the National Academy of Sciences USA* 87:1955–1959.

Stifani, S., R. George, and W. J. Schneider. 1988. Solubalization and chacterisation of the chicken oocyte vitellogenin receptor. *Biochemical Journal* 250:467–475.

Stodola, K. W., E. T. Linder, D. A. Buehler, K. E. Franzreb, D. H. Kim, and R. J. Cooper. 2010. Relative influence of male and female care in determining nestling mass in a migratory songbird. *Journal of Avian Biology* 41:515–522.

Stoleson, S. H., and S. R. Beissinger. 1997. Hatching asynchrony, brood reduction, and food limitation in a neotropical parrot. *Ecological Monographs* 67:131–154.

———. 1999. Egg viability as a constraint on hatching synchrony at high ambient temperatures. *Journal of Animal Ecology* 68:951–962.

Strader, A. D., and J. D. Buntin. 2003. Changes in agouti-related peptide during the ring dove breeding cycle in relation to prolactin and parental hyperphagia. *Journal of Neuroendocrinology* 15:1046–1053.

Strasser, R., and H. Schwabl. 2004. Yolk testosterone organises behavior and male plumage coloration in house sparrows (*Passer domesticus*). *Behavioral Ecology and Sociobiology* 56:491–497.

Styrsky, J. D., K. P. Eckerle, and C. F. Thompson. 1999. Fitness-related consequences of egg mass in nestling house wrens. *Proceedings of the Royal Society B: Biological Sciences* 266:1253–1258.

Surai, P. F. 2002. *Natural Antioxidants in Avian Nutrition and Reproduction*. Nottingham, UK: Nottingham University Press.

Surai, P. F., and B. K. Speake. 1998. Distribution of carotenoids from the yolk to the tissues of the developing embryo. *Journal of Nutritional Biochemistry* 9:645–651.

Svensson, E. 1995. Avian reproductive timing: When should parents be prudent? *Animal Behaviour* 49:1569–1575.

———. 1997. Natural selection on avian breeding time: Causality, fecundity-dependent, and fecundity-independent selection. *Evolution* 51:1276–1283.

Svensson, E., and J.-A. Nilsson. 1997. The trade-off between molt and parental care: A sexual conflict in the blue tit? *Behavioral Ecology* 8:92–98.

Swaddle, J. P., and M. S. Witter. 1997. Food availability and primary feather molt in European starlings, *Sturnus vulgaris*. *Canadian Journal of Zoology* 75:948–953.

Swaddle, J. P., M. S. Witter, I. C. Cuthill, A. Budden, and P. McCowen. 1996. Plumage condition affects flight performance in common starlings: Implications for developmental homeostasis, abrasion and molt. *Journal of Avian Biology* 27:103–111.

Swanson, D. L. 1990. Seasonal variation of vascular oxygen transport in the Dark-Eyed Junco. *Condor* 92:62–66.

Székely, T., I. Karsai, and T. D. Williams. 1994. Determination of clutch-size in the Kentish Plover *Charadrius alexandrinus*. *Ibis* 136:341–348.

Tabibzadeh, C., I. Rozenboim, J. L. Silsby, G. R. Pitts, D. N. Foster, and M. E. El Halawani. 1995. Modulation of ovarian cytochrome P450–17 alpha-hydroxylase and cytochrome aromatase messenger ribonucleic acid by prolactin in the domestic turkey. *Biology of Reproduction* 52:600–608.

Tanvez, A., N. Beguin, O. Chastel, A. Lacroix, and G. Leboucher. 2004. Sexually attractive phrases increase yolk androgens deposition in Canaries (*Serinus canaria*). *General and Comparative Endocrinology* 138:113–120.

Tarugi, P., G. Ballarini, B. Pinotti, A. Franchini, E. Ottaviani, and S. Calandra. 1988. Secretion of apoB- and apoA-I-containing lipoproteins by chick kidney. *Journal of Lipid Research* 39:731–743.

Tatar, M., A. Bartke, and A. Antebi. 2003. The endocrine regulation of aging by insulin-like signals. *Science* 299:1346–1351.

Tatar, M., A. Kopelman, D. Epstein, M.-P. Tu, C.-M. Yin, and R. S. Garofalo. 2001. A mutant *Drosophila* insulin receptor homolog that extends life-span and impairs neuroendocrine function. *Science* 292:107–110.

Thomas, D. W., J. Blondel, P. Perret, M. M. Lambrechts, and J. R. Speakman. 2001. Energetic and fitness costs of mismatching resource supply and demand in seasonally breeding birds. *Science* 291:2598–2600.

Thomas, D. W., P. Bourgault, B. Shipley, P. Perret, and J. Blondel. 2010. Context-dependent changes in the weighting of environmental cues that initiate breeding in a temperate passerine, the Corsican Blue Tit (*Cyanistes caeruleus*). *Auk* 127:129–139.

Thomas, G., T. Szekely, and J. Reynolds. 2007. Sexual conflict and the evolution of breeding systems in shorebirds. *Advances in the Study of Behavior* 37:279–342.

Thomson, D. L., P. Monaghan, and R. W. Furness. 1998. The demands of incubation and avian clutch size. *Biological Reviews* 73:293–304.

Tinbergen, J. M., and C. Both. 1999. Is clutch size individually optimized? *Behavioral Ecology* 10:504–509.

Tinbergen, J. M., and S. Daan. 1990. Family planning in the great tit (*Parus major*): Optimal clutch size as integration of parent and offspring fitness. *Behaviour* 114:161–190.

Tinbergen, J. M., and M. W. Dietz. 1994. Parental energy expenditure during brood rearing in the Great Tit (*Parus major*) in relation to body mass, temperature, food availability and clutch size. *Functional Ecology* 8:563–572.

Tinbergen, J. M., and J. J. Sanz. 2004. Strong evidence for selection for larger brood size in a great tit population. *Behavioral Ecology* 15:525–533.

Tinbergen, J. M., and J. B. Williams. 2002. Energetics of incubation. In *Avian Incubation*, ed. D. C. Deeming, 299–313. Oxford: Oxford University Press.

Tobler, M., M. Granbom, and M. I. Sandell. 2007. Maternal androgens in the pied flycatcher: Timing of breeding and within-female consistency. *Oecologia* 151:731–740.

Török, J., R. Hargitai, G. Hegyi, Z. Matus, G. Michl, P. Péczely, B. Rosivall, et al. 2007. Carotenoids in the egg yolks of collared flycatchers (*Ficedula albicollis*) in relation to parental quality, environmental factors and laying order. *Behavioral Ecology and Sociobiology* 61:541–550.

Travers, M., M. L. Clinchy, R. Boonstra, L. Zanette, and T. D. Williams. 2010. Indirect predator effects on clutch size and the cost of egg production. *Ecology Letters* 13:980–988.

Trivers, R. L., and D. E. Willard. 1973. Natural selection of parental ability to vary the sex ratio of offspring. *Science* 179:90–91.

Tschirren, B., H. Richner, and H. Schwabl. 2004. Ectoparasite-modulated deposition of maternal androgens in great tit eggs. *Proceedings of the Royal Society B: Biological Sciences* 271:1371–1375.

Tschirren, B., J. Sendecka, T. G. G. Groothuis, L. Gustafsson, and B. Doligez. 2009. Heritable variation in maternal yolk hormone yransfer in a wild bird population. *American Naturalist* 174:557–564.

Tsutsui, K. 2009. A new key neurohormone controlling reproduction, gonadotropin-inhibitory hormone (GnIH): Biosynthesis, mode of action and functional significance. *Progress in Neurobiology* 88:76–88.

Tsutsui, K., G. E. Bentley, T. Ubuka, E. Saigoh, H. Yin, T. Osugi, K. Inoue, et al. 2007. The general and comparative biology of gonadotropin-inhibitory hormone (GnIH). *General and Comparative Endocrinology* 153:365–370.

Tyler, C. R., J. P. Sumpter, and R. M. Handford. 1990. The dynamics of vitellogenin sequestration into vitellogenic ovarian follicles of the rainbow trout, *Salmo gairdneri*. *Fish Physiology and Biochemistry* 8:211–219.

Ubuka, T., S. Kim, Y.-C. Huang, J. Reid, J. Jiang, T. Osugi, V. S. Chowdhury, et al. 2008. Gonadotropin-inhibitory hormone neurons interact directly with gonadotropin-releasing hormone-I and -II neurons in European starling brain. *Endocrinology* 149:268–278.

Ubuka, T., K. Ukena, P. J. Sharp, G. E. Bentley, and K. Tsutsui. 2006. Gonadotropin-inhibitory hormone inhibits gonadal development and maintenance by decreasing gonadotropin synthesis and release in male quail. *Endocrinology* 147:1187–1194.

Uller, T. 2008. Developmental plasticity and the evolution of parental effects. *Trends in Ecology and Evolution* 23:432–438.

Uller, T., and A. V. Badyaev. 2009. Evolution of "determinants" in sex-determination: A novel hypothesis for the origin of environmental contingencies in avian sex bias. *Seminars in Cell and Developmental Biology* 20:304–312.

Uller, T., J. Eklöf, and S. Andersson. 2005. Female egg investment in relation to male sexual traits and the potential for transgenerational effects in sexual selection. *Behavioral Ecology and Sociobiology* 57:584–590.

van Balen, J. H. 1973. A comparative study of the breeding ecology of the Great Tit (*Parus major*) in different habitats. *Ardea* 61:1–93.

van de Pol, M., T. Bakker, D. J. Saaltink, and S. Verhulst. 2006. Rearing conditions determine offspring survival independent of egg quality: A cross-foster experiment with Oystercatchers *Haematopus ostralegus*. *Ibis* 148:203–210.

van de Pol, M. D., D. Heg, L. W. Bruinzeel, B. Kuijper, and S. Verhulst. 2006. Experimental evidence for a causal effect of pair-bond duration on reproductive performance in oystercatchers (*Haematopus ostralegus*). *Behavioral Ecology* 17:982–991.

van der Jeugd, H. P., and R. McCleery. 2002. Effects of spatial autocorrelation, natal philopatry and phenotypic plasticity on the heritability of laying date. *Journal of Evolutionary Biology* 15:380–387.

van Noordwijk, A. J., and G. de Jong. 1986. Acquisition and allocation of resources: Their influence on variation in life history tactics. *American Naturalist* 128:137–142.

van Noordwijk, A. J., R. H. McCleery, and C. M. Perrins. 1995. Selection for the timing of great tit (*Parus major*) breeding in relation to caterpillar growth and temperature. *Journal of Animal Ecology* 64:451–458.

van Noordwijk, A. J., J. H. van Balen, and W. Scharloo. 1981. Genetic variation in the timing of reproduction in the Great Tit. *Oecologia* 49:158–166.

Vanderkist, B. A., T. D. Williams, D. F. Bertram, L. Lougheed, and J. P. Ryder. 2000. Indirect, physiological assessment of reproductive state and breeding chronology in free-living birds: An example in the Marbled Murrelet (*Brachyramphus marmoratus*). *Functional Ecology* 14:758–765.

Vander Werf, E. 1992. Lack's clutch size hypothesis: An examination of the evidence using meta-analysis. *Ecology* 73:1699–1705.

Veasey, J. S., D. C. Houston, and N. B. Metcalfe. 2008. A hidden cost of reproduction: The trade-off between clutch size and escape take-off speed in female zebra finches. *Journal of Animal Ecology* 70:20–24.

Veiga, J. P. 1992. Hatching asynchrony in the House Sparrow: A test of the egg-viability hypothesis. *American Naturalist* 139:669–675.

Veiga, J. P., J. Vinuela, P. J. Cordero, J. M. Aparicio, and V. Polo. 2004. Experimentally increased testosterone affects social rank and primary sex ratio in the spotless starling. *Hormones and Behavior* 46:47–53.

Verbeek, N. A. M., and J. L. Morgan. 1980. Removal of primary remiges and its effect on the flying ability of Glaucous-winged Gulls. *Condor* 82:224–226.

Verboven, N., P. Monaghan, D. M. Evans, H. Schwabl, N. Evans, C. Whitelaw, and R. G. Nager. 2003. Maternal condition, yolk androgens and offspring performance: A supplemental feeding experiment in the lesser blackbacked gull (*Larus fuscus*). *Proceedings of the Royal Society B: Biological Sciences* 270:2223–2232.

Verboven, N., J. M. Tinbergen, and S. Verhulst. 2001. Food, reproductive success and multiple breeding in the Great Tit, *Parus major*. *Ardea* 89:387–405.

Verboven, N., and M. E. Visser. 1998. Seasonal variation in local recruitment of great tits: The importance of being early. *Oikos* 81:511–524.

Verhulst, S., and R. A. Hut. 1996. Post-fledgling care, multiple breeding and the costs of reproduction in the great tit. *Animal Behaviour* 51:957–966.

Verhulst, S., and J.-A. Nilsson. 2008. The timing of birds' breeding seasons: A review of experiments that manipulated timing of breeding. *Philosophical Transactions of the Royal Society London B* 363:399–410.

Verhulst, S., and J. M. Tinbergen. 1997. Clutch size and parental effort in the Great Tit (*Parus major*). *Ardea* 85:111–126.

———. 2001. Variation in food supply, time of breeding, and energy expenditure in birds. *Science* 294:471a.

Verhulst, S., J. M. Tinbergen, and S. Daan. 1997. Multiple breeding in the Great Tit. A trade-off between successive reproductive attempts? *Functional Ecology* 11:714–722.

Verhulst, S., J. H. van Balen, and J. M. Tinbergen. 1995. Seasonal decline in reproductive success of the Great Tit: Variation in time or quality. *Ecology* 76:2392–2403.

Vézina, F., A. Gustowska, K. M. Jalvingh, O. Chastel, and T. Piersma. 2009. Hormonal correlates and thermoregulatory consequences of molting on metabolic rate in a northerly wintering shorebird. *Physiological and Biochemical Zoology* 82:129–142.

Vézina, F., and K. Salvante. 2010. Behavioral and physiological flexibility are used by birds to manage energy and support investment in the early stages of reproduction. *Current Zoology* 56:767–792.

Vézina, F., J. R. Speakman, and T. D. Williams. 2006. Individually-variable energy management strategies in relation to energetic costs of egg production. *Ecology* 87:2447–2458.

Vézina, F., and T. D. Williams. 2002. Metabolic costs of egg production in the European starling (*Sturnus vulgaris*). *Physiological and Biochemical Zoology* 75:377–385.

———. 2003. Plasticity in body composition in breeding birds: What drives the metabolic costs of egg production? *Physiological and Biochemical Zoology* 76:713–730.

———. 2005a. Interaction between organ mass and citrate synthase activity as an indicator of tissue maximal oxidative capacity in breeding European Starlings: Implications for metabolic rate and organ mass relationships. *Functional Ecology* 19:119–128.

———. 2005b. The metabolic cost of egg production is repeatable. *Journal of Experimental Biology* 208:2533–2538.

Viney, M. E., E. M. Riley, and K. L. Buchanan. 2005. Optimal immune responses: Immunocompetence revisted. *Trends in Ecology and Evolution* 20:665–669.

Vinuela, J. 1997. Adaptation vs. constraint: Intraclutch eggmass variation in birds. *Journal of Animal Ecology* 66:781–792.

Visser, M. E. 2008. Keeping up with a warming world: Assessing the rate of adaptation to climate change. *Proceedings of the Royal Society B: Biological Sciences* 275:649–659.

Visser, M. E., C. Both, and M. M. Lambrechts. 2004. Global climate change leads to mistimed avian reproduction. *Advances in Ecological Research* 35:89–110.

Visser, M. E., S. P. Caro, K. van Oers, S. V. Schaper, and B. Helm. 2010. Phenology, seasonal timing and circannual rhythms: Towards a unified framework. *Philosophical Transactions of the Royal Society B: Biological Sciences* 365:3113–3127.

Visser, M. E., L.J.M. Holleman, and S. P. Caro. 2009. Temperature has a causal effect on avian timing of reproduction. *Proceedings of the Royal Society B: Biological Sciences* 276:2323–2331.

Visser, M. E., L.J.M. Holleman, and P. Gienapp. 2006. Shifts in caterpillar biomass phenology due to climate change and its impact on the breeding biology of an insectivorous bird. *Oecologia* 147:164–172.

Visser, M. E., and M. M. Lambrechts. 1999. Information constraints in the timing of reproduction in temperate zone birds: Great and Blue Tits. In *Proceedings of the 22nd International Ornithological Congress*, ed. N. J. Adams and R. H. Slotow, 249–264. Durban, Johannesburg: BirdLife South Africa.

Visser, M. E., and C. M. Lessells. 2001. The costs of egg production and incubation in the great tit (*Parus major*). *Proceedings of the Royal Society B: Biological Sciences* 268:1271–1277.

Visser, M. E., S. V. Schaper, L. J. M. Holleman, A. Dawson, P. Sharp, P. Gienapp, and S. P. Caro. 2011. Genetic variation in cue sensitivity involved in avian timing of reproduction. *Functional Ecology 25: 868-877.*

Visser, M. E., B. Silverin, M. M. Lambrechts, and J. M. Tinbergen. 2002. No evidence for tree phenology as a cue for the timing of reproduction in tits *Parus* spp. *Avian Science* 2:77–86.

Visser, M. E., A. J. van Noordwijk, J. M. Tinbergen, and C. M. Lessells. 1998. Warmer springs lead to mistimed reproduction in great tits (*Parus major*). *Proceedings of the Royal Society B: Biological Sciences* 265:1867–1870.

Vleck, C. M. 2002. Hormonal control of incubation behaviour. In *Avian Incubation*, ed. D. C. Deeming, 54–62. Oxford: Oxford University Press.

Vleck, C. M., and T. L. Bucher. 1998. Energy metabolism, gas exchange and ventilation. In *Avian Growth and Development*, ed. J. M. Starck and R. E. Ricklefs, 89–116. New York: Oxford University Press.

Vleck, C. M., L. L. Ross, D. Vleck, and T. L. Bucher. 2000. Prolactin and parental behavior in Adélie penguins: Effects of absence from nest, incubation length, and nest failure. *Hormones and Behavior* 38:149–158.

Vleck, C. M., and D. Vleck. 2002. Physiological condition and reproductive consequences in Adelie penguins. *Integrative and Comparative Biology* 42:76–83.

von Brömssen, A., and C. Jansson. 1980. Effects of food addition to Willow Tit *Parus montanus* and Crested Tit *P. cristatus* at the time of breeding. *Ornis Scandinavica* 11:173–178.

von Engelhardt, N., C. Carere, C. Dijkstra, and T. G. Groothuis. 2006. Sex-specific effects of yolk testosterone on survival, begging and growth of zebra finches. *Proceedings of the Royal Society B: Biological Sciences* 273:65–70.

von Engelhardt, N., C. Dijkstra, S. Daan, and T. G. G. Groothuis. 2004. Effects of 17-beta estradiol treatment of female zebra finches on offspring sex ratio and survival. *Hormones and Behavior* 45:306–313.

von Schalburg, K. R., M. L. Rise, G. D. Brown, W. S. Davidson, and B. F. Koop. 2005. A comprehensive survey of the genes involved in maturation and development of the rainbow trout ovary. *Biology of Reproduction* 72:687–699.

Wagner, E. C., J. S. Prevolsek, K. E. Wynne-Edwards, and T. D. Williams. 2008. Hematological changes associated with egg production: Estrogen-dependence and repeatability. *Journal of Experimental Biology* 211:400–408.

Wagner, E. C., C. A. Stables, and T. D. Williams. 2008. Hematological changes associated with egg production: Direct evidence for changes in erythropoiesis but a lack of resource-dependence. *Journal of Experimental Biology* 211:2960–2968.

Wagner, E. C., and T. D. Williams. 2007. Experimental (anti-estrogen mediated) reduction in egg size negatively affects offspring growth and survival. *Physiological and Biochemical Zoology* 80:293–305.

Wagner, P. D. 1996. Determinants of maximal oxygen transport and utilization. *Annual Review of Physiology* 58:21–50.

Wallace, R. A. 1985. Vitellogenesis and oocyte growth in nonmammalian vertebrates. In *Developmental Biology, Oogenesis*, ed. L. W. Browder, 127–177. New York: Plenum.

Wallander, J., and M. Andersson. 2002. Clutch size limitation in waders: Experimental test in redshank *Tringa totanus*. *Oecologia* 130:391–395.

Walsberg, G. E. 1978. The heat budget of incubating white-crowned sparrows (*Zonotrichia leucophrys oriantha*) in Oregon. *Physiological Zoology* 51:92–103.

———. 1983. Avian ecological energetics. In *Avian Biology*, ed. D. S. Farner and J. R. King, 161–220. New York: Academic Press.

———. 2003. How useful is energy balance as a overall index of stress in animals? *Hormones and Behavior* 43:16–17.

Walzem, R. L. 1996. Lipoproteins and the laying hen: Form follows function. *Poultry and Avian Biology Review* 7:31–64.

Walzem, R. L., R. J. Hansen, D. L. Williams, and R. L. Hamilton. 1999. Estrogen induction of VLDLy assembly in egg-laying hens. *Journal of Nutrition* 129:467S–472S.

Wang, J. M., and S. R. Beissinger. 2009. Variation in the onset of incubation and its influence on avian hatching success and asynchrony. *Animal Behaviour* 78:601–613.

———. 2011. Partial incubation in birds: Its occurrence, function, and quantification. *Auk* 128:454–466.

Wang, S., D. E. Smith, and D. L. Williams. 1983. Purification of avian vitellogenin III: Comparison with vitellogenins I and II. *Biochemistry* 22:6206–6212.

Wang, S.-Y., and D. L. Williams. 1983. Differential responsiveness of avian vitellogenin I and vitellogenin II during primary and secondary stimulation with estrogen. *Biochemical and Biophysical Research Communications* 112:1049–1055.

Ward, S. 1996. Energy expenditure of female barn swallows *Hirundo rustica* during egg formation. *Physiological Zoology* 69:930–951.

Watanabe, T., T. Yamamura, M. Watanabe, S. Yasuo, N. Nakao, A. Dawson, S. Ebihara, et al. 2007. Hypothalamic expression of thyroid-activating and -inactivating enzyme genes in relation to photorefractoriness in birds and mammals. *American Journal of Physiology Regulatory, Integrative, and Comparative Physiology* 292:R568–R572.

Watson, M. D., G. J. Robertson, and F. Cooke. 1993. Egg-laying time and laying interval in the Common Eider. *Condor* 95:869–878.

Weatherhead, P. J., R. D. Montgomerie, and S. B. McRae. 1991. Egg-laying times of American Robins. *Auk* 108:965–967.

Webb, D. R. 1987. Thermal tolerance of avian embryos: A review. *Condor* 89:874–898.

Weggler, M. 2006. Constraints on, and determinants of, the annual number of breeding attempts in the multi-brooded Black Redstart *Phoenicurus ochruros*. *Ibis* 148:273–284.

Weidinger, K., and M. Kral. 2007. Climatic effects on arrival and laying dates in a long-distance migrant, the Collared Flycatcher *Ficedula albicollis*. *Ibis* 149:836–847.

West, S. A., and B. C. Sheldon. 2002. Constraints in the evolution of sex ratio adjustment. *Science* 295:1685–1688.

Wheelwright, N. T., K. A. Tice, and C. R. Freeman-Gallant. 2003. Postfledging parental care in Savannah sparrows: Sex, size and survival. *Animal Behaviour* 65:435–443.

Wiersma, P., C. Selman, J. R. Speakman, and S. Verhulst. 2004. Birds sacrifice oxidative protection for reproduction. *Proceedings of the Royal Society B: Biological Sciences* 271:S360–S363.

Wiggins, D. A., T. Part, and L. Gustafsson. 1994. Correlates of clutch desertion by female Collared Flycatchers *Ficedula albicollis*. *Journal of Avian Biology* 25:93–97.

Wikelski, M., M. Hau, W. D. Robinson, and J. C. Wingfield. 2003. Reproductive seasonality of seven neotropical passerine species. *Condor* 105:683–695.

Wikelski, M., R. W. Kays, N. J. Kasdin, K. Thorup, J. A. Smith, and G. W. Swenson, Jr. 2007. Going wild: What a global small-animal tracking system could do for experimental biologists. *Journal of Experimental Biology* 210:181–186.

Wilkin, T., A. Gosler, D. Garant, S. Reynolds, and B. Sheldon. 2009. Calcium effects on life-history traits in a wild population of the great tit (*Parus major*): Analysis of long-term data at several spatial scales. *Oecologia* 159:463–472.

Williams, B. R., and D. L. J. Kilgore. 1992. Ontogenetic modification of the hypercapnic response ventilatory response in the zebra finch. *Respiration Physiology* 90:125–134.

Williams, G. C. 1966. Natural selection, the costs of reproduction, and a refinement of Lack's principle. *American Naturalist* 100:687–692.

Williams, J. B. 1996. Energetics of avian incubation. In *Avian Energetics and Nutritional Ecology*, ed. C. Carey, 375–416. New York: Chapman and Hall.

Williams, S. E., E. E. Bolitho, and S. Fox. 2003. Climate change in Australian tropical rainforests: An impending environmental catastrophe. *Proceedings of the Royal Society B: Biological Sciences* 270:1887–1892.

Williams, T. D. 1994. Intraspecific variation in egg-size and egg composition in birds: Effects on offspring fitness. *Biological Reviews* 68:35–59.

———. 1995. *The Penguins: Bird Families of the World*. Oxford: Oxford University Press.

———. 1996. Variation in reproductive effort in female zebra finches (*Taeniopygia guttata*) in relation to nutrient-specific dietary supplements during egg-laying. *Physiological Zoology* 65:1255–1275.

———. 1998. Avian reproduction, overview. In *Encyclopedia of Reproduction*, ed. E. Knobil and J. D. Neil, 325–336. New York: Academic Press.

———. 1999. Parental and first generation effects of exogenous 17beta-estradiol on reproductive performance of female zebra finches (*Taeniopygia guttata*). *Hormones and Behavior* 35:135–143.

———. 2000. Experimental (tamoxifen-induced) manipulation of female reproduction in zebra finches (*Taeniopygia guttata*). *Physiological and Biochemical Zoology* 73:566–573.

———. 2001. Experimental manipulation of female reproduction reveals an intraspecific egg-size:clutch size trade-off. *Proceedings of the Royal Society B: Biological Sciences* 268:423–428.

———. 2005. Mechanisms underlying costs of egg production. *BioScience* 55:39–48.

———. 2008. Individual variation in endocrine systems: Moving beyond the "tyranny of the Golden Mean." *Philosophical Transactions of the Royal Society B* 363:1687–1698.

Williams, T. D., and C. A. Ames. 2004. Top-down regression of the avian oviduct during late oviposition in a small passerine bird. *Journal of Experimental Biology* 207:263–268.

Williams, T. D., C. E. Ames, Y. Kiparissis, and K. E. Wynne-Edwards. 2005. Laying sequence-specific variation in yolk estrogen levels, and relationship with plasma estrogen in female zebra finches (*Taeniopygia guttata*). *Proceedings of the Royal Society B: Biological Sciences* 272:173–177.

Williams, T. D., W. O. Challenger, J. K. Christians, M. Evanson, O. Love, and F. Vézina. 2004. What causes the decrease in haematocrit during egg production? *Functional Ecology* 18:330–336.

Williams, T. D., and E. G. Cooch. 1996. Egg-size, temperature and laying sequence: Why do Snow Geese lay big eggs when its cold? *Functional Ecology* 10:112–118.

Williams, T. D., W. L. Hill, and R. L. Walzem. 2001. Egg size variation: Mechanisms and hormonal control In *Avian Endocrinology*, ed. A. S. Dawson and C. M. Chatruverdi, 205–217. New Delhi: Narosa Publishing House.

Williams, T. D., A. S. Kitaysky, and F. Vézina. 2004. Individual variation in plasma estradiol-17β and androgen levels during egg formation in the European starling *Sturnus vulgaris*: Implications for regulation of yolk steroids. *General and Comparative Endocrinology* 136:346–352.

Williams, T. D., M. J.J.E. Loonen, and F. Cooke. 1994. Fitness consequences of parental behaviour in relation to offspring number in a precocial species: The Lesser Snow Goose. *Auk* 111:563–572.

Williams, T. D., and C. J. Martyniuk. 2000. Tissue mass dynamics during egg-production in female Zebra Finches (*Taeniopygia guttata*): Dietary and hormonal manipulations. *Journal of Avian Biology* 31:87–95.

Williams, T. D., and M. Miller. 2003. Individual and resource-dependent variation in the ability to lay supranormal clutches in response to egg-removal. *Auk* 120:481–489.

Williams, T. D., and F. Vézina. 2001. Reproductive energy expenditure, intraspecific variation and fitness. *Current Ornithology* 16:355–405.

Williams, T. D., F. Vézina, and J. R. Speakman. 2009. Individually-variable energy management during egg production is repeatable across breeding attempts. *Journal of Experimental Biology* 212:1101–1105.

Willie, J., M. Travers, and T. D. Williams. 2010. Female zebra finches (*Taeniopygia guttata*) are chronically, but not cumulatively, "anemic" during repeated egg-laying in response to experimental nest predation. *Physiological and Biochemical Zoology* 83:119–126.

Wilson, A. J., A. Charmantier, and J. D. Hadfield. 2008. Evolutionary genetics of ageing in the wild: Empirical patterns and future perspectives. *Functional Ecology* 22:431–442.

Wilson, A. J., and D. H. Nussey. 2010. What is individual quality? An evolutionary perspective. *Trends in Ecology and Evolution* 25:207–214.

Wilson, A. J., D. Réale, M. N. Clements, M. M. Morrissey, E. Postma, C. A. Walling, L.E.B. Kruuk, et al. 2010. An ecologist's guide to the animal model. *Journal of Animal Ecology* 79:13–26.

Wilson, S., D. R. Norris, A. G. Wilson, and P. Arcese. 2007. Breeding expeience and population density affect the ability of a songbird to respond to future climate variation. *Proceedings of the Royal Society B: Biological Sciences* 274:2669–2675.

Wingfield, J. C. 1983. Environmental and endocrine control of avian reproduction: An ecological approach. In *Avian Endocrinology: Environmental and Ecological Perspectives*, ed. S. Mikami, K. Hiomma, and M. Wada, 265–288. Tokyo, Japan Scientific Societies Press.

———. 1994. Environmental and endocrine control of reproduction in the song sparrow, *Melospiza melodia*. 1. Temporal organization of the breeding cycle. *General and Comparative Endocrinology* 56:406–416.

Wingfield, J. C., and D. S. Farner. 1978. The endocrinology of a natural breeding population of the white-crowned sparrow (*Zonotrichia leucophrys pugetensis*). *Physiological Zoology* 51:188–205.

———. 1993. Endocrinology of reproduction in wild species. In *Avian Biology*, ed. D. S. Farner, J. R. King, and K. C. Parkes, 163–327. New York: Academic Press.

Wingfield, J. C., B. K. Follett, K. S. Matt, and D. S. Farner. 1980. Effect of day length on plasma FSH and LH in castrated and intact white-crowned sparrows. *General and Comparative Endocrinology* 42:464–470.

Wingfield, J. C., and A. R. Goldsmith. 1990. Plasma levels of prolactin and gonadal steroids in relation to multiple-brooding and renesting in free-living populations of the song sparrow, *Melospiza melodia. Hormones and Behavior* 24:89–103.

Wingfield, J. C., T. P. Hahn, D. L. Maney, S. J. Schoech, M. Wada, and M. L. Morton. 2003. Effects of temperature on photoperiodically induced reproductive development, circulating plasma luteinizing hormone and thyroid hormones, body mass, fat deposition and molt in mountain white-crowned sparrows, *Zonotrichia leucophrys oriantha. General and Comparative Endocrinology* 131:143–158.

Wingfield, J. C., T. P. Hahn, M. Wada, L. B. Astheimer, and S. J. Schoech. 1996. Interrelationship of daylength and temperature on the control of gonadal development, body mass, and fat score in white-crowned sparrows, *Zonotrichia leucophrys gambelii. General and Comparative Endocrinology* 101:242–255.

Wingfield, J. C., D. L. Maney, C. W. Breuner, J. D. Jacobs, S. Lynn, M. Ramenofsky, and R. D. Richardson. 1998. Ecological bases of hormone–behavior interactions: The "emergency life history stage." *American Zoologist* 38:191–206.

Wingfield, J. C., and L. M. Romero. 2000. Adrenocortical responses to stress and their modulation in free-living vertebrates. In *Coping with the Environment: Neural and Endocrine Mechanisms*, ed. B. S. McEwen, 211–236, Vol. 4 of *The Endocrine System*, ed. H. M. Goodman, Section 7 of *Handbook of Physiology*. Oxford: Oxford University Press.

Wingfield, J. C., M. E. Visser, and T. D. Williams. 2008. Introduction: Integration of ecology and endocrinology in avian reproduction: a new synthesis. *Philosophical Transactions of the Royal Society B* 363:581–1588.

Winkler, D. W., and P. E. Allen. 1995. Effects of handicapping on female condition and reproduction in Tree Swallows (*Tachycineta bicolor*). *Auk* 112:737–747.

Winkler, D. W., and J. R. Walters. 1983. The determination of clutch size in precocial birds. *Current Ornithology* 1:33–68.

Woodburn, R.J.W., and C. M. Perrins. 1997. Weight change and the body reserves of female blue tits, *Parus caeruleus*, during the breeding season. *Journal of Zoology* 243:789–802.

Wood-Gush, D.G.M., and A. B. Gilbert. 1973. Some hormones involved in the nesting behaviour of hens. *Animal Behaviour* 21:98–103.

Woods, D. C., M. J. Haugen, and A. L. Johnson. 2007. Actions of epidermal growth factor receptor/mitogen-activated protein kinase and protein kinase C signaling in granulosa cells from *Gallus gallus* are dependent upon stage of differentiation. *Biology of Reproduction* 77:61–70.

Woods, D. C., and A. L. Johnson. 2005. Regulation of follicle-stimulating hormone-receptor messenger RNA in hen granulosa cells relative to follicle selection. *Biology of Reproduction* 72:643–650.

Wright, J., C. Both, P. A. Cotton, and D. M. Bryant. 1998. Quality vs. quantity: Energetic and nutritional trade-offs in parental provisioning strategies. *Journal of Animal Ecology* 67:620–634.

Wright, J., and I. Cuthill. 1989. Manipulation of sex differences in parental care. *Behavioral Ecology and Sociobiology* 25:171–181.

Yin, H., K. Ukena, T. Ubuka, and K. Tsutsui. 2005. A novel G protein-coupled receptor for gonadotropin-inhibitory hormone in the Japanese quail (*Coturnix japonica*): Identification, expression and binding activity. *Journal of Endocrinology* 184:257–266.

Yohannes, E., M. Valcu, R. W. Lee, and B. Kempenaers. 2010. Resource use for reproduction depends on spring arrival time and wintering area in an arctic breeding shorebird. *Journal of Avian Biology* 41:580–590.

Yom-Tov, Y., and J. Wright. 1993. Effect of heating nest boxes on egg laying in the Blue Tit (*Parus caeruleus*). *Auk* 110:95–99.

Yoshimura, T., S. Yasuo, M. Watanabe, M. Ligo, T. Yamamura, K. Hirunagi, and S. Ebihara. 2003. Light-induced hormone conversion of T4 and T3 regulates photoperiodic response of gonads in birds. *Nature* 426:178–181.

You, S., L. K. Foster, J. L. Silsby, M. E. El Halawani, and D. N. Foster. 1995. Sequence analysis of the turkey LH beta subunit and its regulation by gonadotrophin-releasing hormone and prolactin in cultured pituitary cells. *Journal of Molecular Endocrinology* 14:117–129.

Young, R. L., and A. V. Badyaev. 2004. Evolution of sex-biased maternal effects in birds: 1. Sex-specific resource allocation among simultaneously growing follicles. *Journal of Evolutionary Biology* 17:1335–1366.

Youngren, O. M., M. E. El Halawani, J. L. Silsby, and R. E. Phillips. 1991. Intracranial prolactin perfusion induces incubation behavior in turkey hens. *Biology of Reproduction* 44:425–431.

Yu, J. Y.-L., and R. R. Marquardt. 1973. Development, cellular growth and function of the avian oviduct. *Biology of Reproduction* 8:283–298.

Zadworny, D., K. Shimada, H. Ishida, and K. Sato. 1989. Gonadotropin-stimulated estradiol production in small ovarian follicles of the hen is suppressed by physiological concentrations of prolactin in vitro. *General and Comparative Endocrinology* 74:468–473.

Zanette, L., M. Clinchy, and J. N. M. Smith. 2006. Food and predators affect egg production in song sparrows. *Ecology* 87:2459–2467.

Zera, A. J. 2005. Intermediary metabolism and life history trade-offs: Lipid metabolism in lines of the wing-polymorphic cricket, *Gryllus firmus*, selected for flight capability vs. early age reproduction. *Integrative and Comparative Biology* 45:511–524.

Zera, A. J., and L. G. Harshman. 2001. The physiology of life-history trade-offs in animals. *Annual Review of Ecology and Systematics* 32:95–126.

Zera, A. J., L. G. Harshman, and T. D. Williams. 2007. Evolutionary endocrinology: The developing synthesis between endocrinology and evolutionary genetics. *Annual Review of Ecology, Evolution, and Systematics* 38:793–817.

Zera, A. J., and Z. Zhao. 2006. Intermediary metabolism and life-history trade-offs: Differential metabolism of amino acids underlies the dispersal-reproduction trade-off in a wing-polymorphic cricket. *American Naturalist* 167:889–900.

Index